STUDIES IN INTERFACE SCIENCE

Surface Activity in Drug Action

STUDIES IN INTERFACE SCIENCE

SERIES EDITORS
D. Möbius and R. Miller

Vol. 1 Dynamics of Adsorption at Liquid Interfaces. *Theory, Experiment, Application.* By S.S. Dukhin, G. Kretzschmar and R. Miller
Vol. 2 An Introduction to Dynamics of Colloids. By J.K.G. Dhont
Vol. 3 Interfacial Tensiometry. By A.I. Rusanov and V.A. Prokhorov
Vol. 4 New Developments in Construction and Functions of Organic Thin Films. Edited by T. Kajiyama and M. Aizawa
Vol. 5 Foam and Foam Films. By D. Exerowa and P.M. Kruglyakov
Vol. 6 Drops and Bubbles in Interfacial Research. Edited by D. Möbius and R. Miller
Vol. 7 Proteins at Liquid Interfaces. Edited by D. Möbius and R. Miller
Vol. 8 Dynamic Surface Tensiometry in Medicine. By V.M. Kazakov, O.V. Sinyachenko, V.B. Fainerman, U. Pison and R. Miller
Vol. 9 Hydrophile-Lipophile Balance of Surfactants and Solid Particles. *Physicochemical Aspects and Applications.* By P.M. Kruglyakov
Vol. 10 Particles at Fluid Interfaces and Membranes. *Attachment of Colloid Particles and Proteins to Interfaces and Formation of Two-Dimensional Arrays.* By P.A. Kralchevsky and K. Nagayama
Vol. 11 Novel Methods to Study Interfacial Layers. By D. Möbius and R. Miller
Vol. 12 Colloid and Surface Chemistry. By E.D. Shchukin, A.V. Pertsov, E.A. Amelina and A.S. Zelenev
Vol. 13 Surfactants: Chemistry, Interfacial Properties, Applications. Edited by V.B. Fainerman, D. Möbius and R. Miller
Vol. 14 Complex Wave Dynamics on Thin Films. By H.-C. Chang and E.A. Demekhin
Vol. 15 Ultrasound for Characterizing Colloids. *Particle Sizing, Zeta Potential, Rheology.* By A.S. Dukhin and P.J. Goetz
Vol. 16 Organized Monolayers and Assemblies: Structure, Processes and Function. Edited by D. Möbius and R. Miller
Vol. 17 Introduction to Molecular-Microsimulation of Colloidal Dispersions. By A. Satoh
Vol. 18 Transport Mediated by Electrified Interfaces: Studies in the linear, non-linear and far from equilibrium regimes. By R.C. Srivastava and R.P. Rastogi
Vol. 19 Stable Gas-in-Liquid Emulsions: Production in Natural Waters and Artificial Media Second Edition By J.S. D'Arrigo
Vol. 20 Interfacial Separation of Particles. By S. Lu, R.J. Pugh and E. Forssberg
Vol. 21 Surface Activity in Drug Action. By R.C. Srivastava and A.N. Nagappa

Surface Activity in Drug Action

R.C. Srivastava

Professor Emeritus, Chemistry Group
Birla Institute of Technology and Science
Pilani – 333031 (Rajasthan) India

A.N. Nagappa

Professor, Pharmaceutical Sciences Group
Birla Institute of Technology and Science
Pilani – 333031 (Rajasthan) India

2005
ELSEVIER

Amsterdam – Boston – Heidelberg – London – New York – Oxford
Paris – San Diego – San Francisco – Singapore – Sydney – Tokyo

ELSEVIER B.V.
Radarweg 29
P.O. Box 211, 1000 AE
Amsterdam, The Netherlands

ELSEVIER Inc.
525 B Street, Suite 1900
San Diego, CA 92101-4495
USA

ELSEVIER Ltd
The Boulevard, Langford Lane
Kidlington, Oxford OX5 1GB
UK

ELSEVIER Ltd
84 Theobalds Road
London WC1X 8RR
UK

First edition 2005

Library of Congress Cataloging in Publication Data
A catalog record is available from the Library of Congress.

British Library Cataloguing in Publication Data
A catalogue record is available from the British Library.

ISBN: 0-444-51715-4
ISSN: 1383-7303

♾ The paper used in this publication meets the requirements of ANSI/NISO Z39.48–1992 (Permanence of Paper).
Printed in The Netherlands.

To
K.P. and V.

PREFACE

Surface activity is of ubiquitous presence in living systems. Take any body fluid or cell soup, its surface tension is always less than that of water. Most of the bimolecules, proteins, lipids etc. are surface active in nature. Molecules of surface-active nature are crucial to living matter and its organization. Formation of biological cell is as a matter of fact, a consequence of surface activity. Surface activity in living systems is a matter of evolution i.e., it is need based and therefore should have a crucial role to play in biological actions.

With this thought in mind the investigations recorded in this monograph were started.

Since formation of cell membranes and location of receptor proteins in the lipid bilayer are a consequence of surface activity, it is logical to expect that the drugs acting by altering the permeability of cell membranes after interacting with them may also be surface active in nature. In fact they are, and there is enough circumstantial evidence to indicate that there may exist some crucial step common to the mechanism of action of all surface-active drugs. Surface-active drugs are likely to accumulate at the interface and form films at the site of action modifying access of relevant molecules to the action sites. Our investigations on a wide variety of drugs belonging to different pharmacological categories have revealed that the modification in the access of relevant molecules to the site of action is an important step common to the mechanism of the surface-active drugs and makes significant contribution to drug action. In fact these studies have led us to propose "a liquid membrane hypothesis of drug action" for surface-active drugs. Chapters 1 to 7 contains an account of the hypothesis. Chapter 8 contains a general account of the application of surface activity in therapeutics; this chapter has been added for the sake of completeness of the monograph.

The work recorded in this monograph has been funded by several National Funding Agencies namely the Council of Scientific and Industrial Research (CSIR), the Department of Science and Technology (DST), Government of India, All India Council for Technical Education and the University Grants Commission. The support received from different funding agencies is gratefully acknowledged.

A number of colleagues and associates have participated in the research recorded in this monograph. Some of the prominent names are Drs. S.B. Bhise, C.V.S. Subrahmanyam, D.B. Raju, A.K. Das and A.N. Nagappa; the present co-author. Especial thanks are due to Dr. S.B. Bhise who was the first to work on this problem with conviction, for his doctoral degree. I (RCS) as senior author would also like to offer especial thanks to Dr. A.N. Nagappa who suggested that a monograph be written on our work on liquid membranes in drug action.

This monograph has been written during the tenure of the first author (RCS) as an Emeritus Fellow of the University Grants Commission at the Birla Institute of Technology and Science (BITS) Pilani Rajasthan India. The support from the University Grants Commission, New Delhi and the kind hospitality of the BITS as host organization are gratefully acknowledged, particularly to the Vice-Chancellor Dr. S. Venkateswaran and Dr. L.K. Maheshwari, Director BITS for their affectionate treatment.

Thanks are due to Mr. Ramesh Sharma for word processing and scanning the figures of the manuscript and Mr. K.N. Sharma for artwork.

The inspirational force from two little angels, Krishnapriya(KP) and Vaishnavi(V) kept our zeal undamped. We dedicate this work to them: Krishnapriya is the grand daughter of RCS and Vaishnavi is daughter of ANN. According to Indian traditions, student is like a son to his teacher hence ANN's daughter, Vaishnavi, is also like a grand daughter to RCS.

Table of Contents

Chapter 1

Introduction and scope

1. INTRODUCTION

Formation of cell membranes and location of receptor proteins in lipid bilayers is a consequence of surface activity. It is, therefore, logical to expect that the drugs acting by altering the permeability cell membranes after interacting with them are likely to the surface active in nature. This is because the lipid bilayers with receptors in them represent the interface and the drugs interacting with them will not reach the interface unless they are surface active in nature.

A wide variety of drugs are, in fact, known to be surface active in nature [1-7]. This activity does not appear to be a fortuitous coincidence. In a number of cases excellent correlations between surface activity and biological effects have been demonstrated [8-17]. A typical correlation between surface activity and clinical activity in the case of antipsychotics is shown in Fig 1.

While investigating the actions of drugs like reserpine, prenylamine, chlorpromazine, propranalol etc., which inhibit catecholamine transport, it has been concluded [18] "irrespective of chemical structure the surface activity of psychotropic drugs mainly determines their potency to affect all kinds of membranes especially that of catecholamines storing particles".

Since structural requirements of surface activity are often similar to those for interaction of drugs with receptor sites [19], the correlations between surface activity and biological effects appear to indicate that there might exist a common mode of action for all surface active drugs or there may be at least one crucial step common to the mechanism of all surface active drugs. What can this common mode/crucial steps be?

In view of the liquid membrane hypothesis, which we will describe briefly in the next paragraph, it was suspected that the liquid membranes generated at the site of action of the respective drugs, acting as a barrier to the transcript of relevant permeants, might be an important step common to the mechanism of all surface-active drugs.

The liquid membrane hypothesis [20, 21, 22] was originally propounded to account for enhanced salt rejection in reverse osmosis due to addition of very small amounts, of the order of few ppm, of surfactants like polyvinyl methyl ether to saline feed. According to the hypothesis when a surfactant a added to an aqueous phase, the surfactant layer which forms spontaneously at the interface acts as a liquid membrane and modifies transport across the phase boundary. The hypothesis further postulated that as the concentration of the surfactant is increased the interface gets progressively covered with the surfactant layer liquid membrane and at the critical micelle concentration (CMC) of the surfactant coverage of the interface with the liquid membrane is complete. Experimental evidence from our laboratory [23-25] furnished additional support in favor of progressive coverage of the interface with the liquid membrane.

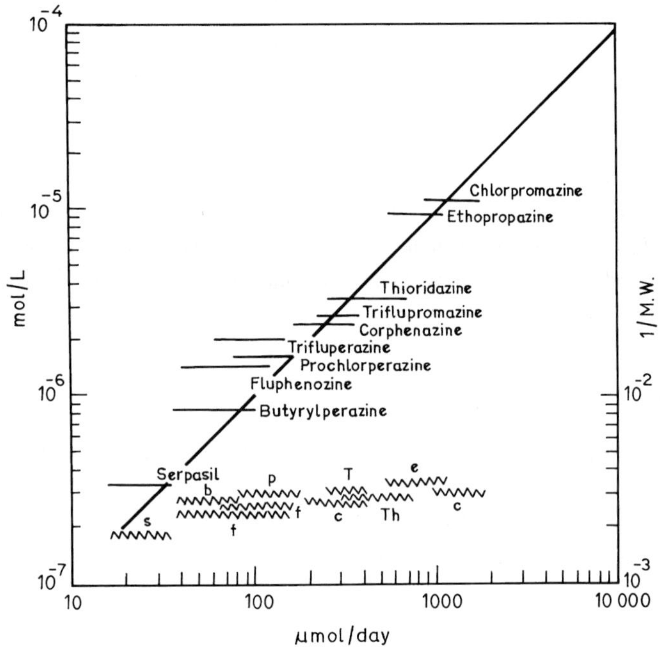

Fig.1. Demonstrating correlation between surface activity and clinical activity. The concentration of drug that lowers the surface tension by 4 dynes/cm is on the ordinate. The abscissa represents the average dose range used to control acute paranoid schizophrenia by one group of physicians on one hospital ward. The solid black bars indicate the surface activity and the wavy lines represent the inverse of molecular weight. The straight line is the theoretical line for an exact 1:1 correlation. It is seen that a hundred fold increases in surface activity is approximately associated with a hundred fold decrease in daily oral dose in micromoles (taken form Ref. 8)

Since molecules of surface-active nature are crucial to living matter and its organization [26], biological implications of the liquid membrane hypothesis have been investigated. These investigations have revealed [25] that liquid membrane bilayers generated on a hydrophobic supporting membrane in accordance with Kesting's liquid membrane hypothesis are capable of acting as mimetic system for biological membrane. In the experimental studies on the role of liquid membranes in the action of surface-active drugs, the liquid membrane bilayer system has been utilized. Therefore, in chapter 5,we will give a consolidated account of the liquid membrane bilayer systems.

Prompted by the conception that the liquid membranes generated at the site of action of respective drugs acting as a barrier to the transport of relevant permeants, might be an important step common to the mechanism of action of all surface-active drugs. A number of investigations as to the role of liquid membrane phenomenon in the mechanism of action of surface-active drugs have been undertaken. For this study structurally dissimilar drugs of different pharmacological categories were chosen. Most of these drugs are antagonistic in action i.e. they act by reducing permeability of relevant permeants to the site of action. The results of these investigations have proved quite revealing. Not only do they explain several observed biological effects by the drugs but they also throw light on the nature and orientation of receptors. This monograph presents a consolidation account of these investigations.

The liquid membrane generated by the drug itself, acting as a barrier modifying access of relevant permeants to the receptors, is a new facet of drug action. If this concept is viewed in the light of existing theories of drug action particularly occupancy theory [27, 28] and rate theory [29, 30] a more rational biophysical explanation for the action of such drugs which act by modifying permeability of cell membranes, emerges. This forms the central theme of this monograph.

REFERENCES

[1] G. Zografi in A. Osol and J.E. Hoove (eds), Remington's Pharmaceutical Science, Mack Publishing Company, 1975, pp 297.

[2] A.T. Florence, Adv. Colloid Interface Sci., 2(1968) 115.

[3] P.S. Guth and M.A. Spirtes, Int. Rev. Neurobiol, 7 (1964) 231.

[4] A. Felmiester, J. Pharm. Sci., 61(1972) 151.

[5] D. Attwood and J. Gibson, J. Pharm. Pharmcol., 30 (1978) 176.

[6] D. Attwood, J. Pharm. Pharmacol. , 24(1972) 751.

[7] D. Attwood, J. Pharm. Pharmacol., 28(1976) 407.

[8] P.M. Seeman and H.S. Bialy, Biochem. Pharmacol., 12(1963) 1181.

[9] J.M. Ritchie and P. Greengard, Annu. Rev. Pharmacol., 6(1966) 405.

[10] F.A. Vilallonga and E.W. Phillips, J. Pharm. Sci., 69(1980) 102.

[11] N.T. Pryanishnikova, Farmakol. Toxicol., (Moscow) 36(1973) 195.

[12] D. Hellenbrechet, B. Lemmer, G. Weithold and H. Grobecker, Naunyn–schmiedeberg's Arch. Pharmacol., 277 (1973) 211.

[13] J.M.A. Sitsen and J.A. Fresen, Pharm. Weekbl, 108(1973) 1053.

[14] K. Thoma and K. Albert, Pharm. Acta Helv., 54(1979) 324.

[15] A.Gesher and A.Li wan Po, J. Pharm. Pharmcol., 30(1978)353.

[16] J.H.Schulman and E.K.Rideal, Nature, 44(1939) 100.

[17] J.C.Skou,Acta Pharmacol.Toxicol., 10(1954)280.

[18] D. Palm, H. Grobecker and I.J. Bak, in H.J. Shumann and G. Kroneberg (eds) "Bayer Symposium II, New Aspects of Storage and Release Mechanism of Catecholamines" Springer Verlag Berlin,1970, pp 188-198.

[19] Attwood, A.T. Florence and J.I.N. Gillan, J. Pharm. Sci., 63(1974) 988.

[20] R.E. Kesting, A. Vincent and J. Eberlin, OSW R&D Report No. 117, Aug 1964.

[21] R.E. Kesting, "Reverse Osmosis Process Using Surfactant Feed Additive" OSW Patent Application, SAL 830, No.3 1965.

[22] R.E. Kesting, W.J. Subcasky and J.D. Paton, J. Colloid Interface Sci., 28 (1968) 156.

[23] R.C. Srivastava and Saroj Yadav, J. Non-equilib Thermodyn., 4 (1979) 219.

[24] R.C. Srivastava and Saroj Yadav, J. Colloid Interface Sci., 69 (1979) 280.

[25] R.C. Srivastava, Liquid Membrane Phenomena: Biological Implications, Indian Society for Surface Science and Technology, Jadavpur University, Kolkata, 2002, pp 275.

[26] C. Tanford, The Hydrophobic effect: Formation of Micelles and Biological Membrane, John Wiley and Sons, New York, 1980.

[27] A.J. Clark, The Mode of Action of Drugs on Cells, E. Arnold Co., London, 1933.

[28] J.H. Gaddum, Pharmacol. Rev., 9 (1957) 211.

[29] W.D.M. Paton, Proc. Roy. Soc., B154 (1961) 21.

[30] W.D.M. Paton and H.P. Rang, Adv. Drug Res., 3 (1966) 57.

Chapter 2

Surface activity of drugs

Since excellent reviews [1-4] on the surface activity of drugs are already available, the survey attempted in this chapter is not intended to be a duplication of earlier efforts; instead, an effort has been made to indicate from the reports on surface activity of drugs, the possibility of surface activity and hence, liquid membrane formation by the drugs alone or in association with membrane lipids contributing to the mechanism of their action.

2.1. Analgesics

Analgesics are classified into inhibitors of cycloxygenase (COX) and centrally acting based on their site and mechanism of action. The centrally acting drugs are more hydrophobic than COX inhibitors. The surface activity of dextropropoxyphene, [5] codienindione, oxycodone [6] antipyrine and its derivatives [7], analgin, amidopyrine [8] has been reported. Analgesic action of five narcotic compounds has been observed to correlate with their surface activity [9].

At high doses, aspirin, sodium salicylate, and the newer non-steroidal anti inflammatory drugs (NSAIDs) inhibit non-prostaglandin (PG)-dependent processes, such as the activity of a variety of enzymes, proteoglycan synthesis by chondrocytes, transmembrane ion fluxes, and chemo-attractant binding. These effects are most likely due to the capacity of aspirin-like drugs to insert into the lipid bilayer of plasma membranes, where they disrupt normal signaling events and protein-protein interactions [10].

NSAIDs associate with zwitterionic phospholipids. This intermolecular association may be the mechanism by which NSAIDs attenuate the hydrophobic barrier properties of the upper gastro intestinal tract. Pre-associating a number of NSAIDs with exogenous zwitterionic phospholipids prevented this increase in surface wettability of the mucus gel layer and protected rats against the injurious gastro intestinal tract side-effects of these drugs, while enhancing their lipid permeability, antipyretic and anti-inflammatory activity [11].

Non-PG-dependent effects of NSAIDs include (a) physical effects of the acidic molecules on surface mucosal cell membranes and mucus, (b) oxyradical production, (c) cytotoxic effects on parietal cells, and (d) inhibitory effects on mucus synthesis, mitochondrial ATP production, cyclic nucleotide production, and a range of other cellular metabolic effects influencing mucosal metabolism and cellular regeneration [12]. Association of fenoprofen sodium molecules to micelles starts at concentrations between 1.0 and 1.2×10^{-1} M.

The CMC values were determined by photon correlation spectroscopy and transmission electron microscopy, which support the assumption that fenoprofen sodium forms disc like micelles [13].

The surfactant behaviour of NSAIDs solutions of the acetic and propionic classes was analyzed in the form of salts to disclose the solubilization property of their aqueous solutions towards a lipid probe. The solubilization values for diclofenac was 35 mM, and for naproxen, sulindac, ketoprofen, indoprofen were reported to be between 100-160 mM [14]. NSAIDs, such as ibuprofen, are amphiphilic substances capable of self-association in aqueous solutions and are able to be absorbed on to the polymers through hydrophobic and electrostatic bonds [15].

The molecular characteristics of the neuropeptide substance P (SP), its agonist, and antagonists were investigated at the air/water interface and when bound to lipid monolayers and bilayers. Measurement of the Gibbs adsorption isotherm showed that the surface areas of SP and its agonist at biologically relevant concentrations were distinctly larger than those of the antagonists [16]. The surface activity of the peptides correlated with the respective binding affinities to lipid membranes [17].

2.2. Antimicorbials

There are many citations confirming the surface activity of antibacterial and antibiotics [18, 19]. Adsorption of drugs and proteins into membrane surfaces and their behavior at interfaces as well as interactions with lipids are of interest in relation to cell membrane organization and functions [20]. The CMC of actinomycin D, penicillin G, streptomycin and sodium fusidate are reported in literature [5,21]. Polyene antibiotics such as nystatin A contain hydrophobic and polar parts in their structure and can be considered as surfactants [22]. Gramicidin is known to form ion channels due to its surface activity [23]. Interactions of adriamycin, cytochrome C, and serum albumin with lipid monolayers containing poly (ethylene glycol)-ceramide were reported by Zhao et al. [24].

Certain phospholipids are known to enhance the antibacterial activity of β lactum antibiotics by enhancing the permeability of bacterial membranes to antibiotics [25]. Most of the compounds classified as virucidal are surfactants in fact. The potential role of detergents including a new antiseptic myramistin as local application via condoms for prevention of HIV infections is under consideration [26]. The CMC of novobiocin, mithramycin, variamycin, erythromycin, oleandomycin and lincomycin were determined by changes in the isotherms of the surface tension and in the maximum absorption of rodamine to the antibiotic concentrations [27].

Omega-acryloyl anionic surfactants, whose polar heads are derived from amino acids, have been telomerized to prepare and to verify whether the antiviral activity is influenced by the degree of polymerization of the polyanions. The oligomeric polyanions were evaluated for their activity against human immunodeficiency virus (HIV-1 or HIV-2) and various other RNA and DNA viruses. With regard to their anti-HIV activity, a minimum number of anionic groups were necessary to achieve an inhibitory effect [28]. The tetracycline enhanced antimicrobial activities on gram-negative bacteria in the presence of surface-active substances were mainly due to increased permeability and association with cell membrane of the antibiotic [29].

High bactericidal activity against poly-resistant were observed in non-bactericidal concentrations of antibiotics with different modes of action, i.e. penicillins, tetracyclines, aminoglycosides, macrolides in the presence of the cationic surface-active substances [30].

Antibacterial activity of a series of alkyl gallates (3,4,5-trihydroxybenzoates) against gram-positive bacteria, especially methicillin resistant *Staphylococcus aureus* strains was evaluated. The bactericidal activity of medium chain alkyl gallates was noted in combination with their ability to disrupt the cell membrane of associated function especially as surface-active agents and to inhibit the respiratory electron transport [31]. Studies have suggested that there is a positive correlation between the antibacterial properties and the surface activity of various organic amine-fluoride molecules [32].

New fluoroalkyl end capped co-oligomers which are highly surface active, containing dimethyl (octyl) ammonium segments [33], N-vinyl-2-pyrrolidone acrylic acid co-oligomers [34], 4-vinyl pyridinium chloride oligomers [35] allyl and diallyl-ammonium chloride oligomers [36] were not only able to reduce the surface tension of water but also exhibited a high antibacterial activity. Fluorinated self-assembled molecular aggregates containing carboxyl and sulfa groups were suggested to interact with positively charged HIV-1 to exhibit a potent anti-HIV-1 activity in vitro. In contrast, fluoroalkyl end-capped oligomers containing cationic segments exhibited not only the unique surface-active properties imparted by fluorine but also high antibacterial activity [37].

It has been reported that cationic surfactants bear antibacterial activity [38]. Surfaces bearing carbohydrate units have been modified in a two-step process to incorporate functionalities (lipophilic with polycationic units) that bear antibacterial activity. The effectiveness of these modified surfaces for antibacterial action against a series of gram-positive and gram-negative bacteria are reported [39]. The antifungal activity of polyene antibiotics depends at least in part on it's binding to a sterol moiety, primarily ergosterol, that is present in the membrane of sensitive fungi, by virtue of their interaction with sterols of cell membrane, leading to channel formation. [40].

Antifungal activity of octyl gallate is primarily due to its surface-active property, similar to alkanols. Thus, the fungicidal activity of gallates was distinctly increased for every additional CH_2 group [41]. SMAP-29 an amphipathic helix is a cathelicidin-derived peptide deduced from sheep leukocytes is a potent antibacterial and antifungal peptide [42]. In an effort to understand the role of the rigid polyene backbone, a sterol recognition site of macrolide antibiotics, modifications of C20 to C33 of amphotericin B was reported to have reduced the antifungal activity [43].

Several lines of evidences suggest that ciprofloxacin (CPX) could have, like other amphiphilic compounds, surface active properties. Thus, it appears that CPX might induce changes on neutral phospholipids. Surface adsorption–insertion on inner/outer phospholipids monolayers of the cytoplasmic bacterial membrane is the first step before reaching the protein efflux pump. Recent findings suggest that CPX is able to adsorb on the phospholipids surface. Moreover, some experiments using black lipid membrane bilayers suggest that CPX would be able to form pores [44].

The complex formed by the interaction of the structurally similar surface-active penicillin drugs, cloxacillin and dicloxacillin, and human serum albumin (HSA) was studied using static light scattering. A maximum in the size of the HSA-cloxacillin complex was found corresponding to the binding of approximately 2100 penicillin molecules per HSA molecule [45]. Quaternary ammonium surfactants are effective antimicrobial agents used in a number of domains such as cosmetics, common antiseptics, sanitizers in hospitals and disinfectants for contact lenses. The efficacy of such agents is conditioned by the amphiphilic nature of the molecule and consequently by its surfactant properties, e.g. ammonium gemini fluorosurfactants analogue. These products possess properties such as reduction of surface tension and a ready attraction for negatively charged surfaces like bacteria and fungi [46].

Defensin A is an inducible antibacterial protein isolated from the larvae of *Phormia terranovae* that interacts with membrane cells by forming ion-conducting pores. Defensin A adsorbs at the air-water interface from an aqueous solution and is able to spread as a monolayers [47]. Cecropins are a group of anti-bacterial, surface active cationic peptides that have an amphipathic N-terminal segment, and a largely hydrophobic C-terminal segment and normally form a helix-hinge-helix structure [48]. Mastoparan M is an amphipathic tetradecapeptide toxin isolated from the venom of the hornet. The biological activities of mastoparans include stimulation of phospholipase A2, phospholipase C, GTP-binding protein and cytotoxic activity against HL60 cells as well as binding to the phospholipid bilayers. Mastoparan M and its analogues are thought to cause the formation of ion conducting channels in lipid membranes so leading to cell lysis [49].

To develop novel antibiotic peptides useful as therapeutic drugs, the analogues of peptides were designed to increase net positive charge by lysine substitution but also hydrophobic helix region by leucine substitution from cecropin A. In particular, cecropin A analogue was designed which showed an enhanced antimicrobial and antitumor activity without hemolysis. Confocal microscopy showed that, cecropin A analogue was located in the plasma membrane [50]. Novel acrylic acid co-oligomers containing fluro alkylated end groups were found to be inhibitors of anti HIV-1.They inhibit the cytopathogenesis induced by virus. Excellent correlation exists between antiviral activity and surface activity [51]. Altering the carbohydrate binding properties of surfactant protein D, e.g., by replacing its carbohydrate recognition domain with that of either mannose binding lectin or conglutinin can increase its activity against influenza A virus [52].

Several surface-active tuberculostatics based on diamino-diphenyl sulfone have been found [53] to be effective *in vitro* at relatively low concentrations. It is suggested [53] that because of their surface activity, these molecules are adsorbed at the bacterial surface with the sulfone portion of the molecule embedded in the cell and polyoxyethylene chains oriented outwards. In another report, antitubercular activity of the drugs has been related to their configuration at air/water interface [54]. Lucanthone, an anti-schistosomiasis drug, and its derivatives, exhibiting structural similarity to phenothiazines are known [55] to be surface active. A relationship between micellar weight of these compounds and the anti-schistosomiasis activity has been discovered [56].

It is also reported [57] that the apparent distribution coefficient of a surface-active compound falls steeply above the CMC, as a result of which there is leveling off of the activity above the CMC. In case of some quaternary ammonium salts, micelle formation has been shown to be a limiting factor in their activity [58]. The CMC values for actinomycin D (1.0×10^{-4}), penicillin G (2.5×10^{-1}) streptomycin (9.0×10^{-5}) and sodium fusidate (3.6×10^{-1}) are reported in literature [5].

2.3. Drugs acting on autonomic nervous system

The autonomic nervous system (ANS) is also called the visceral, vegetative, or involuntary nervous system, widely distributed through out the body and regulates autonomic functions, which occur without conscious control. On the efferent side it consists of sympathetic and parasympathetic as major divisions. There are varieties of categories of drugs that alter the synthesis, metabolism, blockade of transport, attachment at post synaptic receptor, prevention of release, mimicry of endogenous neurotransmitters [59]. Many Drugs acting on ANS are reported to be surface active [60-63].

Surface activity of series of β-blockers has been reported [64]. In another study of β-blockers, properties like effect on myocardial conduction velocity and local anesthesia have been shown to correlate with surface activity and hydrophobicity [61]. Penetration of β-blockers, propranolol, oxprenolol, metaprolol, and nadolol, into model membrane of dimyristoylphosphatidylcholine monolayers using a film balance indicated that incorporation of these drugs is proportional to their lipophilicities [65]. Surface activity of atenolol, metoprolol and propronlol was reported by Nagappa et al [66].

Choline like compounds has been reported to be surface active [67-71]. In case of derivatives of papaverine, spasmolytic activity is shown to be proportional to surface activity of these compounds [72]. Curare like activity is reported in case of a series of polymethyelene-bis-trimethyl ammonium compounds which have been shown to exhibit surface activity [73]. A micellar pattern of association was established for compounds of series of anti acetylcholine drugs based on the diphenylmethane nucleus. The drugs investigated included adiphenine hydrochloride, piperidolate hydrochloride, benztropine mesylate, orphenadrine hydrochloride, chlorphenoxamine hydrochloride, lachesine hydrochloride, poldine methylsulphate, pipenzolate bromide, clidinium bromide, benzilonium bromide and ambutonium bromide for which CMC and aggregation numbers have been determined [74].

The self-association of the anti-acetylcholine drugs, propantheline bromide, methantheline bromide and methixene hydrochloride in aqueous solution, has been examined by surface tension, light scattering and conductometric methods and apparent CMC were reported [75]. The CMC values for adiphenine-HCl (8.2×10^{-2}), chlorphenoxamine-HCl (4.5×10^{-2}) orphenadrine-HCl (9.6×10^{-2}) and penthianate methobromide (2.2×10^{-1}) in molar concentrations are reported in literature [5].

2.4. Antihistamines

Histamine is a hydrophilic molecule distributed in almost all mammalian tissues. The receptors of histamine include H_1, H_2 and H_3 with subtypes. Antihistaminic drugs are mainly used to control allergy (H_1 antagonists) and gastric acid secretions (H_2 antagonists). Surface activity of antihistamines and their interactions with dipalmitoyl lecithin monolayers have also been reported [76, 77].

Olopatadine, an effective topical ocular human conjunctival mast cell stabilizer/ antihistaminic antiallergic drug, was evaluated and compared to selected classical antihistamines for their interaction with model and natural membranes to ascertain potential functional consequences of such interactions. Olopatadine's restricted interaction with membrane phospholipids limits the degree of membrane perturbation and release of intracellular constituents, including histamine and hemoglobin, which is believed to contribute to olopatadine's topical ocular comfort and patient acceptance [78]. Ketotifen, an antiallergic drug with H_1 antihistaminic and mast-cell stabilizing properties, is a basic amphiphilic drug. Enhanced skin permeation of cationic drug ketotifen through excised guinea pig dorsal skin by surfactants with different electric charges was studied. Analysis of the retention of ketotifen in the skin suggested that sodium dodecylsulfate-induced increase in the transfer of hydrophilic ketotifen to the skin is the main reason for the marked increase in skin permeation [79,80]. The CMC values of bromodiphenylhydramine hydrochloride (5.4×10^{-2}), chlorcyclazine hydrochloride (1.27×10^{-1}), diphehydramine hydrochloride (9.0×10^{-2}), diphenylpyraline hydrochloride (4.0×10^{-2}), thenyldiamine hydrochloride (1.0×10^{-1}) and tripelnnamine hydrochloride (1.2×10^{-1}) are reported in molar concentrations in literature [5, 81].

2.5. Drugs affecting renal and cardiovascular function

Drugs affecting renal and cardiovascular function include diuretics, drugs acting on renin angiotensin system, drugs used for the treatment of myocardial ischemia, antihypertensive agents, drugs used for the treatments of heart failure, antiarrythmic agents and drugs for treatments of hypercholestremia and dyslipdiemia. These drugs belong to wide variety of chemical entities comprising hydrophilic and hydrophobic group and hence likely to be surface active.

Surface activity and critical micelle concentrations are reported for diuretic drugs, furosemide and triamterene [82]. The drugs generate a liquid membrane on a supporting membrane. Transport of chloride, sodium, and potassium ions through the liquid membranes generated by the drugs was studied [82]. Transport of furosemide through the liquid membrane generated by diphenylhydantoin, in series with a supporting membrane, has been studied. The data indicate that the reported reduced response to furosemide in the presence of diphenylhydantoin may be due to the impediment of the transport of furosemide by the liquid membrane generated by diphenylhydantoin [83]. Reserpine was shown to be surface active and generate a liquid membrane. Transport of adrenaline, noradrenaline, dopamine, 5-hydroxytryptamine, glutamic acid, and gamma-aminobutyric acid in the presence of the reserpine liquid membrane was studied [84].

The lipid bilayer serves as a structural and dynamic matrix into which numerous functional proteins are embedded. A lipophilic drug may produce some of its effects by perturbing the lipid bilayer. This event, which is manifested as a change in membrane dynamics, may modulate the function of one or more membrane-associated proteins. The elucidation of drug interactions with membranes and their effects on membrane organization is a recent approach to molecular-level mechanisms of action (or side effects) of the drugs [85].

Besides its action on angiotensin AT_1 and AT_2 receptors, it is known that angiotensin II also interacts with the lipid bilayer of biological membranes. Losartan an AT1 antagonist is known for its interactions with the phospholipids bilayer component of membranes. Losartan binding sight appears to be located within the transmembrane regions III, IV, V, VI and VII domains of the AT_1 G protien Coupled Receptor of the (AT I) receptor [86].

The role of surface activity in the mechanism of action of calcium channel blocker (CCB), amlodipine, in the treatment of atherosclerosis has been investigated. Atherogenic low-density lipoproteins (LDL) are characterized by elevations in cholesterol content and increased electro negativity. These are the factors that contribute to aggregation and foam cell formation. A study was designed to test the effect of the positively charged CCB amlodipine on the aggregation properties of oxidized LDL. By contrast, drugs lacking a formal positive charge, including CCBs (felodipine, nifedipine, diltiazem, verapamil) and an angiotensin-converting enzyme-inhibitor (ramiprilate) had no effect on the column binding of the modified electronegative lipids [87].

Enantiomers separations in bulk solutions are possible by liquid membranes. Krieg et al., reported enrichment of chlorthalidone enantiomers by an aqueous bulk liquid membrane containing b-cyclodextrin [88]. Carvedilol is a multiple action antihypertensive drug that has been shown to protect cell membranes from lipid peroxidative damages. Studies on the physical and structural effects of Carvedilol on lipid bilayers by fluorescence techniques, differential scanning calorimetry have demonstrated carvedilol's high affinity for lipid bilayer membranes [89].

The partition coefficient, surface activity and membrane fluidizing/disordering effects of CH-103, a beta-adrenergic receptor antagonist, were compared to those of propranolol and practolol as reference compounds [90]. Several structurally similar pyrazine derivatives, tetramethylpyrazinc, triethylpyrazine and tetraethylpyrazine have been found to inhibit plasmalemma-associated biological activities of various tissues, including ion channels and membrane receptors in a given order of potency that increases with increasing bulkiness and hydrophobicity of these drugs [91].

Studies on the role of liquid membrane phenomenon in biological actions of angiotensin converting enzyme (ACE) inhibitors, captopril and lisnopril have confirmed the surface activity of these drugs. Data on the transport of the relevant permeants in presence of the liquid membrane formed by ACE inhibitors indicate that liquid membrane phenomenon is likely to play a significant role in the action of ACE inhibitors [92].

Antiarrythmic drugs, namely quinidine, disopyramide, procainamide and propranolol, have been shown to generate liquid membranes in series with a supporting membrane. Transports across the liquid membrane generated by these drugs indicate that the transport of sodium ions is impeded which is relevant to the antiarrythmic action of all the four structurally dissimilar drugs [93].

The liquid membrane phenomenon in the actions of digitalis glycosides (digitoxin, digoxin and ouabain) has been studied. Formation of liquid membranes, in series with a supporting membrane, by digitalis alone and by digitalis in association with lecithin and cholesterol has been demonstrated. The results obtained on the transport of relevant permeants, viz. sodium, potassium and calcium ions and dopamine, adrenaline, noradrenaline and serotonin, in the presence of the liquid membrane generated by digitalis in association with lecithin and cholesterol indicate that the liquid membrane barrier to transport may have a relevance to the biological actions of digitalis [94].

2.6. Drugs acting on central nervous system

Brain is an assembly of interrelated neural systems that regulate their own and each other's activity in a dynamic complex fashion. The elucidation of the sites and mechanisms of drugs acting on central nervous system (CNS) demand an understanding of cellular and molecular biology of the brain. Although knowledge of anatomy, physiology, and chemistry of the nervous system is far from complete, the acceleration of interdisciplinary research on CNS has led to remarkable progress. Drugs that act upon CNS influence the lives of everyone every day. These agents are invaluable therapeutically because they can produce specific physiological and psychological effects. Drugs that can affect the CNS can selectively relieve pain, reduce fever suppress disordered movement, induce sleep or arousal, reduce the desire to eat, or allay the tendency to vomit [95].

CNS is rich in lipids, and hence drugs acting on CNS are usually hydrophobic with hydrophilic groups embedded in them and are likely to be surface active. A detailed account on surface activity of CNS acting drugs can be found in literature [4, 5, 96]. Surface activity of drugs acting on CNS will be discussed in the following order viz. anesthetics (general and local), antidepressants, antiepileptics, antipsychotics, and minor tranquillizers (hypnotics and sedatives).

2.6.1. General anesthetics

Many clinical and experimental data have revealed, that hydrophobic anesthetic agents, influence membrane lipid bilayer fluidity [97–99] and membrane-associated proteins [100]. Specifically influenced by halothane, enflurane, isoflurane and desflurane are the ion channels, such as GABA, glycine channels, and neuronal background K^+ channels [101–107]. Effect of the volatile anesthetics has also been demonstrated for cardiac and skeletal muscle tissue [108-109], halothane and isoflurane decrease calmodulin and troponin C affinity for Ca^{2+}. Furthermore, it has been found, that anesthetic agents reduce gap junction conductance [110]. Some experiments showed that ethanol and halothane interact in different ways to the clustering ability of human leukocyte antigen (HLA) class I and II molecules involve protein or protein–lipid interaction [100]. Interestingly, the adhesion of of bacteria to epithelial cells decreases when cells are exposed to halothane [111]. The accumulated experimental data support the concept that volatile anesthetics affect cell-surface receptors and the cell membranes in general.

Barbiturates are known to be surface active; the CMC of thiopental, which is used as intravenous anesthetic agent, has been reported to the 7.0×10^{-3} *M* [112].

Monte Carlo computer simulation of shows that the gel-to-fluid transition of the lipid membrane, manifested in the formation of dynamically coexisting domains of gel and fluid lipids, is strongly influenced by the presence of anesthetics. The calculations reveal, that anesthetics have a high affinity to the fluctuating domain interfaces that are dominated by kink-like lipid-chain conformations. This leads to formation of more interfaces and to a locally high concentration of anesthetics in the interfacial regions, which is much larger than the concentration in the membrane. Important membrane components like cholesterol, which also has been shown to be interfacial active, are found to decrease the absorption of anesthetics and to squeeze out anesthetics from the interfaces [113].

2.6.2. Local anesthetics

Some local anesthetics (LA) behave as a surfactant because of the amphiphilic nature of anesthetic molecules, especially; tetracaine HCl and dibucaine HCl have the CMC equal to 128 and 79 mMol kg^{-1} respectively [114-115]. The strong correlation between hydrophobicity and the affinity for human α -acid glycoprotein is in accord with the two series of local anesthetics, the linear alkyl amino homologs of lidocaine and the piperidine ring-containing homologs of mepivacaine. A linear relationship between dissociation and the octanol buffer partition coefficient of the neutral drug species was observed [116].

A correlation between LA activity and surface tension in a series of esters has been reported [117]. Interaction of LA drugs with lipids has been widely investigated [118-121]. For lipids extracted from nervous tissues, penetration of the drugs into the lipid monolayers has been shown to correlate with their nerve blocking potency. In another study [122], interaction of a series of LA with monolayers of dipalmitoyl lecithin indicated that the minimum blocking concentration of LA lowered the surface tension of the lecithin /water interface approximately to the same extent.

2.6.3. Antidepressants

Drugs with demonstrated efficacy in a broad range of severe psychiatric disorders have been developed since 1950, leading to the development of subspecialty of psychopharmacology. The treatment of depression relies on a varied group of antidepressant therapeutic agents. The premier agents to be used successfully were tricyclic antidepressants (TCA), which elicit a wide range of nueropharmacological effects in addition to their presumed primary action of inhibiting norepinephrine and, variably, serotonin transport into nerve endings. This results in enhanced concentrations of norepinephrine in the synaptic cleft leading to sustained monoaminergic transmission. Inhibitors of monoamineoxidase (MAO) that increase the brain concentrations of monoamines have also been used in treatment of depression [123].

Surface activity of a series of TCA related to imipramine has been reported [124]. A correlation between surface activity of these drugs and their toxicity to chang liver cultures [125,126] and human liver has also been mentioned [127]. It was observed that depressant drugs accumulate at the air - aqueous interface and their pharmacological actions correlate with the surface activity of these drugs [128]. The CMC values for amitryptalline HCl (3.6×10^{-2}), butryptaline HCl (4.2×10^{-2}), cloimipramine HCl (9.0×10^{-5}) desimipramine HCl

(4.9×10^{-2}) imipramine HCl (4.7×10^{-2}) and nortryptaline HCl (2.3×10^{-2}) in molar concentrations are reported in literature [5, 112, 129].

The surface-active drugs, chlorpromazine and imipramine and synthetic surfactant triton X have been tested on large unilamellar vesicles composed of phosphatidylcholine, sphingomyelin, and cholesterol in different proportions. All there molecules behave qualitatively in a similar way, irrespective of bilayer composition: they induce leakage at concentrations well below their CMC and solubilization near the CMC [130].

The suppression of imipramine HCl (IMP)-induced hemolysis by native cyclodextrins (CD) is quantitatively correlated with the surface tension of the solution. The modified beta-CDs are more or less adsorbed on to the air-water interface and occupy larger areas than the wider rim of β-CD. The surface tension data at low concentrations of CD in the presence of 3mM IMP allow one to estimate the 1:1 binding constants of IMP with CDs [131]. A study was aimed at whether or not the antidepressant zimelidine, which is an amphiphilic cationic compound, can induce generalized lipidosis in rats. The results show that zimelidine induces generalized lipidosis in rats although of mild degree when compared with some other amphiphilic cationic drugs [132].

Interaction of tricyclic drug analogs with synaptic plasma membranes were undertaken to study structure-mechanism relationships in inhibition of neuronal Na^+/K^+-ATPase activity. Na^+/K^+-ATPase IC_{50} values decrease linearly with increasing octanol/water partition coefficients (log-log plot) for a series of dimethylethylamine-containing drugs (i.e., chlorpromazine, amitriptyline, imipramine, doxepin, and diphenhydramine), emphasizing the role of surface activity in inhibition [133].

Octanol and dodecane partition coefficients, surface activity and adsorbability to activated charcoal were determined for six tricyclic psychotropic drugs with N-dimethylalkyl side chains. Surface activity correlated well with the partition coefficients, and all drugs obeyed the langmuir adsorption isotherm. A correlation between the reciprocal of the death time of gold fish exposed to drugs and partition coefficients was observed [134].

Amitryptaline is a TCA antidepressant belonging to the first generation of antidepressant drugs, which suffer from several drawbacks, such as anticholinergic, cardiovascular, and antiarrhythmic side effects. The presence of the alkyl amine side chain on TCA molecules confers on them a "surfactant-like" behavior, which may be manifested in the formation of aggregates in aqueous solution. With an aim to enhance bioavailability and reduction in toxicity, conductivity and static fluorescence measurements have been carried out for mixed micelles of dodecyltrimethylammonium bromide and amitryptaline [135].

2.6.4. Hypnotics, sedative and antianxiety agents

A wide variety of agents have the capacity to depress the function of the CNS such that calming or drowsiness (sedation) is produced. The CNS depressants that are used as hypnotics, sedatives and antianxiety agents include benzodaizepines (BZP), barbiturates and as well as sedative- hypnotic agents of diverse chemical structure (paraldehyde, chloral hydrate)[136].

Butylbarbituric acid [137], pentobarbital, quinalbarbital [138] and monoalkyl, dialkyl barbituric acids [139] have been reported to be surface active. A linear correlation between protein binding properties and surface activity with apparent partition coefficient has been observed [140]. Adsorption free energy of barbiturates with phospholipid monolayers has been shown [141] to correlate with their nerve blocking potencies. Changes in ion channels and membrane bound enzymes as a result of drug –lipid interactions have been indicated [142-143] to be involved in the mechanisms of action of barbiturates.

Formation of liquid membranes, in series with a supporting membrane, by barbiturates alone and by barbiturates in association with lecithin and cholesterol has been demonstrated. Data on the transport of relevant permeants, viz. γ-aminobutyric acid, glycine, aspartic acid, serotonin and noradrenaline, in the presence of the liquid membrane generated by barbiturates in association with lecithin and cholesterol have been obtained. The data indicate that modification in the transport of these permeants due to the liquid membrane barrier may have a bearing on the mode of action of barbiturates [144].

Surface activity of benzodaizepines (BZP) was described by Attwood et al. [145]. The liquid membrane phenomenon in BZP and transport of glycine, GABA, noradrenaline, dopamine and serotonin in the presence of the liquid membranes generated by the BZP in association with lecithin and cholesterol has been studied. The data indicate that modification in permeability in the presence of the liquid membranes is likely to make a significant contribution to several biological actions of the BZP [146]. The CMC of diazepam was found to be $1.0 \times 10^{-4} M$ [146,147].

Bouhleal et al. synthesized and studied amphiphilic properties of glycosyl-1, 4-benzodiazepin-2, 5-diones. The structural variations of the sugar group allowed comparison of amphiphilic data such as CMC, surface tension value and water solubility [148].

2.6.5. *Antiepileptic drugs*

Established mechanisms of action of antisiezure drugs fall into three major categories, viz., drugs stabilizing the conformation of inactivated sodium channel, enhanced gama aminobutyric acid (GABA) synaptic transmission and reduction of Ca^{2+} current by acting on T type Ca^{2+} channels. Phenobarbital was the first synthetic organic agent recognized as having antisiezure activity [149]. Nagappa et al. and Chidambaram and Burgess reported Surface activity of phenobarbital [144, 150].

Voltage-gated sodium channels are the molecular targets for anticonvulsant compounds including phenytoin, carbamazepine, and lamotrigine. Each of these compounds blocks sodium channels with striking voltage dependence, having little effect on resting channels, but exhibiting strong block when the channel is inactivated by prolonged depolarization. A fundamental question concerning the mechanism of action of these drugs is whether they act at a common receptor site on sodium channels in exerting their diverse pharmacological effects or at distinct receptor sites on sodium channels or other targets whose different properties lead to the different pharmacological effects [151]. Three structurally dissimilar antiepileptic drugs, namely, phenytion, carbamezapine and sodium valproate are known to stablize biological membranes after interacting with them [152,153]. They are known to contain both hydrophilic and hydrophobic moieties in their structure [154].

2.6.6. Antipsychotic drugs

Clinically effective antipsychotic agents include tricyclic phenothiazines, thixanthenes, and dibenzepines,as well as butyrophenones and congeners, other hetrocyclics, and experimental benzamides. Virtually all of these drugs block D_2 –dopamine receptors and reduce dopamine nuerotranamission in forebrain; some also interact with D_1-and D_4-dopaminergic, 5-HT_{2A}- and5 -HT_{2c}- serotonergic, and α-adrenergic receptors. Antipsychotic drugs are relatively lipophilic, mainly metabolized by hepatic oxidative mechanisms [155].

Phenothiazines are known to reduce membrane permeability at very low concentrations [156]. Ability of phenothiazines to reduce water uptake by frog muscle [157, 158], inhibition of erythrocyte hemolysis [159-162], inhibition of acetylcholine release [163], inhibition of endogenous amines in various tissues [164-168], inhibition of glycine uptake by brain slices [169], etc., are several examples where membrane permeability is shown to be altered by these drugs.

It is also known [156] that not only phenothiazines but tranquillizers also, irrespective of their chemical nature, lower the surface tension of Ringer solution in close correlation to their clinical potency. It is not surprising, therefore, that possibly adsorption of all the phenothiazines on to tissues cells may be explained by the physical chemistry involved in air-water adsorption. Commenting on the mechanism of alteration of membrane permeability, it is argued [156] that tranquillizers form virtually 'monomolecular films' around cell membrane and reduce trans-membrane permeability to solutes.

Interaction of the tranquillizers with insoluble monolayers of lipids, steric acid, etc., constitutes another proof of their surface activity. Chlorpromazine sulfoxide and trifluoroperazine have been shown [170, 171] to interact with lipid monolayers. Interaction of these drugs with lipids as measured by increase in surface pressure has been shown [170, 171] to correlate with their biological activity. Interaction of orphenadrine hydrochloride, chlorpromazine hydrochloride and reserpine with monomolecular films of cholesterol, phosphoglycerides, sphingomyelins and cerebrosides is also documented [172]. The effect of UV radiation on interaction of a series of phenothiazines with dipalymitoyl lecithin films indicated [173, 174] that the ability of these drugs to interact with lecithin monolayer might be a measure of their *in vivo* membrane penetrating and phototoxic properties.

Structural variation in phenothiazines has been shown [175, 176] to alter surface activity, which is evident from the change of CMC, e.g., promazine requires six times higher amount of drug to produce the same surface tension as chlorpromazine below CMC, and 27 times much higher concentration as compared to trifluoropromazine [156]. This observation hints at the possibility that variation in biological activity can be expressed in the form of altered surface activity, which indicates that surface activity should play a significant role in the mechanism of action of these drugs.

Since phenothiazine-nucleotide interactions are mentioned [177] to be important for its action, the fact [177] that "chlorpromazine can from complexes with adenosine triphosphate and di-and monophosphate having surface tensions lower than those of the drug alone" appears interesting. It is further indicated [178] that orientation of phenothiazines at the air/water interface may reflect qualitative and quantitative differences in their pharmacological action. In the case of phenothiazines, thioxanthene, dibenzocycloheptadiene and dibenzazepines, colloidal association and surface activity were found [179] to be dependent on the chemical structure of the drug. Direct action of phenothiazine derivative on cat heart causing fall in blood pressure is probably [180] because of the surface activity of the compound.

Phenothiazines aggregate in a micelle-like manner, depending on number of Carbon in side chain, N being the order of 6-15 [181-190]. The partition coefficients of phenothiazine drugs (trifluoperazine, triflupromazine, chlorpromazine and promazine) between phosphatidylcholine small unilamellar vesicles and water were determined. Partition coefficients of drugs between lipid bilayer vesicles (liposomes) and water provide fundamental information relating to the drug interactions with biomembranes [191]. The interaction of the tertiary amine drugs chlorpromazine and dibucaine in their cationic form with carboxyl groups at the membrane surface is studied at concentrations relevant to anesthesia. They are shown to determine the drug influence on carboxyl groups at the membrane surface, independently of aqueous concentrations [192]. The effect of trifluoroperazine on the sarcoplasmic reticulum membrane indicated that trifluoroperazine interferes with calcium transport *in situ*, as well as with the role of sarcoplasmic reticulum in contractile activation, which is attributed to its surface activity [193].

Haloperidol is known to form a monolayer on water/air or water/lipid interfaces at very low concentrations [194]. It has been further commented that all neuroleptics act like detergent or soaps, i.e., they are powerful surface tension lowering agents. A striking correlation between neuroleptic potency and surface tension lowering activity has been indicated. Thus, formation of monolayers on biological structures has been suggested to be a mechanism of neruroleptic action [194].

The CMC values for chlorpromazine (1.9×10^{-2}), promazine (3.6×10^{-2}) promethazine (4.4×10^{-2}) thioridazine (5.9×10^{-3}) trifluoperazine (4.2×10^{-5}) trifluopromazine (4.5×10^{-3}) flupenthixol (8.5×10^{-3}) in molar concentrations are reported in literature [5,195-197].

2.7. Miscellaneous

Surface properties of some natural and synthetic polypeptides, steroids, prostaglandins, vitamins, toxins, proton pump inhibitors and anticancer drugs were investigated at the air-water interface. The interaction of these substances with lipid monolayers was studied. The experimental data for monolayers were compared with action of these substances at the biological membranes. It was shown that aggregation and/or conformational states of above substances at the air-water interface mainly depend on the mechanical contacts between molecules. The same aggregation and/or conformational states of the substances under study were observed in mixed substance-lipid monolayers.

2.7.1. Surface activity of proteins

Pulmonary surfactant maintains low minimum surface tension at the alveolar air-liquid interface to prevent alveolar collapse at the end of inspiration-expiration cycle. It is composed of approximately 90% by mass of lipids and 5% of surfactant proteins A, B, C, and D [198]. It is believed that surfactant protein (SP) B interacts as a covalently linked dimer with surfactant phospholipids to enable phospholipids monolayers to form at the alveolar fluid-airway interface by accelerating the rate of phospholipid adsorption. A recent report speculates that SP-B and SP-C may, *in vitro*; together contribute to pore formation in lipid bilayers [199].

Microorganisms produce wide variety of surface-active agents. These bioemulsifiers can be classified into low-molecular weight molecules that lower the interfacial tension effectively and high molecular weight polymers that bind tightly to surfaces. These surfactants produced by wide variety of microorganisms have very different chemical structures and surface properties. Several bioemulsifiers have antibacterial and antifungal activity [200]. Recent studies have reported the introduction of a range of new chemical and biochemical functionalities into the structures of amphiphilic molecules. Assemblies spontaneously formed by these amphiphiles are in many cases highly complex and possess properties not found in systems formed from amphiphiles with simpler structures. In particular, the incorporation of peptides and oligopeptides into the hydrophilic domains of amphiphiles has led to new classes of surfactants that self-assemble into structures that mimic a variety of the functions of natural materials including organic scaffolds of bone; inhibitors of proteins involved in viral infection; chiral polymeric amphiphiles; materials that promote adhesion of cells to surfaces. Amphiphiles functionalized with a range of carbohydrates have also been reported. These amphiphiles assemble into aggregate morphologies that depend strongly on the stereochemistry of the carbohydrate. These assemblies offer the basis of promising approaches for the design of polyvalent potent carbohydrate-based drugs [201].

Alzheimer's disease is defined in part by the intraneuronal accumulation of filaments comprised of the microtubule- associated protein tau *in vitro*, fibrillization of recombinant tau can be induced, by treatment with various agents, including phosphotransferases, polyanionic compounds, and fatty acids. Nonetheless, the mechanism by which fatty acids induce tau fibrillization is unknown. Fatty acids resemble detergents in having hydrophobic alkyl chains and charged (anionic) head groups. Above their critical micelle concentrations (CMCs) in aqueous solution, fatty acids form micelles in which their hydrophobic moieties are sequestered, and their charged head groups are exposed to solvent. These data suggest that anionic surfaces presented as micelles or vesicles can serve to nucleate tau fibrillization [202].

Wimley and White proposed hydrophobicity scale experimentally determined for proteins at membrane interfaces [203]. A quantitative description of the coupling of structure formation to partitioning, which may provide a basis for understanding membrane protein folding and insertion, requires an appropriate free energy scale for partitioning. A complete interfacial hydrophobicity scale that includes the contribution of the peptide bond was therefore determined from the partitioning of two series of small model peptides into the interfaces of neutral (zwitterionic) phospholipid membranes. The partitioning of membrane-active oligopeptides into membrane interfaces promotes the formation of secondary structure.

Aromatic residues are found to be especially favoured at the interface while charged residues and the peptide bonds are disfavoured [204].

The knowledge of the membrane binding properties of the proteins involved in coagulation processes is essential to our understanding of blood coagulation. Marie-France et al. reported the involvement of electrostatic and hydrophobic interactions in factor Va binding to membranes containing acidic phospholipids [204].

Class I and class II hydrophobins are small secreted fungal proteins that play a role in a broad range of processes in the growth and development of filamentous fungi. For instance, they are involved in the formation of aerial structures and in the attachment of hyphae to hydrophobic surfaces. The mechanisms by which hydrophobins fulfill these functions are based on their property to self-assemble at hydrophilic-hydrophobic interfaces into a 10 nm-thin highly amphipathic film. Complementation studies have shown that class I hydrophobins belong to a closely related group of morphogenetic proteins, but that they have evolved to function at specific interfaces. Recent evidence indicates that hydrophobins do not only function by self-assembly. Monomeric hydrophobin has been implicated in cell-wall assembly, but the underlying mechanism is not yet clear. In addition, hydrophobin monomers could act as toxins and elicitors [205].

Characterization of unique amphipathic antimicrobial peptides from venom of the African scorpion *Pandinus imperator* has been carried out. The peptides, designated pandinin 1 and 2, are a helical polycationic peptides, with pandinin 1 belonging to the group of antibacterial peptides previously described from scorpions, frogs and insects, and pandinin 2 to the group of short magainin-type helical peptides from frogs. Both peptides demonstrated high antimicrobial activity against a range of gram positive bacteria [206].

Several bacterial pore-forming toxins, which are surface active in nature, have been reported to utilize lipid rafts to intoxicate cells [207]. Aerolysin and Clostridium septicum alpha toxin bind to glycosyl phosphatidyl inositol-anchored proteins in lipid rafts [208] and C. perfringens epsilon toxin and perfringolysin bind to cholesterol in lipid rafts [209, 210]. It has been proposed that lipid rafts serve as concentrating platforms to promote pore formation of these toxins that form oligomers. The findings that beta toxin preferentially binds to lipid rafts and oligomerizes, and that the characteristics of beta toxin resemble those obtained by these pore-forming toxins suggest that the biological activities of beta toxin depend on the proposed function of lipid rafts in the biological membrane [211].

The association behavior of insulin molecules is a complex phenomenon. Till now the molecular mechanism and the structural basis have not been clearly illustrated. Despentapeptide insulin (DPI) is widely studied as a representative of monomer insulin derivatives. Based on the structural comparisons and analyses of 2Zn insulin, DPI and other insulin derivatives, it was suggested that [212] the binding interaction with its receptor molecule should take place mainly on an amphipathic surface of the insulin molecule. When insulin molecule comes close to its receptor, the C-terminus of B-chain will move away from its previous position. So the hydrophobic surface covered by it will be exposed. The result of B29-A1 linked insulin [213] also showed that the linkage of B29 and A1 limited the movement of C-terminus of B-chain, and the hydrophobic surface could not be exposed. This caused the loss of potency [213].

Haematopoietic cells have long been defined as round, nonpolar cells that show uniform distribution of cell surface-associated molecules. Chemical cross-linking and fluorescence resonance energy transfer methods have allowed the visualization of certain glycosyl phosphatidyl inositol-anchored surface-active proteins in lipid rafts. The lipid micro domain resident surface-active proteins, flotillins-reggies, form pre assembled platforms in haematopoietic cells [214].

α-Crystallin, a surface-active major structural protein of the eye lens plays a prominent role in the maintenance of the eye lens transparency and its refractive properties. α- crystallin prevents the aggregation of non-native proteins by providing appropriately placed hydrophobic surfaces and a structural transition above 30°C enhances the protective ability by increasing or reorganizing these hydrophobic surfaces. This suggests a sequential exposure of hydrophobic target binding sites as a function of temperature, with low temperatures binding sites being subset of the total number of sites available at elevated temperatures. This process of enhanced exposure of hydrophobic surfaces at elevated temperatures involves conformational changes in α-crystallin. The non-specificity of α-crystallin, due to common target protein binding surfaces, may be advantageous for its role in binding diverse aggregation-prone molecules in the lens and keep them in solution [215].

Peptide antibiotics act by increasing membrane permeability. The formation of channels is one of the molecular mechanisms by which these peptides increase membrane permeability. Gramicidin A and alamethicin channels have been extensively studied. The channel formed by gramicidin A β-helix allows water and ion passage [216]. Experimentally, antimicrobial peptides have been shown to aggregarate in the membrane [217, 218], as well as in aqueous [217,219-221] or aqueous-organic phase [222]. The detergent-like effect of peptide antibiotics has been widely documented [223-225].

Diphtheria toxin [226], botulinum neurotoxin [227] and tetanus toxin [228] are proteins that are similar in origin and macrostructure. It has been shown that all the three toxins from channels in bilayer lipid membranes (BLM), and that the amino terminus of the heavy chain in the structure of these toxins possesses a channel-forming domain. It was discovered that channels could be formed only in the presence of pH gradient across the BLM, the toxin fragment being present on the acidic side of the membrane, and that reversing the pH gradient effectively blocked channel formation.

Because of the similarity of gross structure of both insulin and vasopressin with the clostridial toxins, the possibility of channel formation by vasopressin and insulin in the lipid bilayers cannot be ruled out and, hence, merits investigation. Transport studies on liquid membrane bilayers generated by a lecithin-cholesterol mixture in the presence of insulin/vasopressin have been carried out [229] with this object in view. Experiments on hydraulic permeability and on solute permeability of relevant permeants through lecithin-cholesterol liquid membrane bilayers in the presence of insulin/vasopressin on a supporting membrane have also been conducted. The data show trends comparable to those reported in the studies demonstrating formation of channels in bilayer lipid membranes by toxins [230-232].

2.7.2. Anticancer Drugs

It has been shown by Ligo et al. that the 1-hexycarbamoyl -5-flurouracil synthesized by Ozaki et al. is more active against various tumors in mice and less toxic to host animals than its parent drug 5-flurouracil. Ligo et al., have tested the activities of these drugs on Lewis lung carcinoma and B16 melanoma [233,234]. Srivatsava et al. studied liquid membrane phenomena in 5-fluorouracil and its two derivatives: 1-(2-tetrahydrofuryl)-5-fluorouracil and 1-hexyl-carbamoyl l-5-fluorouracil [235]. It is evident from the CMC values that and 1-hexyl-carbamoyl-5-fluorouracil is more hydrophobic and more surface active than its parent compound 5-fluorouracil [235].

The interaction of a number of positively charged anti-tumor drugs with cardiolipin-containing model membranes have been investigated using 31P nuclear magnetic resonance, differential scanning calorimetry and monolayer techniques. Measurements of surface pressure and surface potential of cardiolipin monolayers at the air/water interface as well as conformational analysis of the various drug-cardiolipin recombinants showed that the ellipticines are deeply embedded in the acyl chain region of the bilayer, while the anthracyclines and ethidium bromide are preferentially localized in the interface. All drugs share an important electrostatic interaction with the negatively charged phosphates of cardiolipin [236].

Synthesis of a novel series of amphiphilic glycosylated spin-traps derived from alpha-phenyl-N-tert-butyl nitrone (PBN) and an initial characterization of their anti-caspase-3 activity was reported. Preliminary investigation of their anti-apoptosis effect showed that they dramatically inhibit the activity of caspase-3 in cultured neuronal cells following induction of apoptosis by hydrogen peroxide [237].

The drug cisplatin has broad antineoplastic activity against advanced testicular and ovarian cancers, epithelial malignancies, cancers of the head, neck, bladder, oesophagus and lungs. Peripheral neurotoxicity, ototoxicity and nephrotoxicity are its major side effects. The nonspecific action of this drug on the lipid bilayer architecture of membranes has been studied by following the effects produced on the electrical characteristics of model planar bilayer lipid membranes (BLM). The results confirm that the drug has a strong surface interaction with the zwitterionic polar head groups of the amphipathic phospholipids constituting the BLM. The permeability characteristics of cisplatin through the hydrophobic core are limited. Cisplatin does not fluidize the membrane sufficiently to cause its breakdown but creates small ion conducting defects on the membrane bilayer resulting in a marginal increase in ion conductivity. These results indicate that cisplatin exhibits a non-specific action on the lipid bilayer component of the membrane, which might be partly responsible for its neurotoxic side effects [238].

The temperature and concentration-induced effects of tamoxifen on dipalmitoyl phosphatidylcholine (DPPC) model membranes were investigated by the Fourier Transform-infrared spectroscopic technique [239]. The effect of two anti-cancer agents, vinblastine sulphate and vincristine sulphate on the gel-liquid crystal transition of fully hydrated DPPC has been studied by differential scanning calorimetry (DSC). DSC diagrams were established for various mixtures of DPPC with agent and a fixed (50%) amount of water. It is concluded that, vinblastine sulphate perturbs the hydrated DPPC structure more strongly than vincristine sulphate. This conclusion confirms the idea proposed by Tar-Minassian-Saraga et al. that these anti-mitotic drugs might also affect the functioning of cell membranes [240].

2.7.3. Steroids.

Steroids are known to be surface active [241, 242]. The CMC values for ethinyl estradiol (2.7×10^{-7}), progesterone (9.0×10^{-5}) testosterone propionate (3.87×10^{-6}) in molar concentrations are documented literature [243, 244].

Bile acids are amphipathic end products of cholesterol metabolism. Cholesterol excretion is mediated by bile acids. The water-soluble amphipathic molecules are formed from cholesterol in the hepatocyte. In addition to their role in cholesterol homeostasis, bile acids also are functional detergents that induce bile flow and transport lipids as mixed micelles in the bilary tract and small intestine. Bilary secretion also provides an excretory route for lipophilic steroids and drug metabolites. Bile also has a high concentration of phospholipids, which consist mostly of phosphatidylcholine (PC) and which form mixed micelles with bile acids. These mixed micelles contain amphipathic microdomains that can solubilize cholesterol. Mixed micelle formation also lowers the monomeric activity of bile acids and prevents their destroying the apical membrane of the bilary epithelial cells. IgA, an immunoglobulin, and mucus are secreted into bile, where their role is to prevent bacterial growth and adhesion. Finally, bile contains tocopherol, which may prevent oxidative damage to the bilary and small intestinal epithelium [245].

Spivak et al [246] have developed a simple biologically non-invasive method for determining the critical micellar concentration of bile salts using pure naturally occurring bilirubin IX alpha monoglucuronide, an important bile pigment present in virtually all mammalian biles. The CMC for sodium taurochenodeoxycholate is between 2.5 and 3.0 mM.

Cationic steroid antibiotics have been developed that display broad-spectrum antibacterial activity. These compounds are comprised of steroids appended with amine groups arranged to yield facially amphiphilic morphology. These antibiotics are highly bactericidal, while related compounds effectively permeable to the outer membranes of gram-negative bacteria sensitizing these organisms to hydrophobic antibiotics. Cationic steroid antibiotics exhibit various levels of eukaryote vs. prokaryote cell selectivity, and cell selectivity can be increased via charge recognition of prokaryotic cells. Studies of the mechanism of action of these antibiotics suggest that they share mechanistic aspects with cationic peptide antibiotics [247].

The hepatoprotective effect of β muricholate a surface-active steroid, against cholestasis induced by hydrophobic steroids was studied in rats. The hydrophilic-hydrophobic balance of β muricholate was estimated by surface tension measurements. β muricholate appeared to have a weak affinity for a hydrophobic interface. It generated a lower surface pressure than ursodeoxycholate and much more lower than chenodeoxycholate. The low surface activity of β muricholate could account for its non-toxicity and protective action towards hepatocyte membranes [248].

The design and evaluation of a novel class of DNA delivery agents based on steroid–polyamine conjugates bearing a flexible linker are reported. The hydrophobic regions are based on steroids, i.e. chlolestane and lithocholic acid motifs. The gene transfection activity of the steroid–polyamine conjugates is influenced by the polyamine chain length and steroid structure. Molecular modeling of the relevant amphiphilic molecules revealed low-energy structures in which the polyamine chains are folded rather than stretched. This highlights the significant effect of space filling, i.e. the shape and orientation of the hydrophilic and hydrophobic regions, upon the efficiency of gene transfection [249].

2.7.4. Prostaglandins

The prostaglandins are among the most prevalent autacoids and have been detected in almost every tissue and body fluid; they produce, in minute amounts, a remarkably broad spectrum of effects that embrace practically every biological function. No other autacoids show more numerous and diverse effects than do prostaglandins. The CMC values for prostaglandin E_1 ($3.85\times10^{-8}M$) and prostaglandin $F_{2\alpha}$ ($1.93\times10^{-8}M$) are reported in literature [250]. Transport through liquid membranes generated by lecithin, cholesterol and lecithin-cholesterol mixtures has been studied in the presence of prostaglandins. The data indicate that prostaglandins in association with cholesterol may be responsible for the aqueous pores present in the lipid bilayers controlling passive transport through biomembranes. The data further indicate that the presence of cholesterol in each of the two constituent monolayers of the lipid bilayer is essential for pore formation by prostaglandins [251].

Colacicco and Basu have reported the correlations between molecular structure and surface function of six prostaglandins in a model membrane system. Using spread films at the air/water interface, they determined surface pressure and surface potential of PGs A_1, A_2, E_1, E_2, $F_{1\alpha}$ and $F_{2\alpha}$. All the prostaglandins formed films with low pressure (0 to 9 dynes/cm) and relatively low surface potentials (10 to 250 mV) [252]. Roseman and Yalkowsky have reported the physicochemical properties of prostaglandin $F_{2\alpha}$ (trimethamine salt): solubility behavior, surface properties, and ionization constants [253].

Monomolecular film compression-relaxation behavior was examined for select dinoprost C-15 alkyl esters. Higher homologs of the series such as palmitate and decanoate esters yielded stable expanded monolayers that exhibited minimal relaxation of surface pressure during noncompression. Their limiting molecular areas were consistent with a Hirschfelder model projection in which the prostaglandin moiety assumes a horizontal orientation at the interface with its alkyl ester chain oriented vertical to the surface plane. Shorter chain homologs such as hexanoate, valerate, butyrate, propionate, and acetate also formed expanded monolayers but exhibited increased instability with decreased alkyl chain length, as reflected in their lower surface pressure development during compression and significant relaxation of pressure during noncompression. Such instability can be tied to their increased solubility in the sub-phase solution and higher desorption rate from the interface [254].

2.7.5. Vitamins

Vitamins are organic nutrients that are required in small quantities for a variety of biochemical functions and which, generally, cannot be synthesized by the body and must be supplied by the diet. Vitamins are classified into water soluble and fat-soluble vitamins based on their solubility.

Tandon et al. reported the surface-activity of vitamin E (CMC: $5\times10^{-8}M$) [255]. The liquid membrane phenomenon in vitamin E and hydraulic permeability data have been obtained to demonstrate the existence of the liquid membranes in series with a supporting membrane generated by α-tocopherol and also by the lecithin-cholesterol-α-tocopherol mixtures. Data on the transport of oestrogen, progesterone, cystine, methionine, creatinine and sodium, potassium and calcium ions in the presence of the liquid membrane generated by the lecithin-cholesterol-α-tocopherol mixture have been obtained and discussed in the light of the various syndromes caused by vitamin E deficiency [255].

Nagappa et al. reported The CMC of vitamin D_3 ($8\times10^{-9}M$) [256]. Transport through liquid membranes generated by vitamin D_3 in series with a supporting membrane has been studied. The data on the modification in the permeability of cations, glucose and phosphate has been shown to be consisitent with the reported biological actions of vitamin D_3 [256]. 25-Hydroxycholesterol and 25-hydroxy vitamin D_3 increased the permeability of liposomes to Ca^{2+} measured by the arsenazo III encapsulation technique. This effect was sensitive to the lipid composition of the membrane. Changes that decreased the motional freedom of phospholipid acyl chains decreased Ca^{2+} permeability. The highest permeability was observed with the zwitterionic phospholipids, phosphatidylcholine and phosphatidylethanolamine, whereas the acidic phospholipids, phosphatidylinositol and phosphatidyl serine, depressed Ca^{2+} permeability. The effect did not appear to be due to iontophoretic properties of the sterols, and it is suggested that perturbation of the membranes by the polar 25-hydroxyl group may play a role in increasing membrane permeability [257].

Intracellular release of free DNA from the vector complex is one of the critical steps limiting the efficiency of non-viral gene delivery. New cationic amphiphilics made from the natural pro-vitamin, lipoic acid, which reversibly binds and releases DNA depending on the redox state of the lipoate moieties are synthesized. The cationic amphiphilic derivatives of natural non-toxic compound lipoic acid provide a new promising class of synthetic vectors for gene delivery [258].

The role of the surface activity of vitamin A has been studied in the light of the liquid membrane hypothesis of drug action. Transport of relevant amino acids such as serine, threonine, arginine, and histidine and various ions such as calcium, sodium, and potassium in the presence of liquid membranes generated by vitamin A has been studied. The data on the modifications in the permeability of relevant amino acids and ions indicate that the liquid membranes generated by vitamin A may also play a significant role in its physiological action [259]. The effects of up to 20mol% incorporation of all-trans-retinol (vitamin A), retinal (vitamin A aldehyde) and retinoic acid (vitamin A acid) on acyl chain order and dynamics in liquid crystalline dipalmitoyl phosphatidyl choline membranes at pH 7.5 were studied by electron spin resonance. All three retinoids restrict acyl chain motion to a similar extent approaching the center of membrane [260].

2.7.6. Proton pump inhibitor

Omeprazole and lansoprazole, the therapeutically important drugs belonging to proton pump inhibitor category are extensively used in the treatment of gastric ulcers. Transport through liquid membranes generated by these drugs in lecithin cholesterol mixture in series with a supporting membrane has been studied. The data obtained show the formation of liquid membrane in series with the supporting membrane. Transport studies of cations, chloride and bicarbonate ions in the presence liquid membranes generated by omeprazole (CMC: 0.978×10^{-6} *M*) and Lansoprazole (CMC: 0.712×10^{-6} *M*) indicate the modification in the permeability of various permeants [261].

The fact that such a wide variety of drugs are surface active in nature and a correlation between surface activity and biological activity is indicated hints at the possibility that a common mode of action may be operative in the mechanism of action of these drugs. For surface-active substances, reduction of surface tension is accompanied by formation of a surface layer, which is complete at or above CMC. Hence, the surface-active drugs are expected to form a "liquid membrane" interposed between the biological membrane and its relevant permeant. Several factors, e.g., orientation of the surface active drug with respect to biological membrane, active interaction of the drug with biomembrane, nature of interaction within the surface active drugs, presence of double layer around the drug liquid membrane, may influence its permeability characteristics. However, one possibility appears quite obvious, i.e., access of the permeant to the 'receptors' located on the biomembrane will certainly be altered by the presence of a liquid membrane generated by the drugs.

The liquid membrane generated by the drugs would alter transport of various other permeants also in addition to the permeants relevant for the specific biological effect of the drug. This, in turn, indicates existence of multiplicity of biological effects for surface-active drugs. Multiple effects have, in fact, been observed in the case of several surface-active drugs [262, 263]. Fig.1 from Florence's excellent review article [1] clearly depicts interrelation between several categories of surface-active drugs.

It has been indicated [264] "aqueous films formed from solutions of surface-active drugs might be sensitive tools with which to study the interactions of the drugs with ions, especially those of biological importance, and with which to obtain a more quantitative insight into the behaviour of the molecules at surfaces." It has also been commented "surface tension reduction is only a symptom of many physicochemical attributes, and much work still remains to be done before surface activity *per se* can be reliable guide to biological activity in a homologues series." The liquid membrane hypothesis of drug action appears to be a positive step in this direction.

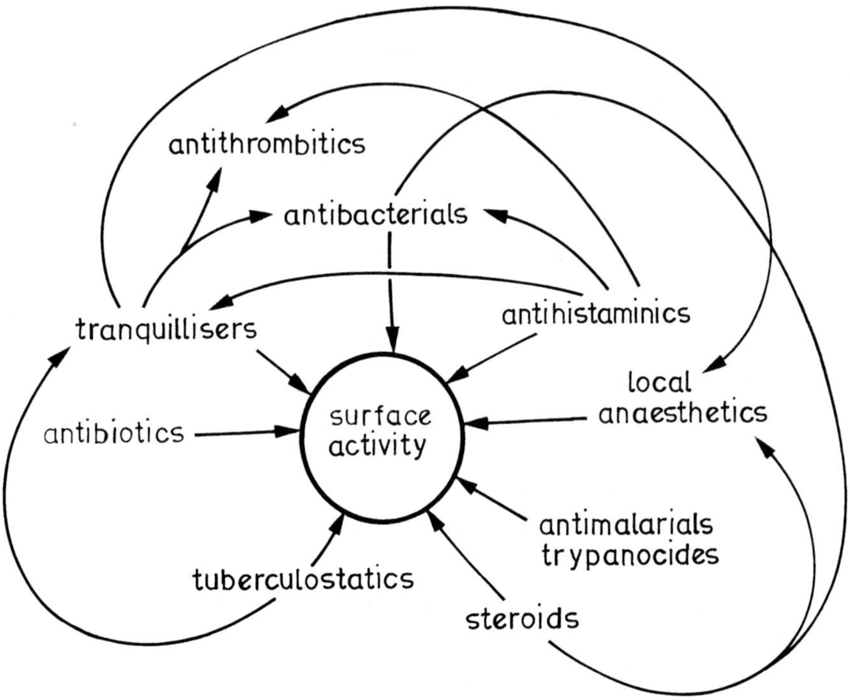

Fig.1. Diagram showing inter-relationship between drugs that have surface activity. Arrows linking different pharmacological types show secondary activity of some of the members of the class, e.g., some steroids have local anaesthetic activity; some tuberculostatics have shown tranquillizing activity (Taken from Ref. 1).

Before we state the hypothesis in Chapter 4, a brief presentation of the existing relevant theories of drug action particularly, "occupancy theory" and "rate theory" are given in Chapter 3. This is desirable because the proposed "liquid membrane hypothesis of drug action" will be finally discussed in the light of these theories.

REFERENCES

[1] A.T. Florence, Adv. Colloid Interface Sci., 2(1968)115.

[2] P.S. Guth and M. A. Sprites, Int. Rev. Nuerobiol., 7(1964)231.

[3] A. Felmeister, J. Pharm. Sci., 61(1972)151.

[4] S. Schreier, S.V. P. Malheiros and E. de Paula, Biochim. Biophys. Acta, 1508 (2000) 210.

[5] D. Attwood, A.T. Florence Surfactant Systems, Their Chemistry, Pharmacy and Biology, Chapman and Hall, New York, 1983.

[6] B. Silwa Zesz., Nauk.Univ Jageillon, 6(1961)25.

[7] A. Giacalone and D.D. Maggio, Gazz. Chim. Ital., 69(1939)122.

[8] T. Shirirkova, A.A.Trapeznikov and LV Komarova, Farmatsiya (Moscow), 25(1976)3.

[9] N.T. Praynishnikova, Farmakol Tosikol (Moscow), 36(1973)195.

[10] S. Abramson and G. Weissmann, Clin. Exp. Rheumatol, 7(1989)163.

[11] L.M. Lichtenberger, Z.M. Wang, J.J. Romero, C.Ulloa, J.C. Perez, M.N. Giraud and J.C. Barreto, Nat. Med., 1(1995) 602.

[12] K.D. Rainsford, J. Gastroenterol., 163(1989)9.

[13] T. Rades and C.C. Muller-Goymann, Int. J. Pharm., 159(1997) 215.

[14] A. Fini, G. Fazio, G. Feroci, Int. J. Pharm., 126(1995) 95.

[15] R. Rodryguez, C. Alvarez-Lorenzo and A. Concheiro, Eurp. J.Pharm.Sci., 20(2003)429.

[16] A. Seelig, Biochim. Biophys. Acta, 1030(1990) 111.

[17] A. Seelig, Biochemistry, 31(1992) 2897.

[18] N.D. Weiner, F. Hart and G. Zografi, J Pharm. Pharmacol., 17(1965)350.

[19] E. Zissman, C.R. Acad. Sci. Paris, 245(1957) 237.

[20] P.K. Kinnunen, Chem. Phys. Lipids, 57(1991) 375.

[21] M.C. Carey, D.M. Small and J. Lipid Res., 12(1971) 604.

[22] I. Krivoshein and I. Krivorutchenko, Mikrobiol. Z., 59(1997) 87.

[23] A.E. Vallejo and C.A. Gervasi, Bioelectrochemistry, 57(2002) 1.

[24] H. Zhao, P.M. Dubielecka, T. Soderlund, and P.K.J. Kinnunen, Biophys. J., 83(2002) 954.

[25] K.A. Krogfelt, M. Utely, H.C. Kirvan, D.C. Laux and P.S. Cohen, J. Antimicrob. Chemotherp., 46(2000)377.

[26] I. Krivorutchenko, Mikrobiol. Z., 61(1999) 95.

[27] V.V. Maidanov, N.L. Egutkin and I.Nikitin, Antibiotiki., 28(1983)663.

[28] A. Leydet, V. Barragan, B. Boyer, J.L. Montero, J.P. Roque, M. Witvrouw, J. Este, R. Snoeck, G. Andrei and E. De Clercq, J. Med. Chem., 40(1997) 342.

[29] E.P. Anan'eva, G.E Afinogenov and N.P. Elinov, Antibiotiki., 7(1978)605.

[30] G.E. Afinogenov, E.F. Panarin and E.P. Anan'eva, Antibiotiki., 6(1977)511.

[31] K. Fujita and K. Nihei, Bioorg. Med. Chem., 11(2003) 4255.

[32] S. Shani, M. Friedman and D. Steinberg, Int. J. Pharm., 131(1996) 33.

[33] H. Sawada, K. Yonagida, Y. Inaba, M. Sugiya, T. Kawase and T.Tomita, Eur. Polym. J., 37(2001) 1433.

[34] H. Sawada, A. Wake, T. Maekawa, T. Kawase, Y. Hayawake, T. Tomita and M. Baba, J.Flourine Chem., 83(1997) 125.

[35] H. Sawada, Y. Yamashita, T. Kawase, T. Tomita, M. Baba and Y. Hayawake, J. Flourine Chem., 84(1997) 155.

[36] H. Sawada, K. Tanba, T. Tomita, T. Kawase, M. Baba and T. Ide, J. Flourine Chem., 84(1997) 141.

[37] H. Sawada, J. Fluorine Chem., 121(2003) 111.

[38] G. McDonnell and A.D. Russell, Clin. Microbiol. Rev., 12(1999) 147.

[39] T. Abel, J. Lee, I. Cohen, R. Engel, M. Filshtinskaya, A. Melkonian and K. Melkonian, Carbohydrate Research, 337(2002) 2495.

[40] R.A. Demel, F.J.L. Crombag, L.L.M. Van, Deenean and S.C. Kinsky, Biochim. Biophys. Acta, 150(1968) 1.

[41] K.Fujita and I.Kubo, Int.J. Food Microbiol., 79(2002)193.

[42] B.Skerlavaj, M. Benincasa, A. Risso, M.Zanetti and R.Gennaro, FEBS Letters, 463(1999) 58.

[43] N. Rogers, M. E. Selsted, and S. D. Rychnovsky, Bioorg. Med. Chem. Lett., 7(1997)3177.

[44] J. Hernández-Borrell and M. Teresa Montero, Int. J. Pharm., 252(2003) 149.

[45] S. Barbosa, P. Taboada, J. M. Ruso, D. Attwood and V. Mosquera, Colloids and Surfaces A: Physicochem. Eng. Aspects, 224(2003) 251.

[46] L. Massi, F. Guittard, R. Levy, Y. Duccini and S. Geribaldi, Euro. J. Med.Chem., 38(2003)519.

[47] R. Maget-Dana, Ch. Hetru and M. Ptak, Thin Solid Films, 284-285(1996) 841.

[48] W. Wang, D. K. Smith and H. M. Chen, Biochim. Biophys. Acta, 1473(1999) 418.

[49] Ming-Liang, L. Ru-Wen, Q. Jian-Wen, W. Zhi-Jian and W. Tzong-Ming, Int. J. Antimicrob. Agents, 13(2000) 203.

[50] Y. Park, D.G. Lee, J. Seung-Hwan, W. Eun-Rhan, H.G. Jeong, C. Cheol-Hee and H. Kyung-Soo, Biochim. Biophys. Acta, 1645(2003) 172.

[51] H. Swada, T. Kei-ichi, N. Itoh, C. Hosoi, M. Oue, M. Baba, T. Kawase, M. Mitani and M. Nakajima, J. Fluorine Chem., 77(1996) 51.

[52] R. White, E. Crouch, D. Chang and K. L. Hartshorn, Biochem. Biophys. Res. Comm., 286(2001) 206.

[53] B. Eiseman, J. Exptl. Med., 88(1948) 189.

[54] Y. Sakagami and S. Kagi, J. Antibiotics (Tokyo) Ser.A., 19(1966)161.

[55] W. Scholtan, Kolloid Z., 170(1960) 19.

[56] W. Scholtan and R. Gonnert, Prog. Med. Chem., 5(1956) 314.

[57] P.H. Elworthy and A.T. Florence, J. Pharm. Pharmacol., 19(1967)1405.

[58] J.A. Cella, L.A. Harriman, D.N. Eggenberger and H.J. Harwood, J. Am. Chem. Soc., 77(1955) 4264.

[59] B.B. Hoffman, P. Taylor in J.G. Hardman, L.L. Limibird and A. G. Gilman (eds.) Goodman and Gilman's The Pharmacological Basis of Therapeutics, McGraw-Hill, New York, 2001, (pp115-154).

[60] D. Hellenbrecht, B. Lemmer, G. Wiethold and H. Grobecker: Naunyn Schmiedebergs Arch Pharmacol., 277(1973)211.

[61] D. Attwood and S.P. Agarwal, J. Pharm. Pharmacol., 31(1979)392.

[62] R.P. Quintana,: J Pharm Sci., 54(1965)573.

[63] U. Holzgrabe, M. Wagener and J. Gasteiger, J. Mol. Graph., 14(1996)185.

[64] D.N. Elliot, P.H. Eloworthy and D. Attwood, J. Pharm. Pharmacol., 25(1973)188.

[65] S.L. Krill, K.Y. Lau, W.Z. Plachy and S.J. Rehfeld, J Pharm Sci., 87(1998) 751.

[66] A.N. Nagappa, P. L. Kole, P.V. Pandi, K.Giresh, P.K. Rahul and I. Shanmukha, J.Membr. Sci., 211(2003) 349.

[67] K.G. Swan and N.G. White, U.S. Patent, 2, 408, 893(1946).

[68] K.C. Swan and N.G. White, U.S. Patent, 2, 432, 049(1947).

[69] I.P. Komkov and N.I. Vorona, Izv. Vysch, Ucheb. Zaved. Techknol.,14(1971)1369.

[70] N.I. Vorona and G.S. Vasilev, U.S.S.S.R. Patent, 488(1975) 807.

[71] N.I. Vorona and G.S. Vasilev, Izv Vysch. Ucheb. Zaved. Khim. Techknol., 20(1977)1243.

[72] J. Szede and M. Telbisz, Ber. Ungar.Ges., 13(1937)449.

[73] X. Mamaliga and M. Serban, Rev. Chim. Acad. Rep. Popularie Recomane, 2(1957) 325.

[74] D. Attwood, J. Pharm. Pharmacol., 28(1976)407.

[75] D. Attwood, J. Pharm. Pharmacol., 28(1976)762.

[76] D. Attwood, J. Pharm. Pharmacol., 24(1972)751.

[77] D. Attwood and D.K.Udeala, J Pharm Pharmacol., 27(1975)806.

[78] H. Brockman, G. Graff, J. Spellman and J.Yanni, Acta Ophthalmol Scand. Suppl., 230(2000) 10.

[79] U. Breyer-Pfaff, U. Mey, M.D. Green and T. R. Tephly, DMD 28(2000) 869.

[80] S. Kitagawa and A.Ikarashi, Chem. Pharm. Bull., 51(2003)1183.

[81] D. Causon, J. Gettins, J. Gromally, R. Greenwood, N. Natarajan and E. Wyn-Jones, J. Chem. Soc. Faraday Trans., 1177(1981) 143.

[82] S.B. Bhise, D. Bhattacharjee and R.C. Srivastava, J. Pharm. Sci., 73(1984) 1579.

[83] R.C. Srivastava, S.B. Bhise, R. Sood and M.N.A. Rao, Colloids and Surfaces, 19(1986) 83.

[84] S.B. Bhise, P.R. Marwadi, S.S. Mathur and R.C. Srivastava, J. Pharm. Sci., 72(1983) 599.

[85] E. Theodoropoulou and D. Marsh, Biochim. Biophys. Acta, 1461(1999) 135.

[86] A.J. Balmforth, A.J. Lee, F.H. Shepherd, P. Warburton, D. Donnelly and S.G. Ball, Biochem. Soc. Trans., 25(1997) 1041.

[87] J. E. Phillips and R.P. Mason, Atherosclerosis, 168(2003) 239.

[88] H.M. Krieg, J. Lotter, K. Keizer and J.C. Breytenbach, J. Membr. Sci., 167(2000) 33.

[89] C. Hung-Yuan, C. S. Randall, W. W. Holl, P. P. Constantinides, Y. Tian-Li and G. Z. Feuerstein Biochim. Biophys. Acta., 1284(1996) 20.

[90] E. Varga, J. Szollosi, K. Antal, P. Kovacs and J.Z. Szabo, Pharmazie, 54(1999) 380.

[91] N. Song-Qin, I. Majarais, K. Chiu-Yin and R.M. Epand, Eur. J. Mol. Pharmacol., 266(1994) 11.

[92] A.N. Nagappa, R.T. Patil, V. Pandi and K. Ziauddin, Indian J. Biochem. Biophys., 38(2001)412.

[93] S.B. Bhise, C.V.S. Subrahmanyam, R.K. Sharma and R.C. Srivastava, Int. J. Pharm., 28(1986) 145.

[94] V. Singh, B. Malhotra, D.B. Raju, A.N. Nagappa and R.C. Srivastava, Indian J. Biochem. Biophys., 28(1991) 34.

[95] F.E. Bloom in J.G.Hardman, L.L. Limibird and A. G. Gilman (eds.) Goodman and Gilman's The Pharmacological Basis of Therapeutics, McGraw-Hill, New York, 2001, pp293-320.

[96] R.C. Srivastava, Liquid Membrane Phenomena: Biological Implications, Indian Society for Surface Science and Technology, Kolkata, 2002.

[97] Z. Lalchev, Y. Christova, R. Todorov, V. Alexandrov, P. Stoichev and R. Petkov, Appl. Cardiopul. Pathophysiol., 4(1992)315.

[98] W.R. Tobin, H.E. Kaiser, A.M. Groeger and M.J. Lema, In vivo, 14(2000) 157.

[99] S.M. Johnson and A.D. Bangham, Biochim. Biophys. Acta, 193(1969) 92.

[100] M. Dzoljic, L. Bene, Z. Krasznai, S. Damjanovich and B. Van Duijn, J. Photochem. Photobiol. B, 56(2000) 48.

[101] N.P. Franks and W.R. Lieb, Nature, 333(1994) 662.

[102] N.P. Franks and W.R. Lieb, EMBO J., 15(1998) 6854.

[103] C. Ries and E. Puil, J. Nuerophysiol., 81(1999) 1802.

[104] J.E. Sirois, J.J. Pancracio, C. Lynch and D.A. Bayliss, J. Physiol. (Lond.), 512(1998) 851.

[105] J.E. Sirois, Q. Lei, E.M. Talley, C. Lynch and D.A. Bayliss, J. Neurosci., 20(2000) 6347.

[106] B. Wineger, D. Owen, C. Yost, J. Forsayeth and E.Mayeri, Anesthesiology, 85(1996) 889.

[107] R. Nicoll and D. Madisson, Science, 217(1982)1055.

[108] A. Kudoh and A. Matsuki, Anesth. Analg., 91(2000) 440.

[109] T. Blanck, Ital. J., Neurol. Sci., 20(1999) 431.

[110] D. He and J. Burt, Circ. Res., 86(2000) 104.

[111] I. Batai and M. Kerenyi, Acta Anaesthesiol. Scand., 43(1999) 760.

[112] A. Seelig, R. Gottschich and R.M.Devant, Proc. Natl. Acad. Sci.USA, 91(1994) 68.

[113] K. Jorgensen, J.H. Ipsen, O.G. Mouritsen and M.J. Zuckermann, Chem. Phys. Lipids, 65(1993) 205.

[114] H. Satake, H. Matsuki and S. Kaneshina, Colloids Surf. A, 71(1993) 135.

[115] H. Matsuki, S. Hashimoto, S. Kaneshina and M. Yamanaka, Langmuir, 10(1994) 1882.

[116] S. Taheri, I. P. Cogswell, A. Gent and G. R. Strichartz, J. Pharmacol. Expt. Therap., 304(2003) 71.

[117] A. Sekebra and G. Vrba, J. Pharm. Sci., 49(1960) 394.

[118] J.C. Skou, Acta. Pharm. Toxicol., 10(1954) 280.

[119] J.C. Skou, Acta. Pharm. Toxicol., 10(1954) 317.

[120] J.C. Skou, Acta. Pharm. Toxicol., 10(1954) 325.

[121] J.C. Skou, Biochim. Biophys. Acta, 30(1958) 625.

[122] N.L. Gershfeld, J. Phys. Chem., 66(1962) 1923.

[123] R.J. Baldessarini in J.G. Hardman, L.L. Limibird and A.G. Gilman (eds.) Goodman and Gilman's The Pharmacological Basis of Therapeutics, McGraw-Hill, New York, 2001, (pp. 447-483).

[124] D. Attwood and J. Gibson, J. Pharm. Pharmacol., 30(1978) 176.

[125] H. Yasuchara, C.A. Dujovne and I. Ueda, Pharmacology, 18(1979) 95.

[126] H. Yasuchara, C.A. Dujovne, I. Ueda and K. Arakawa, Toxicol. Appl. Pharmacol., 47(1979) 47.

[127] H. Yasuchara, T. Kashara, K. Sakomoto and C.A. Dujovne, Rinsho Yakuri, 10(1970) 121.

[128] A.S. Buchaman, G.M. Laycock and A. Shulman, Eur. J. Pharmacol., 7(1969) 60.

[129] D. Attwood, V. Mosquera and V. Perez-Villar, J. Chem. Soc. Faraday Trans., I 85(1989) 3011.

[130] H. Ahyayauch, M.A. Requero, A. Alonso, M. Bennouna and F.M. Goni, J. Colloid Interface Sci., 256(2002) 284.

[131] N. Funasaki, T. Okuda and S. Neya, J. Pharm. Sci., 90(2001) 1056.

[132] H. Bockhardt and R. Lullmann-Rauch, Acta. Pharmacol. Toxicol., 47(1980) 45.

[133] M.A. Carfagna and B.B.Muhoberac, Mol. Pharmacol., 44(1993) 129.

[134] A. Gescher and A. Li Wan Po., J Pharm. Pharmacol., 30(1978) 353.

[135] A. Rodríguez E.Junquera P. D.Burgo and E.Aicart , J. Colloid Interface Sci., 269(2004) 476.

[136] D.S.Charnery S. J. Mihic and R.A. Harris in J.G. Hardman, L.L. Limibird and A. G. Gilman (eds.), Goodman and Gilman's The Pharmacological Basis of Therapeutics, McGraw-Hill, NewYork, 2001, pp399-427.

[137] R.J. Kuffiner M.T. Bush and L.T.Bricher, J.Am.Chem.Soc., 79(1957) 1587.

[138] J.J.Lewis, Introduction to Pharmacology, Livingstone, Edinburgh, Scotland, 1960.

[139] A. Agiacalone, Gazz.Chim.Utal., 70(1940) 656

[140] M.Bornschein R.Voigt and H.Florstedt, Phamazie, 28(1973) 757.

[141] F.A. Villalonga and E.W. Phillips, J. Pharm. Sci., 69(1980) 102.

[142] P. Seeman, Pharmacol. Rev., 24(1972) 583.

[143] A.G. Lee, Biochim. Biophys. Acta, 455(1976) 102.

[144] A.N. Nagappa, D.B. Raju and R.C. Srivastava, Colloids and Surfaces, 35(1989) 17.

[145] D. Attwood, R. Blundell, V. Mosquera and M. Gracia, J. Colloid Interface Sci., 161(1993) 19.

[146] D.B. Raju, A.N. Nagappa and R.C. Srivastava, Indian J. Biochem. Biophys., 26(1989) 104.

[147] R.C. Srivastava R.K. Sharma and S.B.Bhise, J. Colloid Interface Sci., 93(1983) 72.

[148] D. Bouhlal, P. Gode, G. Goethals, M. Massoui, P. Villa and P. Martin, Carbohydrate. Res., 329(2000) 207.

[149] A.Hauptmann, Med.Wochenschr., 59(1912)1907

[150] N. Chidambaram and D. J. Burgess, J. Pharm. Sci., 2(2000) 1.

[151] G. Liu, V. Yarov-Yarovoy , M. Nobbs , J.J. Clare , T. Scheuer and W.A. Catterall, Neuropharmacology, 44 (2003)413.

[152] S.B. Bhise, C.V.S. Subrahmanyam, A.K. Malhotra and R.C. Srivastava, Int. J. Pharm., 24(1985)297.

[153] W.C. Bowman and M.J. Rand, Textbook of Pharmacology, Blackwell Scientific Publications, Oxfords, 1980, p18.30.

[154] E.I. Isaacson and J.N. Dalgado, in Burger's Medicinal Chemistry, part III, M.E. Wollf (Ed.), John Wiely and Sons, NewYork, 1981, p854.

[155] R.J. Baldessarini and F.I. Tarazi in J.G. Hardman, L.L. Limibird and A. G. Gilman (eds.) Goodman and Gilman's The Pharmacological Basis of Therapeutics, McGraw-Hill, New York, 2001, pp485-520.

[156] P.M. Seeman and H.S. Baily, Biochem.Pharmacol., 12(1963) 1181.

[157] E.T. Eckhardt and W.M. Govier, Proc.Soc.Biol. (N.Y.), 97(1958) 124.

[158] B.N. Halpem and J. Reuse, C.R. Acad. Sci. (Paris), 228(1949) 514.

[159] H. Chaplin, H. Crawford, M. Cutbush and P.L. Mollision, J. Clin. Path., 5(1952) 91.

[160] B.N. Halpem, B. Dreyfus and G. Bourdon, Presse. Med., 58(1950) 1151.

[161] A.R. Freeman and M.A. Spirtes, Biochem. Pharmacol., 12(1963) 47.

[162] O. Schales, Proc.Soc.Biol.(N.Y.), 83(1953) 593.

[163] P.S. Guth and V.P. Whittaker, Pharmacologist, 2(1960) 93.

[164] H.J. Dengler and E.O. Titus, Biochem.Pharmacol., 8(1961) 64

[165] R.S. Satcey, Brit.J.Pharmacol., 16(1961) 284.

[166] A. Plestscher and K.F. Grey, in W.D.M. Paton (Ed.), Pharmacological Analysis of Central Nervous Action, Macmillan, New York, 1962, p.75.

[167] J. Axelord, L.G. Whitby and G. Hertting, Science, 133(1961) 383.

[168] J. Axelord, L.G. Whitby, G. Hertting and T.L. Kopin, Circulat.Res., 9(1961) 715.

[169] O. Lindon, J.H. Quastel and S. Svec, Canad.J.Biochem., 35(1957) 1145.

[170] G. Zografi, D.E Auslander and P.L. Lytell, J.Pharm.Sci., 53(1964)573.

[171] G. Zografi and D.E Auslander, J.Pharm.Sci., 54(1965) 1313.

[172] L.L.M. Van Deenen and R.A. Demel, Biochem.Biophys.Acta, 94(1964) 314.

[173] A. Felmeister and R. Schaubman, J. Pharm. Sci., 57(1968) 178.

[174] A. Felmeister and R. Schaubman, J. Pharm. Sci., 58(1969) 64.

[175] K. Thoma and M. Arning, Arch.Pharm.(Weinhein Ger.), 309(1976) 851.

[176] G. Zografi and M.V. Munski, J. Pharm. Sci., 59(1970) 819.

[177] I. Blei, Arch.Biochem.Biophys., 109(1965) 321.

[178] F. Villalonga, E. Fried and J.A. Izquierdo, Arch. Int. Pharmacodyn. Ther., 130(1961) 260.

[179] K. Thoma and K. Albert, Pharm.Acta.Helv., 54(1979) 334.

[180] H. Chaplin, H. Crawford, M. Cutbush and P.L. Mollison, J. Clin.Pathol., 5(1952) 91.

[181] D. Attwood, A.T. Florence and J.M.N. Gillian, J. Pharm. Sci., 63(1974) 988.

[182] D. Attwood and R. Natarajan, J. Pharm. Pharmacol., 33(1981) 136.

[183] A.D. Atherton and B.W. Barry, J. Colloid Interface Sci., 106(1985) 479.

[184] E. Wajnberg, M. Tabak, P.A. Nussenzveig, C.M. Lopes and S.R. Eouro, Biochem. Biophys. Acta, 944(1988) 185.

[185] D. Attwood, V. Mosquera and V. Perez-Villar, J. Chem. Soc., Faraday Trans. I, 85(1989) 3011.

[186] D. Attwood, D. Doughty, V. Mosuera and V. Perej-Villar, J. Colloid Interface Sci., 141(1991) 316.

[187] W. Wanbahdi, H. Mwakibete, D.M. Bloor, R. Palepu and E. Wyn-Jones, J. Phys. Chem., 96(1992) 918.

[188] V. Perez-Villar, M.E. Vazquez-Iglesias and A. Degeyer, J. Phys. Chem., 97(1993) 5149.

[189] D. Attwood, V. Mosquera, J.L. Lopez-Fontan. M. Garcia and F. Sarmiento, J. Colloid Interface Sci., 184(1996) 658.

[190] D. Attwood, E. Boitard, J.P. Dubes and H. Tachoire, J. Phys. Chem., 101(1997) 9586.

[191] M. Prez-Rodrigues, G. Prieto, C. Rega, L.M. Varela, F. Sarmiento and V. Mosquera Langmuir, 14(1998) 4422.

[192] S. Takegami, K. Kitamura, T. Kitade, A. Kitagawa, and K. Kawamura, Chem. Pharm. Bull., 51(2003) 1056.

[193] S.R. Louro, C. Anteneodo and E.Wajnberg, Biophys Chem., 74(1998) 35.

[194] P.A. Janssen, Int. J. Neuropsychiat. S., (1967) 10.

[195] D. Attwood, V. Mosquera and J.L. Lopez-Villar, J.Chem.Soc.Faraday Trans. I, 85(1989) 3011.

[196] A.D. Atherton and B.W. Barry, J.Colloid Interface Sci., 106(1985) 479.

[197] S.V. P. Malheiros, E. de Paula and N.C.Meirelles, Biophys. Chem., 83(2000) 89.

[198] M.R. Kaser and G. G. Skouteris, Peptides, 18(1997) 1441.

[199] G. Devendra and R.G. Spragg, Respir. Res., 3(2002) 19.

[200] E.Z. Ron and E.Roesenberg, Environmental Microbilogy, 3(2001) 229.

[201] L. Yan-Yeung and N. L. Abbott, Curr. opin. Colloid and Interface Sci., 7(2002) 267.

[202] C.N. Chirita, M. Necula and J. Kuret, J. Biol. Chem., 278(2003) 25644.

[203] W.C. Wimley and S.H. White., Nature Struct. Biol., 3(1996) 842.

[204] M.F.Lecompte, G. Bouix, and K. G. Mann, J. Biol.Chem., 269(1994) 1905.

[205] H.A.B. Wösten, Annu. Rev. Microbiol., 55(2001) 625.

[206] G. Corzo, P. Escoubas, E. Villegas, K. J. Barnham, W. He, R.S. Norton and T. Nakajima, Biochem. J., 359(2001) 35.

[207] W. Schraw, Y. Li, M.S. McClain, F. G. Van der Goot, and T. L. Cover, J. Biol. Chem., 277(2002) 34642.

[208] A.A. Waheed, Y. Shimada, H. F. Heijnen, M. Nakamura, M. Inomata, M.Hayashi, S. Iwashita, J.W. Slot, and Y. Ohno-Iwashita, Proc. Natl. Acad. Sci., U.S.A., 98(2001) 4926.

[209] S. Miyata, J. Minami, E. Tamai, O. Matsushita, S. Shimamoto, and A. Okabe, J. Biol. Chem., 277(2002) 39463.

[210] M. Nagahama, S. Hayashi, S. Morimitsu, and J. Sakurai, J. Biol. Chem.,278 (2003)36934.

[211] D.C. Liang, W.R. Chong, J.P. Zhang and Z.L. Wan, Sci. China B., 35(1992) 547.

[212] U. Derewenda, Z. Derewenda, E. J. Dodson, C.G. Bimgx and J. Markcissen, J. Mol. Biol., 220(1991) 425.

[213] L. Rajendran, M. Masilamani, S. Solomon, R. Tikkanen, C.A.O. Stuermer, H. Plattner and H. Illges, Proc. Natl.Acad.Sci., USA, 100(2003) 8241.

[214] M. Delaye and A. Tardieu, Nature, 302(1983) 415.

[215] V. Srinivas, Siddhartha, A. Datta, T. Ramakrishna, and Ch. Mohan Rao, Mol. Vis., 7 (2001) 114.

[216] G.A. Wooley and B.A. Wallace, J. Membr. Biol., 129(1992) 109.

[217] G. Schwartz, S. Stankowski and V. Rizzo, Biochim. Biophys. Acta, 861(1986) 141.

[218] M. Ge and J.H. Freed, Biphys. J., 76(1999) 264.

[219] W.R. Veatch and E.R. Blout, Biochemistry, 13(1974) 5257.

[220] H.H. Paradise, Biochim, Biophys. Res. Commun., 88(1979) 810.

[221] R. Urruita, R.A. Cruciani, J.L. Barker and B. Kachar, FEBS Lett. 247(1999) 17.

[222] H.S. Mcjapirab, J.S. Hyde and J.B. Feix, Biochemistry, 32(1993) 11895.

[223] Z. Oren and Y. Shai, Biopolymers, 47(1998) 451.

[224] B. Bechinger, Biochim. Biophys. Acta, 1462(1999) 157.

[225] S.L. Keller, S.M. Grunner and K. Gawrish, Biochim. Bipphys. Acta, 1278(1996) 241.

[226] D.M. Gill, in J.T. Jelgaszewicz and T. Wadstrom (Eds.), Bacterial Toxins and Cell Membrane, Academic Press London, 1978, pp.291-332.

[227] L.L. Simpson, Pharmacol. Rev., 33(1981) 155.

[228] J.P. Robinson and J.H. Nash, Mol. Cell Biochem., 48(1982) 33.

[229] V. Singh, D.B. Raju, C. Ravindran and R.C. Srivastava, J. Colloid Interface Sci., 139(1990) 337.

[230] B.L. Kagan, A. Finkelstein and M. Colombini, Proc. Natl. Acad. Sci., USA, 78(1981) 4950.

[231] J.J. Donovan, M. Simon, R. Draper and M. Montal, Proc. Natl. Acad. Sci., USA, 78(1981) 172.

[232] D.H. Hoch, M.R. Mira, B.E. Ehyilich, A, Finkelstein, B.R. Dasgupta and L.L Simpson, Proc. Natl. Acad. Sci., USA, 82(1985) 1692.

[233] M. Ligo, A. Hoshi, A. Nakamura and K. Kunetani, J.Pharm.Dyn., 1(1978) 49.

[234] S. Ozaki, Y. Ike, H. Mizuno, K. Ishikawa .Mori, Bull. Chem. Soc., Jpn., 50(1977) 2406.

[235] R.C. Srivastava, M. Shankar, A. Tandon and A. N. Nagappa, Int. J. Pharm., 38(1987) 239.

[236] K. Nicolay, A.M. Sautereau, J.F. Tocanne, R. Brasseur, P. Huart, J.M. Ruysschaert and D. B. Kruijff, Biochim. Biophys. Acta, 940(1988) 197.

[237] G. Durand, A. Polidori, J.P. Salles and B. Pucci, Bioorg. Med. Chem. Lett., 13(2003) 859.

[238] K.U. Maheswari, T. Ramachandran and D. Rajaji, Biochim Biophys Acta., 1463(2000)230.

[239] H. Boyar and F. Severcan, J. Mol. Struct. , 408-409(1997) 265.

[240] L. Ter-Minassian-Saraga, G. Madelmont, C. Hort-Legrand and S. Metral, Biochem. Pharmacol., 30(1981) 411.

[241] H.U. Bergmeyer and W. Dirscher, Biochem.Z., 323(1952)104.

[242] E.N. Wilnier, Biol.Rev., 36(1968) 368.

[243] R.C. Srivastava, R.K.Sharma and S.B.Bhise, Colloids and Surfaces, 14(1985) 1.

[244] A.N. Nagappa, K.U. Bhaskar, S. Shankar, S. Mitra and R.C. Srivastava, Indian J Biochem. Biophys., 25(1988) 350.

[245] A.F. Hofmann, News Physiol. Sci., 14(1999) 24.

[246] W. Spivak, C. Morrison, D. Devinuto and W. Yuey, Biochem J. , 252(1988) 275.

[247] P.B. Savage, C. Li, U. Taotafa, B. Ding and Q. Guan, FEMS Microbiology Letters 217(2002) 1.

[248] X. M. Zhao, A. M. Montet and J. C. Montet, J. Nutr. Biochem., 4(1993) 105.

[249] T. Fujiwara, S. Hasegawa, N. Hirashima, M. Nakanishi and T. Ohwada, Biochim. Biophys. Acta, 1468(2000) 396.

[250] R.C. Srivatava, A. Tandon, A.N. Nagappa and R.K. Sharma, J. Colloid Interface Sci., 117(1987) 375.

[251] R.C. Srivastava, D.B. Raju, V. Singh and S. Upadhyay, J. Membr. Sci., 58(1991) 211.

[252] G. Colacicco and M.K. Basu, Prostaglandins, 16(1978) 189.

[253] T.J. Roseman and S.H. Yalkowsky, J. Pharm. Sci., 62(1973) 1680.

[254] B.E. Sims and G.R. Derr., J. Pharm. Sci., 66(1977)823.

[255] A. Tandon, A. N. Nagappa and R. C. Srivastava, Int. J. Pharm., 32(1986) 39.

[256] A.N. Nagappa, Kzeeyauddin, P.V. Pandi, R.T. Patil, K. Girish and R.C.Srivatava, J. Surface. Sci., Technol., 18(2002) 43.

[257] R. P. Holmes and N. L. Yoss, Biochim. Biophys. Acta, 770(1984) 15.

[258] M. Balakirev, G. Schoehn and J. Chroboczek, Chem. Biol., 7(2000) 813.

[259] A.N. Nagappa, P.L. Kole, P.V. Pandi, K. Zeeyauddin, R.T. Patil, and I. Shanmukha, J. Colloid Interface Sci., 271(2004) 416.

[260] S.R. Wassall, T. M. Phelps, M. R. Albrecht, C.A. Langsford and W. Stillwell, Biochim. Biophys. Acta, 939(1988) 393.

[261] A.N. Nagappa, P.V. Pandi, P.K.R. Mishra, K. Girish and I. Shanmukh, Indian J. Biochem. Biophys., 39(2002) 406.

[262] W.W. Douglos, in A.G. Gilman, L.S. Goodman and A. Gilman, (Eds.). The Pharmacological Basis o Therapeutics, , Macmillanm, New York, 1980, p.202.

[263] R.J. Baldesarini, in A.G. Gilman, L.S. Goodman and A. Gilman, (Eds.). The Pharmacological Basis o Therapeutics, Macmillanm, New York, 1980, p.404.

[264] K.J. Mysels, J. Phys. Chem., 68(1964) 3441.

Chapter 3

Theories of drug action:

Before we take up theories of drug action let us introduce the definitions of some of the commonly used terms.

3.1. Commonly used terms

Definitions of some of the commonly used terms in the context of the theories of drug action to be presented are in section 3.2 as follows:

3.1.1. Receptor

The first discussions of the concept of a receptor were presented by Langley [1] while studying the action of atropine and pilocarpine on salivary flow in cats. Langley referred it to the receptive substance of muscle that received the stimulus and transferred it to the contractile material. Paul Ehrlich, however, first used the term "receptor", to describe hypothetical specific chemical groupings of "side chains" on cells upon which chemotherapeutic agents were postulated to act.

Many disease processes, such as myasthenia gravis and diabetes involve the modification of the number of receptors present in a target organ or abnormalities in the structure or function of a receptor. In addition, since cell growth and differentiation is under strict receptor control, it is suspected that modifications of growth factors or growth factor receptors are involved in tumor genesis. Knowledge of the nature of the receptor and its functions can suggest treatments for the disease. More importantly for the practicing clinician, understanding of the receptor involved in the disease can help with the diagnosis and treatment of unusual cases.

Most drugs interact with specific receptors, which may be the same as the site for a neurotransmitter or hormone. It may also be a site on an ion channel, enzyme, or other cellular constituent. The effective use of a new drug requires knowledge of its pharmacokinetics and sites of action. In particular, knowledge of the sites of action of a drug (i.e., the type of receptor) can help one to predict possible adverse drug interactions.

Most forms of communication between cells are mediated by receptor-ligand interactions. For example, the movement of skeletal muscles is entirely dependent upon the interaction of acetylcholine with the acetylcholine receptor at the neuromuscular junction. The control of heart rate is mediated by central nervous system neurotransmitter receptors and receptors in the autonomic nervous system. In addition, all hormone action are mediated by either membrane bound receptors on the cell surface or soluble receptors in the cytoplasm

The term receptor has been used operationally to denote any cellular macromolecule to which a drug binds to initiate its effects. Among most important drug receptors are cellular proteins, whose normal function is to act as receptors for endogenous regulatory ligands – particularly neurotransmitters, growth factors and hormones. The function of such receptors consists of the binding the appropriate ligand and, in response, propagating its regulatory signal in the target cell. A receptor by definition exists in at least two conformational states, active and inactive [2].

In order to define a specific receptor, three criteria should be satisfied: saturability, specificity, and reversibility:

Saturability**:** A finite number of receptors per cell (or per weight of tissue or protein) should be present as revealed by a saturable binding curve. By adding increasing amounts of drug, the number of drug molecules bound should form a plateau at the number of binding sites present.

Specificity: The drug should have structurally complementary to the receptor. This can be demonstrated by using a series of drugs varying slightly in chemical structure and showing that affinity differs with differing chemical structure. Also, if the drug is optically active, then the two isomers may have markedly different affinities.

Reversibility: The drug should bind to the receptor and then dissociate in its nonmetabolized form. This property distinguishes receptor-drug interactions from enzyme-substrate interactions

3.1.2. Antagonism

Drugs, which act by combining with receptors, can be classified as: (i) full agonist, (ii) partial agonists (iii) neutral antagonist and (iv) partial inverse agonist (v) full inverse agonist.

If two drugs bind to the same receptors at the same site than there exists competition among molecules for the binding site. In such situation the binding of the ligand to the receptor leading to a conformational changes of the receptor would complicate our understanding regarding nature of drug-receptor interactions. The theoretical structural modeling and functional studies of mutant proteins are helpful in our understanding regarding dynamics of protein structure and protein drug interactions. In brief, full agonist is a drug that has a higher affinity for the active conformation than for the inactive conformation and hence will drive the equilibrium to the active state and thereby activates the receptor. Such a drug will be an agonist. A full agonist is sufficiently selective for the active confirmation and therefore, at a saturating concentration, it will drive the receptor essentially to the active state. A drug, which binds to the same site on receptor, but with moderately greater affinity for active conformation than inactive conformation is called partial agonist. Neutral antagonist is a drug, that binds with equal affinity to either conformation and therefore, will not alter the activation equilibrium. Partial inverse agonists are those with moderately greater affinity for inactive conformation of the receptor. It acts by reducing the action exerted by the agonist when an agonist and antagonist are administered together. Full inverse agonists are those with preferential affinity for inactive conformation of the receptor and act by reducing the action exerted by the agonist when an agonist and antagonist are administered together.

Most of the biological actions caused either by agonists or antagonists are mediated through membranes and hence, interaction of agonists or antagonists with the membrane components, i.e., receptors is essential. Because of the complexity of biomembranes, nature and details of the interaction and the mechanism of the consequent response is far from being completely understood. In order to gain information on these aspects, it has been customary to investigate the quantum of biological response as a function of the drug concentration at the site of its action. Results of such investigations are usually presented in the form of dose-response curves, various forms of which are summarized below [3].

3.1.3. Dose-response curve

One of the ways of expressing the course of drug action is dose-response curve. If the response is plotted as percentage fraction of the maximal response against the dose on an arithmetic scale, a hyperbolic curve results. Representative curves are shown in Fig. 1.

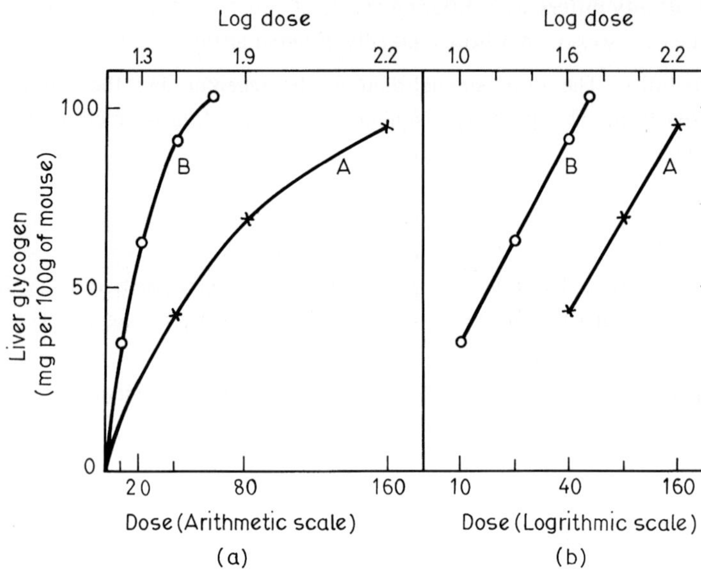

Fig. 1a & 1b. Linear and semi-logarithmic does response curves. Effects of two steroids, 11-dehdrocorticosterone (a) and cortisone (b) on liver glycogen in mice

3.1.4. Log dose-response curve (LDR)

From a practical standpoint, logarithmic dose scale is preferable. In semi-logarithmic presentation a sigmoid curve is obtained. The major portion of these curves being linear, it is much easier to deal with statistical analysis. The drugs that produce the same biological effect by similar mechanism but differ in potency yield parallel line segments (see Fig. 1b). This is very convenient for analysis. Another practical advantage of this presentation is that a wide range of doses can be presented on a single graph conveniently. A representative LDR curve is shown in Fig. 1(b).

3.1.5 Double-reciprocal plot

A third way of representation, inspired by the Line weaver-Burk plots in enzyme kinetics, is also adopted. In this form of representation, reciprocal of response (Δ) is plotted against reciprocal of dose (X). It gives a straight line with positive intercept on Y-axis and negative intercept on X-axis. The intercept on Y-axis gives reciprocal of maximum response, corresponding to infinite dose. Slope of this line represents (K_x/V_{max}) where K_x is dissociation constant of the drug-receptor complex and V_{max} is the maximal response. Thus, by knowing the intercept on Y-axis and the slope of the double-reciprocal plot, dissociation constant of the drug-receptor complex can also be computed. A typical double-reciprocal plot is shown in Fig.2.

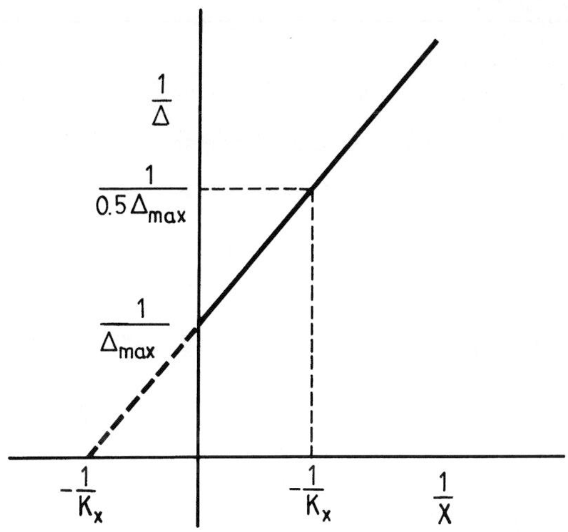

Fig.2. The double reciprocal plot. Δ represents response and X represents dose.

Out of these several types of dose-response curves mentioned above, LDR and double-reciprocal plots are more convenient for understanding the nature of antagonism. Similar to phenomenon of inhibition in enzyme kinetics, antagonism exhibited by drugs is of two types-competitive and non-competitive. Both the types of antagonisms are reflected in the nature of LDR and double-reciprocal plots.

In case of LDR plots, if one compares the curve for an agonist alone with that for a mixture of an agonist and an antagonist, it is observed that there is a right shift in LDR curve for the mixture. This indicates that for obtaining the same quantum of biological response, which is obtained by an agonist alone, a comparatively higher dose of the agonist is needed in the presence of antagonist. In case of competitive antagonism, right-shifted curve is parallel to the curve for the agonist alone and the maximal response is obtainable even in presence of the antagonist, but at a higher concentration of the agonist. Unlike this, in case of noncompetitive antagonism, right-shifted curve is not parallel to the curve for the agonist alone, and the maximal response in presence of the antagonist is not attainable, even at a very high concentration of an agonist.

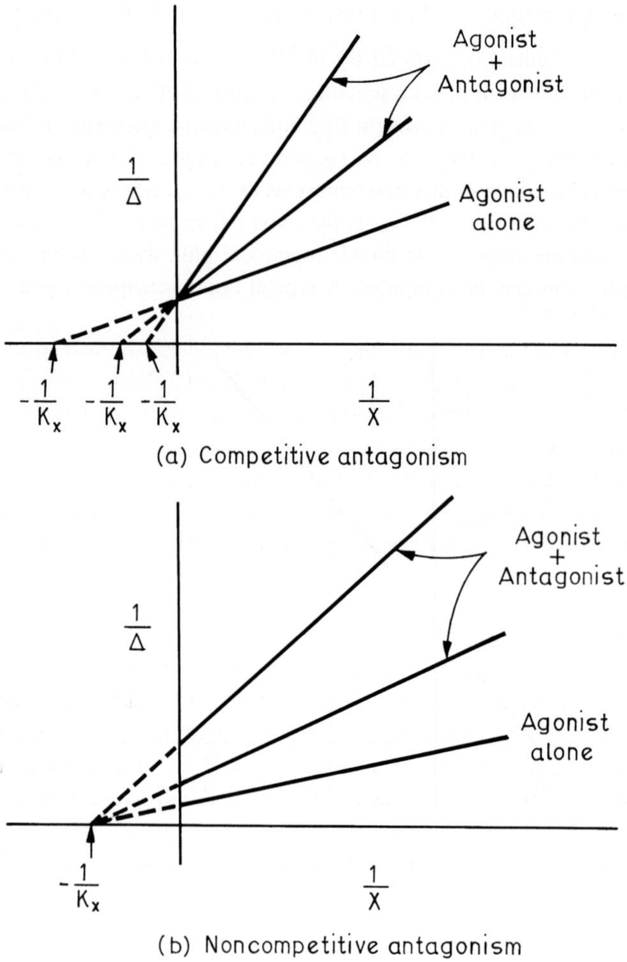

Fig.3. Analysis of antagonism by double reciprocal plots.

The double-reciprocal plots in the case of competitive and non-competitive antagonism are shown in Fig. 3. In case of competitive antagonism, the plots yield straight lines which have the same intercept on Y-axis representing reciprocal of response, while the slope for the agonist-antagonist mixture is greater than that for the agonist alone. This makes the value of extrapolated intercept on X-axis representing reciprocal of dose, greater for the agonist-antagonist mixtures than that for the agonist alone. In case of noncompetitive antagonism, slopes of the plots show the same trend as in case of competitive antagonism, whereas the intercepts do not. Value of the extrapolated intercept on X-axis is the same for all the straight line, while the intercepts on Y-axis are different, the intercept for the agonist being less than the intercept for the agonist-antagonist mixture. Thus, the intercepts on Y-axis lead to the conclusion that in the case of competitive antagonism the maximal response due to the agonist alone corresponding to $(1/X) \rightarrow 0$, is, in principle, attainable in the presence of antagonists, while in the case of noncompetitive antagonism, it is not.

3.1.6. PA_x values [4]

Antagonism of a drug is also expressed quantitatively on p-scale ($p=-log_{10}$). It is expressed as pA_x, where A_x is the molar concentration of the antagonist in the presence of which the potency of the agonist is reduced X times. The pA_x, values, which are used quite often are pA_2 and pA_{10}.

3.2. Theories of drug action [5-9]

Important theories relevant to the discussion presented in the next section are (i) Occupancy Theory, (ii) Rate Theory and (iii) Inactivation Theory. In the occupancy theory, response is stated to be a function of the occupation of receptors by agonist, while in the rate theory; response is considered to be a function of rate of formation/dissociation of the drug receptor complex. Inactivation theory assumes that the receptor ligand complex is an intermediate "active" state that gives rise to an inactive form of the receptor, which is part of a receptor ligand complex. The three theories are summarized below.

3.3. Occupancy theory [5, 6]

Biological responses to drugs are, as a rule, graded; they can be measured on a continuous scale and, as pointed out earlier, there is a systematic relationship between the dose of a drug and the magnitude of the response. Application of the law of mass action to the dose-response relationship was largely done by Clark [5, 6]. An observed biological effect was assumed to be a reflection of the combination of drug molecules with receptors. The magnitude of a response was postulated to be directly proportional to the occupancy of receptors by drug molecules. The maximal response is assumed to be obtained when all the receptors are occupied.

Simple mass law principle enables one to express quantitatively, dependence of biological effect upon dose. If $[X]$ represents concentration the drug at the site, $[R]$ represents the concentration of receptors not occupied by the drug, $[RX]$ represents concentration of drug-receptor complex and Δ represents magnitude of biological response then one can write

$$R + X \underset{k_2}{\overset{k_1}{\leftrightarrows}} RX$$. At the equilibrium

$$[R]\,[X]\,/\,[RX] = k_2/k_1 = K_x \qquad (1)$$

Further, since biological response Δ is assumed to be proportional to the concentration of occupied receptors, we can write,

$$\Delta = k_3\,[RX] \qquad (2)$$

In Eqs. (1) and (2), k_1, k_2 and k_3 are the corresponding rate constants and K_x is the dissociation constant of drug-receptor complex. The total receptor concentration R_T is given by the equation,

$$[R_T] - [R] + [RX] \qquad (3)$$

Substituting the value of R from Eq. (3), Eq. (1) becomes

$$\{[R_T] - [RX]\}\,[X]\,/\,[RX] = K_x \quad (4)$$

which after rearrangement can be written as,

$$[RX]/[R_T] = [X]/\{K_x+[X]\} \quad (5)$$

In view of the fact that maximum biological response, which the system is capable of, is obtained only when all the receptors are occupied, i.e., $\Delta_{max} = k_3[R_T]$, Eq. (5) can be rewritten as

$$\Delta/\Delta_{max} = [RX]/[R_T] = [X]/\{K_x+[X]\} \quad (6)$$

It has been pointed out [10] that Eq. (6) is based on the following implicit assumptions:

(i) An all-or-none stimulus is elicited by the combination of each receptor with an agonist molecule.

(ii) There is summation of these individual stimuli.

(iii) The effect is linearly proportional to the number of stimuli.

(iv) The maximal stimulus occurs when every receptor site is occupied by antagonist molecule.

(v) The drug-receptor complex is formed by readily and rapidly reversible chemical bonds.

(vi) The occupation of one receptor does not affect the tendency of the other receptors to be occupied.

Although explanations of observations related to the response caused by most agonist molecules can be provided on the basis of occupancy theory, the observations related to responses caused by a variety of other agonist molecules need postulation of a few additional concepts which are summarized below.

3.3.1. Affinity

If sets of LDR curves for a series of co-generic drugs of varying potencies interacting with the same receptor are examined, it is observed that these curves do not overlap. These curves indicate that for a particular biological response to be elicited, the most potent agonist drug requires the least concentration. This is expressed by saying that the most potent drug has the highest affinity for the receptors, while the congeners have lesser affinity. From Eq. (5) which can be rearranged to read,

$$[RX]/[R_T] = 1/\{1+K_x/[X]\} \quad (7),$$

it follows that [11] the ratio $[RX]/[R_T]$ increases with concentration of the drug $[X]$, and decreases with the dissociation constant, K_x. of the drug-receptor complex RX. Thus 'affinity' of the drugs to the receptors is proportional to the reciprocal of K_x.

Thus, the more the potency of an agonist, the higher will be its affinity for the receptor and, hence, the lower will be the dose required to elicit a particular quantum of biological response.

If LDR curves for an agonist alone and for a mixture of agonist and competitive antagonist are compared, it can be inferred that LDR for the mixture shows a behavior similar to agonist, but with lesser affinity for the receptor. This indicates that presence of a competitive antagonist alters the effective affinity of the agonist for the receptor.

3.3.2. Efficacy (intrinsic activity)

According to the occupancy assumption, the number of receptors occupied determines the magnitude of a response. Agonist drugs are supposed to differ in their affinity for the receptors and, therefore, different doses are required to achieve the same degree of receptor occupancy and hence, the same response. A molecule of any agonist occupying a given receptor site is assumed to make the same quantal contribution to the overall response as a molecule of any other agonist. It may, however, occur at a higher concentration of the other agonist, if its affinity is low.

Instances are known in which various agonists that apparently act on the same receptor site produce maximal responses of different magnitudes, an observation not accounted for by the theory. Hence, the theory has been modified by introducing the concept of intrinsic activity [12] or efficacy [13]. It is defined as the capacity of a drug to initiate a response once it occupies the receptor sites.

Thus, affinity describes the tendency of the drug to form a stable complex with the receptor, and efficacy describes the biologic effectiveness of the drug-receptor complex. The two properties are considered to be unrelated. Since biological effect would be determined both by the extent of receptor occupancy, i.e., affinity and also by efficacy, it follows that equal biological responses need not imply equal degree of receptor occupancy, and maximal responses may vary from drug to drug.

According to Ariens [11, 12], Eq. (6) should be rewritten as

$$\Delta / \Delta_{max} = \alpha [RX]/ [R_T] \quad (8)$$

where α is termed as the intrinsic activity [12] factor. In Stephenson's alternative framework [13], efficacy *(e)* denotes the capacity of a drug to initiate a response once it occupies receptor sites. The value of parameter *'e'* can range from zero to a large positive number.

In the sequence of events represented by the equations,

$$X + R \underset{k_2}{\overset{k_1}{\rightleftarrows}} RX \overset{k_3}{\longrightarrow} \text{response}$$

the rate constant k_3 is related to efficacy. It is viewed as a measure of the probability that an agonist occupying a receptor will induce a shift to the configuration that provides the stimulus

Stimulus, *'S'* and efficacy, *'e'* can be related by the following equation,

$$S = e [RX]/[R_T] = e[X]/\{K_x + [X]\} \quad (9)$$

Thus, as for receptor occupation, Stephenson's efficacy (*e*) and Arien's intrinsic activity factor (α) are the same. However, the factors differ when the relationship between receptor occupancy and responses is considered.

In Stephenson's framework, biological response is a function of the stimulus, i.e.,

$$\Delta/\Delta_{max} = f(S) = f(e [RX]/[R_T] \quad (10)$$

The relationship between response and stimulus is arbitrarily defined such that $S=1$ when response is half the maximal response produced by a highly active agonist. Eq. (9) can be rearranged to read

$$S = e\,[X]/K_x/\{1+[X]/K_x\} \tag{11}$$

From equation (11) it follows that, for a highly active agonist, where e has a high value; $S \rightarrow e[X]/K_x$.

3.3.3. Spare receptors [14,15]

According to the assumptions made in the occupancy theory, maximal response is attainable only when the agonist drug occupies all the receptors. As a corollary, therefore, when an antagonist is added to the system, at no stage maximal response should be attainable. However, in case of competitive antagonism, this prediction is not observed to be true, i.e., even in the presence of a competitive antagonist the same maximal response is attainable, which is obtained in the absence of an antagonist, but at a higher concentration of the agonist. This discrepancy has been resolved by proposing "spare receptors". It is hypothesized that there are some additional receptors, which become available to the agonist in the presence of an antagonist. It is also stated that in obtaining maximal response due to agonist alone, there is no combination between agonist and the so-called "spare receptors".

In short, for a highly active agonist with a high efficacy the maximal response will be produced by a concentration that does not occupy all the receptors. The receptors, which remain unoccupied, are termed 'spare receptors'.

It has also been suggested [16] that it is better to hypothesize 'spare cells' rather than "spare receptors". An antagonist, being applied for a short time, would block only the superficial cells, and not the deep ones. Experimental evidence [17], in case of α-adrenergic receptors of rat *vas deferens* does not provide evidence for spare receptors. Paton [7, 8] has commented that for occupancy, existence of spare receptors merely seems to be a puzzling extravagance. The "spare receptors", thus, continues to be a hypothetical assumption.

***3.3.4. Rate Theory* [7-9].**

The central idea in this theory is different from that in the occupancy theory. Instead of attributing excitation to the occupation of receptors by drug molecules, it is attributed to the process of occupation -each association between a drug molecule and a receptor providing one quantum of excitation. The magnitude of biological response is proportional to the rate at which drug molecules associate with receptor sites. This rate depends on the concentration of free drug, the concentration of free receptor sites and k_1, and the rate constant for association of drug molecules with receptors.

This theory abandons the occupancy assumption, and adopts the principle of intrinsic activity. Efficacy, in this theory, is no longer an *ad hoc* constant, but is defined by the rate constant k_1, which may differ from drug to drug. The distinction between an agonist and an antagonist is based on the value of k_2 the dissociation rate constant for drug-receptor complex. Drugs with higher values of k_2 are agonists because if k_2 is large, the rate of dissociation of drug-receptor complex will be making free receptor sites available at high rate for new effective collisions with drug molecules. In contrast, if k_2 is small, drug-receptor complex will be more stable; the rate of dissociation will be low making the availability of free receptors to the drug molecules for new association events infrequent. This will, consequently, lead to little or no excitation. Thus, the drugs with low k_2 will act as weak agonists or as antagonists. Antagonism, therefore, implies persistent occupancy of the receptor by the drug-antagonist. Both for agonists and antagonists, potency is determined by the equilibrium dissociation constant k_2/k_1, which describes affinity of the drug for the receptor.

The theory explains why antagonists tend to be bulkier than agonists since it is indicated that, as compared to small molecules, bulky molecules may have more non-specific binding and, hence, a lower dissociation constant. It is also claimed that the theory explains why potent antagonists have a slow onset of action since more potent they are, lower is the dose at which they must be used and, consequently, slower will they equilibrate. Both these arguments apply with equal force to the mass law theory based on the occupancy assumption. Rate theory has been used [7] to predict slopes of LDR curves, which are found to depend on the association and the dissociation rate constants.

Since agonists have high dissociation rates than antagonists, postulation of a large shift in antagonist occupancy during exposure to agonist for a short time appears unreasonable. Yet, it is on such a postulate that a quantitative account of competitive antagonism rests; it is generally accepted that the theory describes experimental results with considerable accuracy [8]. If it is accepted that agonists and antagonists combine with receptors in a mutually-exclusive way, it is expected that the extent of dissociation of antagonist from the receptor taking place during the brief testing period with doses of agonist must be very small. A large response may be produced within a few seconds of adding a dose of agonist, even while almost all the receptors are still occupied by antagonist; this indicates that agonist occupancy responsible for higher response can only be very limited. This argument directly leads to the "spare receptor" hypothesis. The only variation in arguments of rate theory is that the notion of “spare receptors” is replaced by that if spare capacity for more rapid association.

3.3.5. Inactivation theory

Receptor inactivation theory is based on the two state model originally proposed by Katz and Thesleff for ion channels [18]. Kenakinn [19] on his work on the Torpedo nicotinic receptor reported that the multimeric receptor exists in active and inactive states with ligand binding altering the equilibrium between these two states. Receptor inactivation theory reflects a synthesis of both occupancy theory and rate theory providing an alternative consideration for the study of the receptor ligand interaction. Inactivation theory assumes that *RL* complex is an intermediate “active state” that gives rise to an inactive form of the receptor, *R’*, which is part of an *RL* complex termed *R’L* [20].

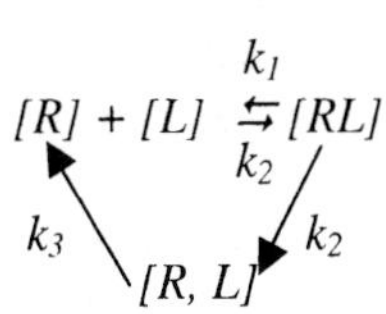

where *R* stands for receptor and *L* for ligand

A growing body of evidence suggests that the number of drug receptors on cell surfaces is not fixed, but is dynamically regulated by circumstances that include exposure to the ligand itself. Because most traditional theories of drug action are based on the assumption of a fixed number of receptors, it is desirable to examine the importance of this regulatory process on the interpretation of dose-effect data. The two major theories of drug action, however, are occupancy theory [5] and rate theory [7-9]. The liquid membrane hypothesis of drug action, which is the central theme of this monograph, will be discussed in the light of these two theories. The liquid membrane hypothesis of drug action is described in the following chapter.

REFERENCES

[1] J.N. Langely, J. Physiol., 23(1898) 240.

[2] J.H. Gaddum, Pharmacol.Rev., 9(1957) 211.

[3] E.M. Ross and T.P. Kenakin, in J.G. Hardman, L.L. Limibird and A.G. Gilman (eds.) Goodman and Gilman's The Pharmacological Basis of Therapeutics, 10th Edn., McGraw-Hill, New York, 2001, (pp31-43).

[4] O. Arunalakashana and H.O. Schild, Brit .J. Pharmacol., 14(1959) 48.

[5] A.J. Clark, The Mode of Action of Drugs on Cells; E.Arnold Co., London(1933).

[6] A.J. Clark, General pharmacology in Handbuch der Experimentellen pharmacologie,Vol IV; Ed. by A.Heffter,Springer-Verlag, Berlin,1937.

[7] W.D.M. Paton, Proc. Roy. Soc., B, 154(1961) 21.

[8] W.D.M. Paton and H.P.Rang, Advan. Drug Res., 3(1966) 57.

[9] W.D.M. Paton, Proc. Roy. Soc. Med., 53(1960) 815.

[10] W.C. Bowman and M.J. Rand, Textbook of Pharmacology, 2nd Ed., Blackwell Scientific Publications, Oxford (1980),PP.39.19.

[11] E.J. Ariens (Ed.), Molecular Pharmacology, Academic Press, New York(1964),pp137.

[12] E.J. Ariens and A.M. Simonis, J. Pharm. Pharmacol., 16(1956) 379.

[13] R.P. Stephenson, Brit.J.Pharmacol.,11(1956) 379.

[14] R.P. Stephenson and R.B. Barlow, in Eds. R. Passmore and J.S. Robson, A companion to Medical Studies, Blackwell Scientific Publications Ltd., F.A. Davis Co., Philadelphia (1970), Chapter3.

[15] E.J. Ariens, J.M. Van Rossum and P.C. Koopman, Arch.Int.Pharmacodyn., 127(1960) 459.

[16] D.R. Waud, Pharmacol.Rev., 49(1968) 49.

[17] J.F. Moran, C.R. Triggle and D.J. Triggle, J. Pharm. Pharmcol., 21(1969) 38.

[18] B. Katz and S. Thesleff, J. Physiol., (Lond.), 138(1957) 63.

[19] T.P. Kenakin, Pharmacologic Analysis of Drug- Receptor Interaction, 3rd ed., Lippincot, Philadelphiaa, PA, 1997.

[20] M. Williams, C. Mehlin and D. Triggle in (Ed.) D.J. Abraham Burgers Medicinal Chemistry and Drug Discovery, , 6th Edn., Vol. 2, 2003 pp.327.

Chapter 4

The liquid membrane hypothesis of drug action

Before we state the liquid membrane hypothesis of drug action we would describe in some detail the liquid membrane hypothesis per se. This sequence of narration we believe will be helpful in understanding the liquid membrane hypothesis of drug action.

4.1. The liquid membrane hypothesis

The liquid membrane hypothesis was propounded by Kesting et al [1-4] in the context of water desalination. The accumulation of surface-active molecules at the interface, when they are added to an aqueous phase, is an expected behaviour. But, nonetheless, there are two novel features of the liquid membrane hypothesis: (i) the surfactant layer which forms spontaneously at the interface acts as a liquid membrane in series with the supporting membrane, and (ii) the progressive coverage of the supporting membrane by the surfactant layer liquid membrane, i.e., as concentration of the surfactant additive is increased, the interface gets progressively covered by the surfactant layer liquid membrane and at the CMC of the surfactant, it is completely covered.

Kesting et al. [3] have investigated the liquid membrane hypothesis utilizing various surfactants, to correlate surfactant concentrations at the cellulose acetate/saline solution interface with transport rates of water and salt from saline-surfactant solutions. The surfactants used by these authors were polyoxyethylenenonylphenols and poly (vinylakyl) ethers. The cellulose acetate membranes used in their experiments were casted from the solution developed by Loeb, Manjikian and McCutcheon [5]. The 0.25 mm thick films thus casted, were placed in stainless steel cell of the type described by Kesting, Brash and Vincent [6]. A 1% aqueous NaCl solution was circulated past the membranes at a rate sufficient to dissipate the boundary layer built up of the salt. The pressure within the test cell was 55± 1 atm at the temperature, 25±0.2°C. The transport rates of the salt (NaCl) and water were estimated as functions of bulk feed additive concentrations. The membranes used in these experiments were pre-heated at 85°C for 30 minutes. Such membranes were chosen for their lack of porosity to eliminate the possibility of pore plugging by the additives, as suggested by Michaels, Bixler and Hodges [7]. The normalized values of water flux (J_1/J_1°) and salt flux (J_2/J_2°), where J_1 and J_2 represent the values of water flux and salt flux respectively and J_1° and J_2° are the values prior to the addition of surfactants, were obtained at various concentrations of the surfactant additive. The data in the particular case of Dowfax *9N-9* are reproduced in Fig 1, which indicate that as the concentration of the surfactant additive is increased a layer of surfactant progressively covers the interface.

The resistance offered by this layer to the transport of both water and the salt, progressively increases upto the CMC; at this concentration the cellulose acetate membrane surface is completely covered by the surfactant layer. The minor changes in the material transport beyond the CMC (Fig. 1) were considered to be secondary effects associated with increasing density within the already fully developed surfactant layer. Because this layer is perm selective and offers resistance to material transport in the same manner as the gel membrane, which it covers without permeating, itself constitutes a membrane, though one in the liquid state.

Considerable amount of variability in the permeability and selectivity of the various membranes was observed. The liquid membranes generated by the surfactants possessing a high hydrophilic/hydrophobic ratio, offered higher resistance to salt flow and lower resistance to water flow than those generated by the surfactants possessing a low hydrophilic/hydrophobic ratio (Table 1, Fig 2). The fact that resistance to water transport decreased with increasing hydrophilic/hydrophobic ratio, supports the 'active site' concept in which a membrane allows water to permeate in proportion to the number of hydrophilic sites with which water can associate by hydrogen bonding.

Table1. The relationship between the CMC and the normalized water and salt transport*.

Surfactant	CMC in 1% aq. NaCl (ppm)	Minimum surfactant concentration at which fully developed liquid membranes exist		Average values of normalized transport rates at the CMC	
		Deduced from J_1/J_1°	Deduced from J_2/J_2°	J_1/J_1°	J_2/J_2°
Dowfax-9N-6	25	20	20	0.75	0.56
Dowfax-9N-9	43	40	40	0.85	0.56
Dowfax-9N-15	50	60	70	0.90	0.47
PVM	6	5	5-6	0.83	0.45

* 1% NaCl feed at 54.4 atm at 25°C; feed velocity, 100cm/s (taken from Ref.3)

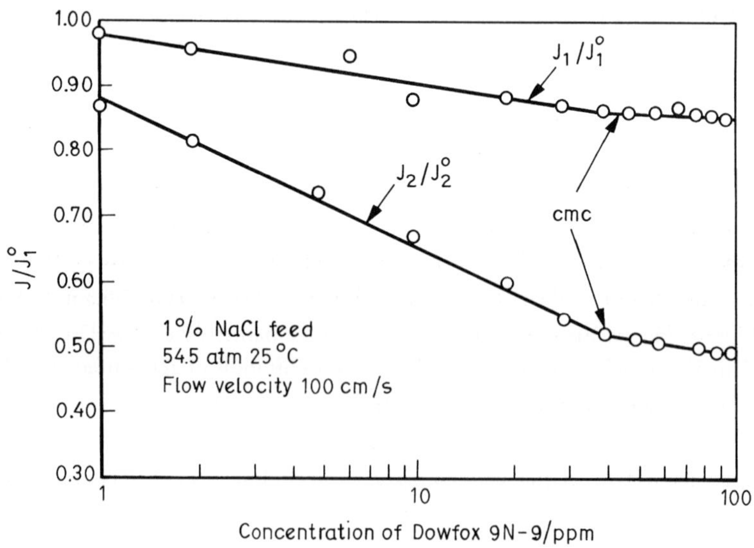

Fig.1 The effect of surfactant concentration on normalized transport rates (Ref. 3)

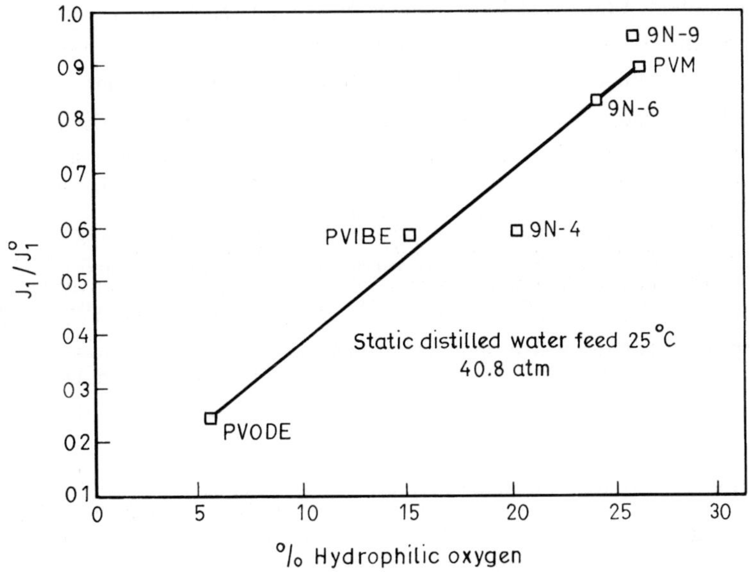

Fig 2. The effect of surfactant hydrophilicity on normalized rates of water transport (Ref. 3)

4.1.1. Further experiments on liquid membrane hypothesis

The experiments conducted by Srivastava and Yadav [8] on hydraulic permeability, electro-osmotic velocity, streaming potential and current in presence of polyvinylmethylether (PVM) have lent additional support to the liquid membrane hypothesis. The electro-osmotic cell [8] used for the transport studies essentially consisted of two compartments separated by a Sartorius cellulose acetate micro-filtrating membrane (average pore size, 0.2 μm). One of the compartments was attached to a pressure head, and the other a capillary for the measurement of volume flux. The solutions in the two compartments of the transport cell were well stirred using a magnetic stirrer. One of the compartments of the transport cell was filled with varying concentrations, ranging from 0 to 12 ppm of the aqueous solutions of PVM, and the other compartment was filled with distilled water and the data on hydraulic permeability, electro-osmotic velocity, streaming potential and current were obtained. Since the value of the CMC of aqueous PVM is 6 ppm [3], the concentration range of 0 to 12 ppm was purposely chosen so that the data are obtained on both the lower and the higher side of the CMC of PVM. The details of the experiments and procedures are available in the original paper [8].

The data on hydraulic permeability, electro-osmotic velocity and streaming current at various concentrations of PVM, obtained by Srivastava and Yadva [8], are reproduced in Fig 3 to 5. Except for the hydraulic data, all the other were found to be in accordance with the equations,

$$(J_v)_{\Delta P=0} = L_{12}\Delta\phi \qquad (1)$$

for electro-osmotic velocity, and

$(I)_{\Delta\phi=0} = L_{21}\, \Delta P$ (2)

for streaming current, derived form the linear phenomenological equations,

$J_v = L_{11}\, \Delta P + L_{12}\, \Delta\phi$ (3)

and

$I = L_{21}\, \Delta P + L_{22}\, \Delta\phi$ (4)

with $L_{12} = L_{21}$ on account of Onsager's theorem, obtained using non-equilibrium thermodynamic treatment [9] of electro-osmotic effects. In Eqs. (3) and (4), J_v stands for volume flux, I stand for flow of electricity; ΔP and $\Delta\phi$ are respectively the pressure difference and the electrical potential difference and the coefficients, L_{ik} are the phenomenological coefficients. The hydraulic permeability data in the presence of PVM (Fig. 3), instead of the equation,

$(J_v)_{\Delta\phi=0} = L_{11}\Delta P$ (5)

derived from eqs.(3), were found to be represented by the equation,

$J_v=L_{11}\, [\Delta P - \Delta P_0\, \{1 - exp\, (-\Delta P/\Delta P_0)\}]$ (6)

where ΔP_0 is the extrapolated intercept of the straight-line part of the curves on the ΔP-axis. When ΔP assumes such high values that the term $exp\, (-\Delta P/\Delta P_0)$ becomes much smaller than unity, the Eq. 6 simplifies to

$J_v = L_{11}\, (\Delta P - \Delta P_0)$ (7)

which represents the straight-line parts of the curves II to IV in Fig. 3. Since water flow through the cellulose acetate microfiltration membrane obeys the linear relationship (5) (Fig. 3, curve I), this kind of non-ideal behaviour (Eq. 6) appears to be on account of the flow through the liquid membrane generated by PVM in series with the cellulose acetate microfiltration membrane as hypothesized by Kesting et al [1-4]. However, the cause for the non-proportional flow (Eq. 6) through PVM liquid membrane remains unexplored. A more definite indication of the formation of the liquid membranes can be had from the gradation in the values of the phenomenological coefficients as the concentration of PVM is increased from 0 t0 12 ppm (Table 2). Values of the coefficient L_{11} (estimated from the slopes of the straight line portions of the curves in (Fig. 3), show a progressive decrease as the concentration of PVM is increased form 0 to 6 ppm (the CMC value). When the concentration of PVM is increased further beyond 6 ppm, the value of L_{11} also decreases but this decrease is much less pronounced than the decrease observed upto 6 ppm – the CMC value for aqueous PVM. Similar trends can be seen in the values of L_{12} and L_{21} (Table 2). These trends are in keeping with the liquid membrane hypothesis of Kesting et al [1-4] and indicate the progressive coverage of the supporting membrane, the cellulose acetate micro-filtration membrane in these experiments [8], as the concentration of the surfactant is increased from 0 to the CMC of the surfactant. The slight decrease in the values of L_{11} etc., beyond the CMC could possibly be, as hypothesized by Kesting et al [3], due to increase in the density of the surfactant layer liquid membrane which at the CMC is fully developed and completely covers the supporting membrane.

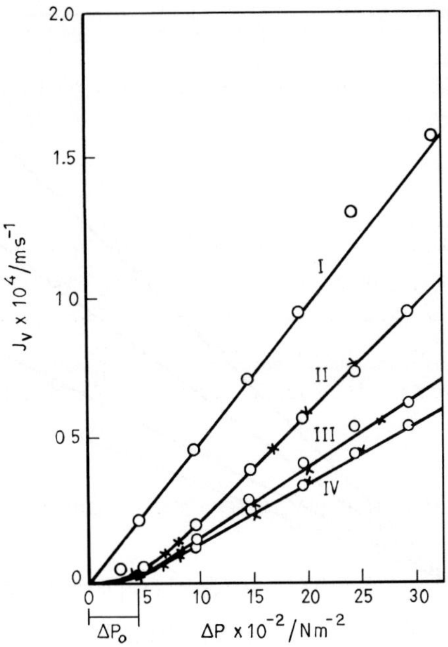

Fig.3. The hydraulic permeability data. Curves I, II, III are for 0, 3, 6 and 12 ppm PVM respectively; O, experimental points; x, theoretical points as predicted by the Eq. (6) (Ref.7).

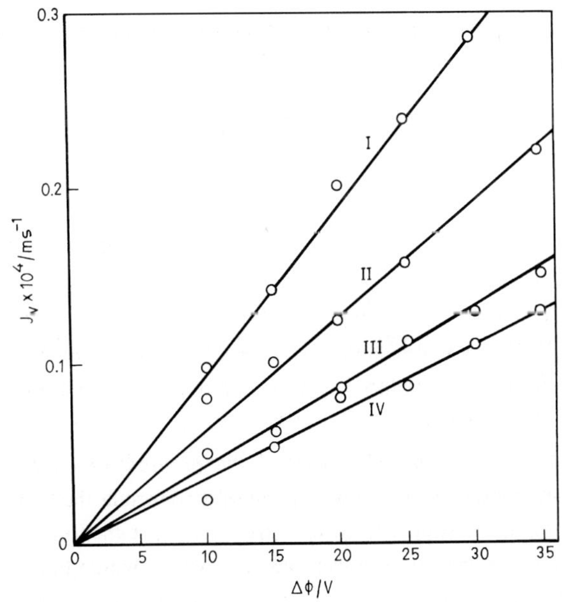

Fig.4. The electro-osmotic velocity data. Curves I, II, III, IV are 0, 3, 6 and 12 ppm of PVM respectively (Ref. 8).

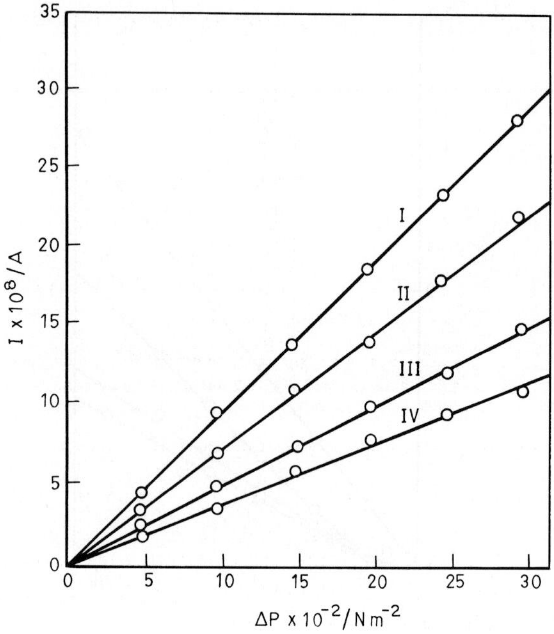

Fig.5. The streaming current rate. Curves I, II, III, IV are for 0, 3, 6 and 12 ppm of PVM respectively (Ref. 8).

Table 2. Values of the phenomenological coefficients L_{11}, L_{12}, L_{21}, L_{22} and the electrical resistance at various concentrations of PVM (Ref. 8).

Concentration of PVM/ppm	0	3	6	12
$L_{11} \times 10^7/m^3N^{-1}s^{-1}$	0.486	0.38	0.246	0.212
$L_{12} \times 10^6/mA\ J^{-1}$	0.936	0.63	0.44	0.37
$L_{21} \times 10^6/mA\ J^{-1}$	0.925	0.62	0.45	0.36
$L_{22} \times 10^2/ohm^{-1}m^{-2}s^{-1}$	2.36	2.33	2.23	2.14
Resistance x 10^{-6}/ohm	0.445	0.45	0.47	0.49

To furnish further evidence in favors of the progressive coverage, the method of analysis for mosaic membranes [10-12] was utilized [8]. Since at the CMC of the surfactant, the supporting membrane is supposed to be fully covered with liquid membrane, one can logically expect that at concentrations lower that the CMC, it would be only partially covered with the liquid membrane. The situation is pictorially depicted in Fig. 6. For such a situation the volume flow per unit area (J_v) through the mosaic membrane would be given by the equation,

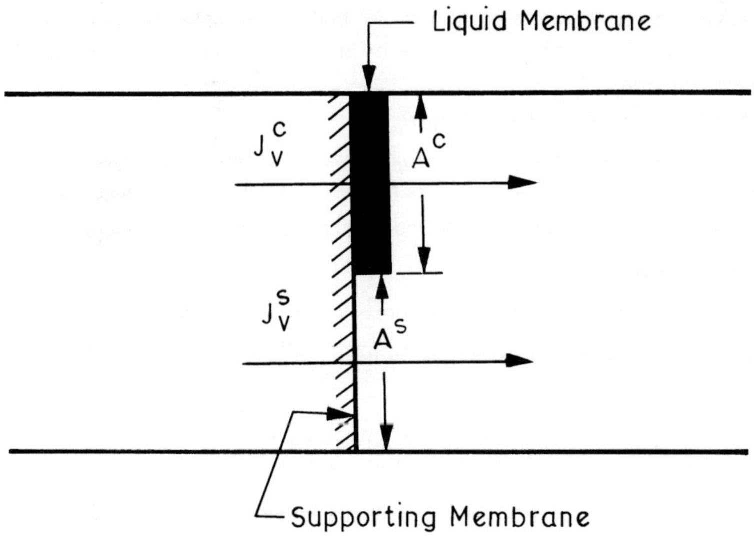

Fig.6. The schematic representation of mosaic membrane formed when the concentration of the surfactant is lower than its critical micelle concentration.

$$J_v (A^s + A^c) = J_v^c A^s + J_v^s A^c \tag{8}$$

where *s* and *c*, respectively, stand for the bare supporting membrane and the composite membrane consisting of the liquid membrane and the supporting membrane in series with each other, and *A* represents the area of the membrane denoted by the superscripts. If we focus attention on the linear region of the curves in Fig. 3 and utilize the linear relationships, Eq. 5 and 7 and, between volume flux and the pressure difference, Eq. 8 can be transformed in to

$$J_v = [[L_{11}^s A^s/(A^s + A^s) + L_{11}^c A^c/(A^s + A^c)]\Delta P - [L_{11}^c A^c/(A^s + A^c)]\Delta P_0 \tag{9}$$

Since at 6 ppm (the CMC value), the supporting membrane is supposed to be fully covered with the surfactant layer liquid membrane, the concept of progressive coverage would imply that when concentration of the surfactant is one-half of its CMC, i.e., at 3 ppm concentration, the fraction of the total area of the supporting membrane covered with the liquid membrane would be equal to one half and when the surfactant concentration is one-third of its CMC, the fraction of the total area covered with the liquid membrane would be equal to one-third and so on. Thus, the slope of the straight line part of the J_v versus ΔP curve for 3 ppm concentration of PVM (Fig 3) should be equal to $(L_{11}^s + L_{11}^c)/2$, where L_{11}^s and L_{11}^c, respectively,, are the values of the slopes corresponding to 0 to 6 ppm concentrations of PVM. Value of the slope, thus computed, comes out to be 0.336×10^{-7}, which matches with the experimental value of L_{11} for 3 ppm concentration of PVM (Table 2). Similar considerations apply to the other phenomenological coefficients.

In cases where variation of J_v with ΔP both in the presence and in the absence of surfactant layer liquid membrane is given by the proportional relationship, Eq.(9) would reduce to the equation,

$$J_v = [L_{11}^s A^s/(A^s + A^c) + L_{11}^c A^c/(A^s + A^c)]\Delta P \quad (10)$$

In general terms, the concept of progressive coverage in such cases would imply that when the concentration of the surfactant is n times its CMC, n being less than or equal to 1, the coefficient L_{11} would be related to the coefficient L_{11}^s and L_{11}^c by the equation,

$$L_{11} = (1 - n)\ L_{11}^s + n\ L_{11}^c \quad (11)$$

where L_{11}^s and L_{11}^c represent the values of L_{11} for 0 and the CMC of the surfactant, respectively. This kind of analysis can be utilized with other phenomenological coefficients also.

4.1.2. Examples of liquid membrane from biologically relevant substances: for example bile salts

Bile salts, which are excellent surfactants, have also been shown [13] to generate liquid membranes at the interface, in accordance with Kesting's hypothesis [1-4]. The experiments were conducted on sodium deoxycholate. The hydraulic permeability data and their analysis in the light of the mosaic membrane model, in the manner described in the previous section, were utilized to demonstrate the formation of a liquid membrane at the Sartorius cellulose acetate micro filtration membrane/aqueous interface. Studies on the simultaneous transport of solute (potassium chloride) and solvent (water) through the liquid membrane generated by sodium deoxycholate in series with the supporting membrane were also conducted [13] and the value of reflection coefficient (σ) and the solute permeability (ω) for the liquid membrane were estimated. For this, the equations [9, 14],

$$J_v = L_p\,(\Delta P - \sigma\Delta\Pi) \quad (12)$$

$$J_s = \omega\Delta\Pi + \bar{C}_s\,(1-\sigma)\,J_v \quad (13)$$

derived from the linear phenomenological equations [9, 14],

$$J_v = L_p\ \Delta P\ +\ L_{PD}\ \Delta\Pi \quad (14)$$

$$J_D = L_{DP}\ \Delta P + L_D\ \Delta\Pi \quad (15)$$

$$\text{with } L_{PD} = L_{DP} \quad (16)$$

between fluxes and forces in an osmometric situation, were utilized. In equations (12) to (15), J_v represents the volume flux, J_D measures the velocity of solute relative to that of solvent, J_s the solute flux, ΔP, the pressure difference, $\Delta\Pi$, the osmotic pressure difference and L_p, and L_{PD} and L_D are the phenomenological coefficients and related to σ and ω by the Eqs. (17) and (18),

$$\sigma = (\Delta P/\Delta\Pi)\ j_{v=0} = -\ L_{PD}/L_P \quad (17)$$

$$\omega = (J_s\,/\,\Delta\Pi)_{Jv=0} =\ \bar{C}_s\,[(L_P L_D - L_{PD} L_{DP})/L_P] \quad (18)$$

$\bar{C}_s$ being average of the solute concentrations in the two compartments of the transport cell. The value of σ lies between 0 and 1. When $\sigma = 1$, the membrane is said to be an ideal semi permeable membrane. The values of σ were estimated using equation (12), according to

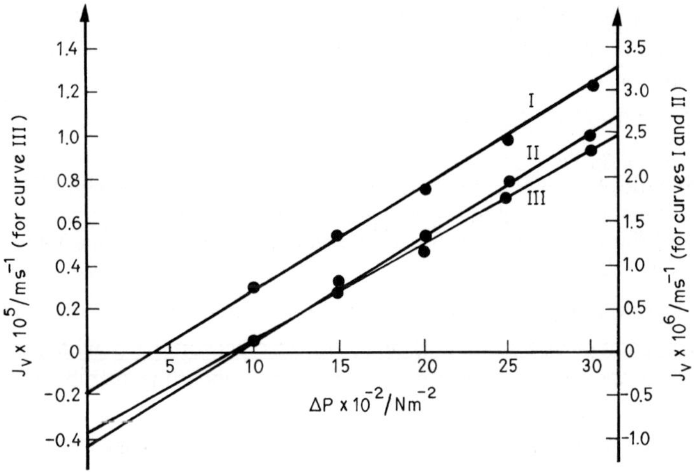

Fig.7. Determination of L_P and σ using Eqn. (12). Curve I, ... $\Delta\Pi = 25.98 \times 10^2\ N\ m^{-2}$. Curve II........... $\Delta\Pi = 10.38 \times 10^2\ N\ m^{-2}$; Currve III... (Blank experiment) $\Delta\Pi - 25.98 \times 10^2\ N\ m^{-2}$ (Ref. 13).

which if $\Delta\Pi$ is held more or less constant, a plot of J_v against ΔP should be a straight line with an intercept equal to $\sigma\Delta\Pi$ on the ΔP axis. The plot of the experimental data [13] utilized to evaluate σ is reproduced in Fig. 7.

The values of ω were estimated using the definition given by the Eq.(18). The details of the experiment are given in the original paper [13]. From the experimentally determined values of σ and ω for the supporting membrane and the composite membrane consisting of the liquid membrane generated by sodium deoxycholate in series with the supporting membrane, the values of σ and ω for liquid membrane were estimated using the analysis summarized below [15, 16].

Writing the phenomenological Eqs. (14) and (15) in the inverted matrix from, i.e.,

$$\Delta P = R_P J_V + R_{PD} J_D \tag{19}$$

$$\Delta\Pi = R_{DP} J_V + R_D J_D \tag{20}$$

with $R_{PD} = R_{DP}$ (21)

where the resistance coefficients R_{ik} are related to the phenomenological coefficients L_{ik} by the Eqs.,

$$R_P = L_D / L_P L_D - L_{PD} L_{DP} \tag{22}$$

$$R_{PD} = - L_{PD} / L_P L_D - L_{PD} L_{DP} \tag{23}$$

$$R_{DP} = - L_{DP} / L_P L_D - L_{PD} L_{DP} \tag{24}$$

$$R_D = L_P / L_P L_D - L_{PD} L_{DP} \tag{25}$$

and invoking the assumptions made by Kedem and Katchalsky [15] in the theory for the permeability of series of composite membranes, the following combination rules connecting the resistance coefficients for the series composite membranes and of its constituent membrane elements can be written as follows:

$$R_P^T = R_P^S + R_P^l \quad (26)$$

$$R_{PD}^T = R_{PD}^S + R_{PD}^l \quad (27)$$

$$R_{DP}^T = R_{DP}^S + R_{DP}^l \quad (28)$$

$$R_D^T = R_D^S + R_D^l \quad (29)$$

where the superscript T, S and *l* stand for the total, i.e., series composite membrane, the supporting membrane and the liquid membrane respectively. From equations, (19) and (20), one can write the following equation for the reflection coefficient for the total composite membrane,

$$\sigma^T = (\Delta P/\Delta \Pi)_{Jv=0} = R_{PD}^T / R_D^T \quad (30)$$

which, using the combination rules (27) and (29), can be correlated with the reflection coefficients for the constituent membrane elements, the supporting membrane and liquid membrane, i.e.,

$$\sigma^T = \sigma^S (R_D^S / R_D^T) + \sigma^l (R_D^l / R_D^T) \quad (31)$$

The coefficients of σ^S and σ^l in Eq.(31) can be seen to be related to each other by the Eq.(32),

$$R_D^S / R_D^T + R_D^l / R_D^T = 1 \quad (32)$$

Using equations, (31) and (32), it is possible to compute the value of σ^l from the experimentally determined values of σ^T and σ^S, because in view of Eq.(18), the value of (R_D^S / R_D^T) can be estimated from the experimentally determined values of ω for the supporting membrane and the total series composite membrane. The value of solute permeability for the liquid membrane (ω^l) can be estimated from the experimentally determined values for the supporting membrane and the total series composite membrane using the relationship (33) derived by Kedem and Katchalsky [15, 16], i.e.,

$$1/\omega^T = 1/\omega^S + 1/\omega^l \quad (33)$$

The values of σ^l and ω^l thus computed for the transport of potassium chloride through the liquid membrane generated by sodium deoxycholate were found to be *0.442* and 0.59×10^{-9} mol s^{-1} N^{-1} respectively.

The fact that bile salts generate liquid membranes, which influence solute transport, appears to have significant biological implications. Bile salts are present in plenty in the intestinal region, which is an important location for absorption and digestion of food. Bile salts are not only capable of creating a liquid membrane barrier for the permeants, but according to some reports, are also capable of influencing the action of enzymes [17]. In fact, there are several reports [18, 19] on the influence of surfactants on enzyme action.

Bile salts are not isolated examples. Molecules of surface-active nature are of wider occurrence in biological systems, and are crucial to organization of living matter. The formation of cell membrane and location of proteins in the lipid bilayer part of the membranes is a consequence of surface activity of the molecules constituting biological membrane. Hence, the liquid membranes mono and bilayer systems generated by the constituents of biological membranes should also be capable of acting mimetic systems for biomembranes. Efforts made in this direction have indicated in favors of such a possibility. The following chapter 5 contains an account of such efforts.

The model studies, as to the role of liquid membrane of drug action, described in Chapter 6, have been conducted on liquid membrane systems. This fact makes chapter 5 relevant and also a pre-requisite to the understanding of the studies contained in Chapter 6.

4.2. The liquid membrane hypothesis of drug action

Having described the liquid membrane hypothesis and its genesis we are now in a position to take up the liquid membrane hypothesis of drug action.

In the previous chapter (chapter 3) we have given an account of the theories of drug action. The two theories, which are most relevant to the present discussion, are "the occupancy theory" and the rate theory.

The two theories though differ in arguments, have a common premise that the observed biological effects are a consequence of interaction of drugs with membrane components. The antagonistic drugs, in general, are stated to interact with the membrane components, and occupy the sites with which the agonist drugs would have interacted to give the desired biological response.

Thus, it can be stated that antagonistic drugs act by creating hindrance in the interaction of agonist drugs with receptor sites. How is this hindrance created? It is contained in the liquid membrane hypothesis of drug action.

The membranes represent an interface. As a corollary, any drug, which acts by modifying the permeability of cell membranes after interacting with them, of necessity, has to be surface active in nature. Since surface-active substances are capable of forming liquid membranes, which can influence mass transfer across the interface (Kesting's hypothesis), the formation of liquid membrane at the site of action could be an important event in the mechanism of action of surface-active drugs. Thus, the central concept in the liquid membrane hypothesis of drug action is that the surface-active drugs may generate a liquid membrane at the site of action, which acts as a barrier modifying the transport of relevant molecules to these sites. This is in addition to the concepts like structural complementarily of the antagonist drugs enabling them to interact with the same receptor sites with which the agonist molecules interact. The liquid membrane generated by the drug itself, acting as barrier modifying access of relevant molecules to the site of action is a new facet of drug action. If this concept is viewed in the light of the "occupancy theory" and the 'rate theory', a more rational biophysical explanation for the action of surface-active drugs, which act by modifying the permeability of cell membranes, emerges.

The reason, why such a possibility has gone unnoticed so far, appears to be due to the fact that passive transport processes have traditionally been considered unimportant for biological actions – transport through the liquid membranes are undoubtedly passive in nature.

It may, however, be clarified that the liquid membrane hypothesis in no way disputes the specific/active interaction between the agonist drugs and their receptors. The liquid membrane formation is an event, which precedes the active interaction. The new point of the hypothesis lies in the assertion that the passive transport through the liquid membrane also makes a significant contribution to drug action.

REFERENCES

[1] R.E. Kesting, A. Vincent and J. Eberlin, OSW, R and D Report 117, August 1964.

[2] R.E. Kesting, "Reverse Osmosis Process using Surfactant Field Additives" OSW Patent Application SAL 830 Nov.3, 1965.

[3] R.E. Kesting, W.J. Subcasky and J.D. Paton, J. Colloid Interface Sci., 28 (1968) 156.

[4] R.E. Kesting, Synthetic Polymeric Membranes – A Structural Perspective, 2nd edition, Wiley Interscience, 1985.

[5] S. Leob, S. Manjikian and L. McCutchenon, "Paper presented at the first International Symposium on Water Desalination", Washington D.C., Oct. 3-9, 1965.

[6] R.E. Kesting, M. Barsh and A. Vincent, J. Appl. Polymer Sci., 9(1965)1873.

[7] A. Michaels, H. Bixler and R. Hodges, Jr., J. Colloid Sci., 20(1965)1034.

[8] R.C. Srivastave and Saroj yadav, J. Colloid Interface Sci., 69(1979) 280.

[9] A. Katchalstry and P.F. Curran, "Non-equilibrium Thermodynamics in Biophysics", Harvard University Press, Cambridge, Mass, 1965.

[10] K.S. Spiegler and O. Kedem, Desalinaton, 1(1966) 311.

[11] T.K. Sherwood, P.L.T. Brain and R.E. Fischer, Ind. Eng. Chem., Fundam, 6(1967) 2.

[12] F.L. Harris, G.B. Humphreys and K.S. Spiegler in P. Meares (Ed), Membrane Separation Process, Chapter 4, Elsevier, Amsterdam, 1976.

[13] R.C. Srivastava and Saroj Yadav, J. Non-equilib. Thermodyn., 4(1979) 219.

[14] O. Kedem and A. Katchalsky, Biochim. Biophys. Acta, 27(1979) 229.

[15] O. Kedem and A. katchalsky, Trans. Faraday Soc., 59(1963)1941.

[16] A. Katchalsky and O. Kedem, Biophys. J., 2(1962) 53.

[17] S.B. Bhise, M.N.A. Rao and R.C. Srivastava, J. Colloid Interface Sci., 78(1980) 563.

[18] W.P. Jencks, "Catalysis in Chemitry and Enzymology", McGrawHill, New York, 1969.

[19] R. Verger and G.H. de Haas, Annu. Rev. Biophys. Bioeng., 5(1976) 77.

[20] R.C. Srivastava, S.B. Bhise and S.S. Mathur, Adv. Colloid Interface Sci., 20(1984) 131.

Chapter 5

Liquid membranes as biomimetic system

Since the studies contained in chapter 6, as to the role of liquid membranes is drug action, have been conducted on the liquid membrane (mono or bilayer) systems, we in this chapter describe, the liquid membranes, how they are formed and their capability to mimetic biologically relevant transport processes. In demonstrating the biomimetic capability of liquid membranes we have, among others, also chosen the biomimetic studies on the action of biological agents like insulin, vasopressin, polyene antibiotics, prostaglandins etc. The studies on such biological agents are expected to be relevant to studies on drug action described in the following chapter.

5.1. Introduction

Biological membranes play a crucial role in almost all cellular phenomena. They consist mainly of lipids and proteins. Although biological membranes are complex and highly variable both in structure and in function, there is a basic construct common to all of them. The gross organizations of proteins and lipids, which are common to all biological membranes have been suggested by Singer and Nicoloson [1]. This is known as fluid mosaic model and is pictorially depicted in Fig.1. It has been shown that the lipid bilayer, which is in a fluid state and in which proteins are incorporated, is the basic matrix of all biological membranes - the lipids are the mortar and the proteins are the bricks. This is the reason why model systems have been prepared for the lipid bilayer part of biomembranes, e.g., BLM [2], in which the desired molecules can be incorporated depending upon which property of the biomembranes one desires to mimic. The attempt in this chapter is to show that the liquid membrane bilayers generated by the membrane constituents (lipids) also possess the potential of being used as mimetic systems for biomembranes. Experiments carried out giving a positive indication in favors of such a possibility are described and discussed in this chapter.

5.2. Liquid membranes from cholesterol, lecithin and lecithin-cholesterol mixtures [3, 4]

5.2.1. Liquid membranes from cholesterol [3]

Cholesterol, which is an important constituent of bio-membranes, though very slightly soluble in water, has been shown [5, 6] to lower considerably the surface tension of water. Cholesterol has a maximum solubility [7,8] of $4.7 \mu M$ in water and the measured surface tension of its saturated solution in water is 33 dyne/cm at 25°C. Its CMC is in the range [7, 8] of 25-40 μM. All this indicates that cholesterol is a very effective surfactant and should be capable of generating a liquid membrane at the interface in accordance with Kesting's liquid membrane hypotheses. The data on hydraulic permeability, electro-osmotic velocity and streaming current have been utilized [3] to establish the existence of liquid membrane phenomena in cholestrol. Experiments have also been designed [3] to demonstrate the formation of bilayers of cholesterol liquid membrane.

Aqueous solutions of cholesterol were prepared using the method described by Gershfeld and Pagano [6]. The necessary weight of cholesterol to attain the desired concentration was dissolved in ethanol and the solution was added with constant stirring to the aqueous phase. The stirring was continued for several days, always for more than 120 hours. The CMC of aqueous cholesterol, as determined from the variation of surface tension with concentration, was found to be 30.08 nM, which falls within the range reported in literature. In the aqueous solutions of cholesterol, prepared using the method described by Gershfeld and Pagano [6], the final concentration of ethanol was never allowed to increase 0.1% by volume because it was experimentally found that a 0.1% solution of ethanol in water did not lower the surface tension of water to any measurable extent.

The all-glass cell [3] used for the transport studies is diagrammed in Fig.2, which has been well labeled to make it self-explanatory. A Sartorius cellulose acetate microfiltration membrane Cat. No. 11107, average pore size, $0.2\mu M$ of thickness, 1×10^{-4}m and area, $2.55\times10^{-5}m^2$, in fact acted as a support for the liquid membrane dividing the transport cell into two compartments C and D. For the measurements of hydraulic permeability, electro-osmotic velocity, streaming potential and current, the two halves, i.e., the compartment C and the compartment D of the transport cell (Fig.2) were filled with the solutions of desired concentrations.

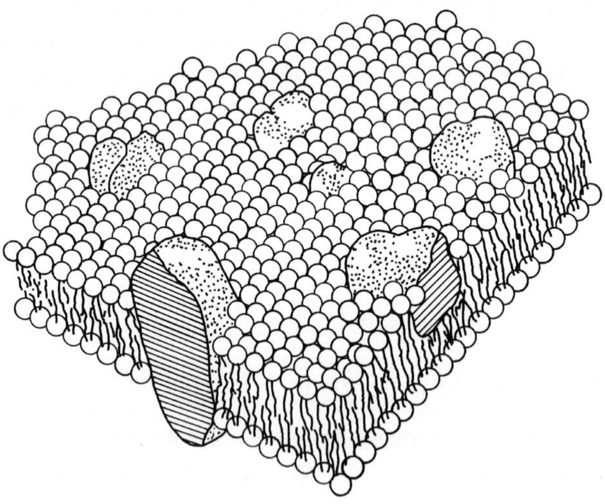

Fig. 1. The lipid-globular protein mosaic model with a lipid matrix (the fluid mosaic model): Schematic three-dimensional and cross-sectional views. The solid bodies with stippled surfaces represent the globular integral proteins, which at long range are randomly distributed in the plane of membrane. At short range, some may from specific aggregates, as shown (Ref. 1).

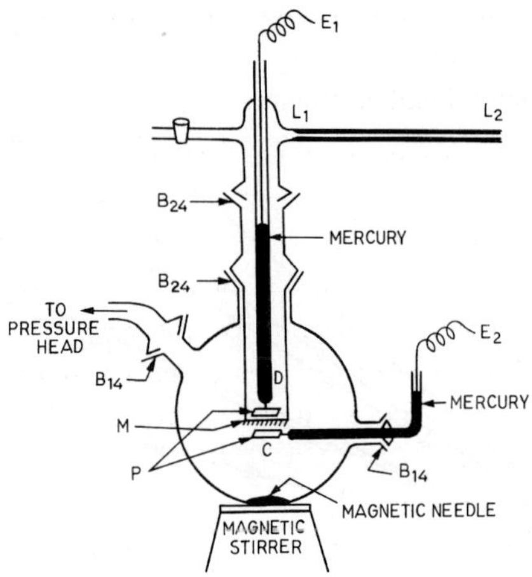

Fig. 2. The transport cell: M-supporting membrane, P-bright platinum electrodes, E_1, E_2-electrode terminals, L_1L_2-capability.

Since the CMC of aqueous cholesterol is 30.08 nM, the concentration range form 0 to 56.4 nM was chosen for the measurement of hydraulic permeability, electro-osmotic velocity, streaming potential and current in order to obtain data on both the lower and the higher side of the CMC of cholesterol. For measurements of hydraulic permeability the electrodes E_1 and E_2 were short-circuited and the volume flow consequent to the various pressure difference (ΔP) applied across the membrane, was noted in the capillary, L_1L_2 (Fig.2). For electro-osmotic velocity measurements, the condition, $\Delta P=0$ was imposed on the system and the volume flow, induced by the electrical potential differences across the membrane, was noted. For measurements of streaming potentials/ current, known pressure differences were applied across the membrane, and when flow in the capillary, L_1L_2 became steady, the electrodes E_1 and E_2 were joined with measuring devices.

The transport data [3] for various concentrations of cholesterol is reproduced in Figs. 3 to 6. The straight line plots are in accordance with the linear equations for hydraulic permeability, electro-osmotic velocity, streaming potential and current derived from the linear phenomenological equations, for electro osmotic effects (see Eqs. (3) and (4) of chap 4). Values of the various phenomenological coefficients, viz., L_{11}, L_{12}, L_{21} and L_{22} occurring in Eqs. (3) and (4) of chapter 4 at various concentration of cholesterol, estimated from the slopes of the straight lies in Fig.3 to 6 are reproduced in Table 1. The validity of Onsager's equality, viz., $L_{12} = L_{21}$, for all concentrations of cholesterol, is obvious from the values recorded in Table 1.

Let us now focus attention on the hydraulic permeability data (Fig.3, Table 1). From the variation of the coefficient L_{11} with cholesterol concentration, plotted in Fig.7, it is obvious that as concentration of cholesterol increase, the resistance to volume flow also increases in a progressive manner and it is maximum when the concentration of cholesterol equals its CMC beyond which it becomes more or less constant. Although , it is by the way, from the nature of curve.

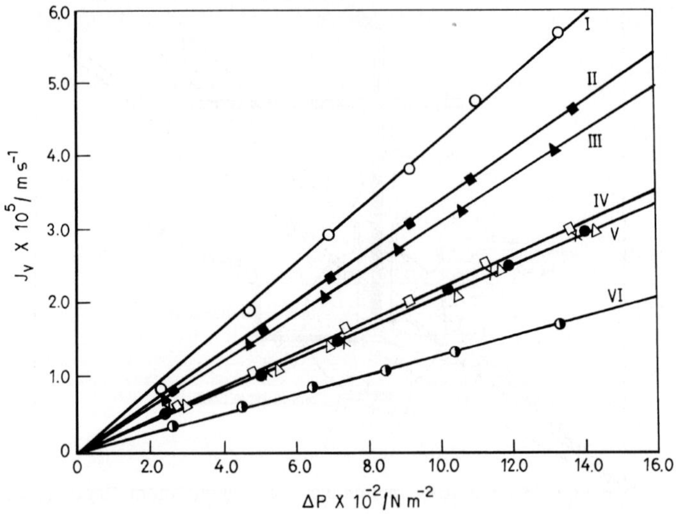

Fig. 3. Hydraulic-permeability data. Curves I-V are for the case when compartment C was filled with cholesterol solutions and compartment D with water. Cholesterol concentration: (O)0, (■)9.4, (▲)15.04, (□)28.2, (Δ)30.08, (●)37.6 and (x)56.4 nM. Curve VI is for the case when both compartments were filled with cholesterol solution of concentration equal to its CMC.

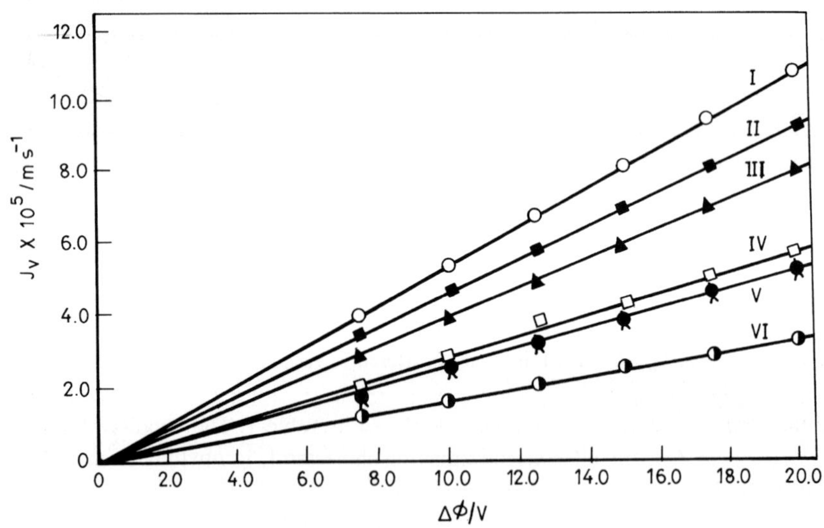

Fig. 4. Electro osmotic velocity data. Curves I-V are for the case when compartment C was filled with cholesterol solutions and compartment D with water. Cholesterol concentration: (O)0, (■)9.4, (▲)15.4, (□)28.2, (Δ)30.08, (●)37.6 and (x)56.4 nM Curve VI is for the case when both compartments were filled with cholesterol solution of concentration equal to its CMC.

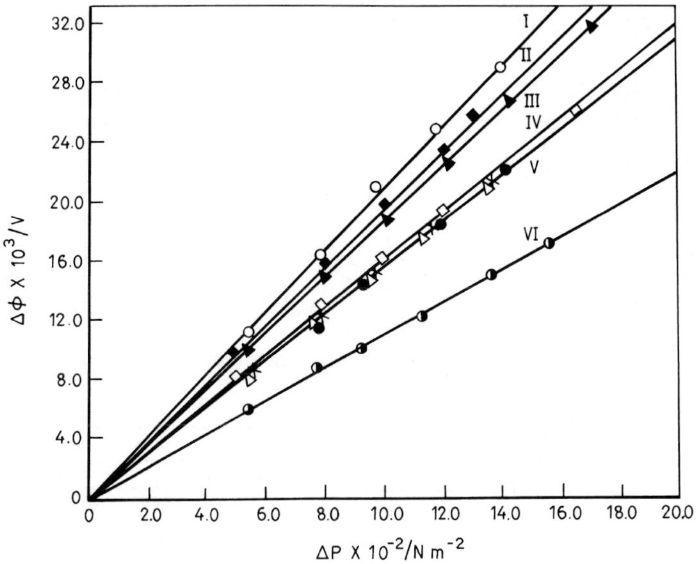

Fig. 5. Streaming potential data. Curves I-V are for the case when compartment C was filled with cholesterol solutions and compartment D with water. Cholesterol concentration: (O)0, (■)9.4, (▲)15.4, (□)28.2, (Δ)30.03, (●)37.6 and (x)56.4 nM. Curve VI is for the case when both compartments were filled with cholesterol solution of concentration equal to its CMC.

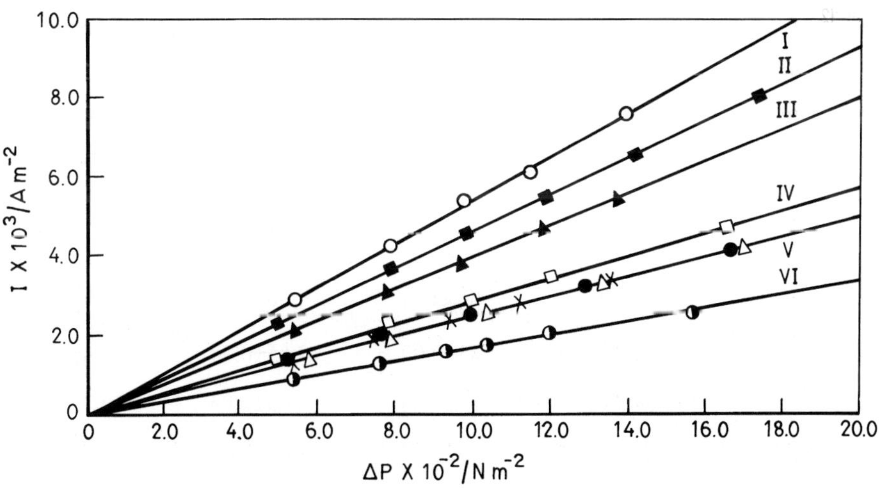

Fig. 6. Streaming current data. Curves I-V are for the case when compartment C was filled with cholesterol solutions and compartment D with water. Cholesterol concentration: (O)0, (■)9.4, (▲)15.4, (□)28.2, (Δ)30.08, (●)37.6 and (x)56.4 nM. Curve VI is for the case when both compartments were filled with cholesterol solution of concentration equal to its CMC.

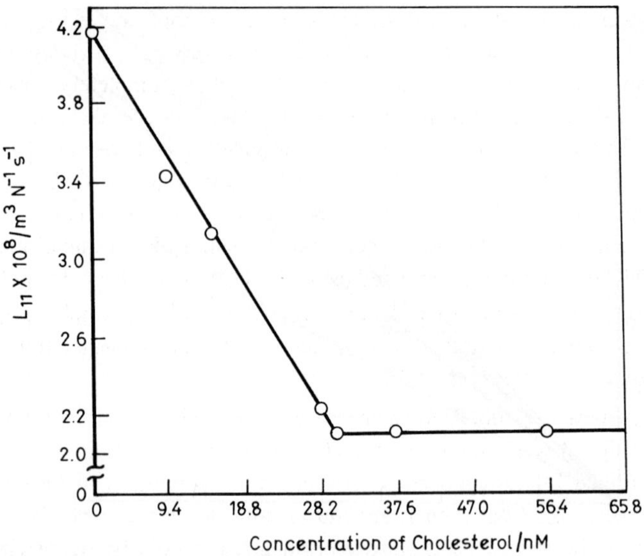

Fig. 7. Variation of L_{11} with the concentration of cholesterol (Ref. 3).

in Fig.7 one is tempted to suggest that the variation of the phenomenological coefficient with concentration can be exploited for the CMC determination of surfactants. The trend in Fig.7 is in accordance with Kesting's liquid membrane hypothesis [9-11].

Table 1. Values of the phenomenological coefficients, L_{ik} at various cholesterol concentrations (Ref.3).

Cholesterol conc. (*nM*)	$L_{11}\times10^8$ ($m^3\ s^{-1}\ N^{-1}$)	$L_{12}\times10^6$ ($mA\ J^{-1}$)	$L_{21}\times10^6$ ($mA\ J^{-1}$)	$L_{22}\times10^6$ ($ohm^{-1}\ m^{-2}$)
0.00	4.17±0.08	5.36±0.03	5.36±0.12	2.62±0.17
9.40	3.43±0.03	4.65±0.05	4.67±0.03	2.31±0.13
15.04	3.14±0.02	3.99±0.03	3.99±0.06	2.07±0.11
28.20	2.22±0.03	2.83±0.02	2.85±0.04	1.78±0.08
30.08*	2.11±0.04	2.62±0.04	2.60±0.04	1.63±0.07
37.60	2.13±0.03	2.63±0.04	2.62±0.04	1.57±0.06
56.40	2.14±0.01	2.61±0.02	2.59±0.01	1.51±0.06
C**	1.34±0.04	1.70±0.03	1.67±0.04	1.26±0.04

* CMC. **Values of L_{ik} for the system both the compartments, C and D (Fig.2) were filled with cholesterol solution of concentration equal to its CMC.

As concentration of the surfactant is increased the supporting membrane – the cellulose acetate microfiltration membrane in this case gets progressively covered with the surfactant layer liquid membrane and at the CMC, it is completely covered. When the concentration of the surfactant increases beyond the CMC almost all of the added surfactant remains in the bulk of the solutions in the form of micelles, and does not go to the interface. This is why the resistance to flow does not increase beyond the CMC of the surfactant. Analysis of the transport data in light of the mosaic membrane mode [12-14] furnishes further evidence in favors of the liquid membrane formation. The value of the coefficient L_{11} for half the CMC of cholesterol computed using Eq. (11) of chapter 4 derived on the basis of the mosaic model comes out to be $(3.14\pm0.03)\times10^{-8}\ m^3\ N^{-1}\ s^{-1}$, which compares favorably with the experimentally determined value (Table 1). Similar considerations apply to other phenomenological coefficients as well.

The transport data obtained in the case when both compartments C and D of the transport cell (Fig. 2) were filled with the cholesterol solution of concentration equal to its CMC, were utilized [3] to demonstrate the formation of bilayers of the cholesterol liquid membrane. Since at the CMC, the supporting membrane gets completely covered with the liquid membrane, the supporting membrane in this case would be sandwiched between the two layers of the liquid membrane generated, one on either side of it. In dealing with a situation like this it is more convenient to utilize the inverse phenomenological equations [15] between thermodynamic forces, X and fluxes, J, i.e.,

$$X_i = \sum_k R_{ik}\ J_k \tag{1}$$

where the resistance coefficients, R_{ik} are related to the coefficients, L_{ik} by

$$R_{11} = (L_{22}/|L|),\ R_{12} = (-L_{12}/|L|),\ R_{21} = (-L_{21}/|L|),\ R_{22} = (L_{11}/|L|) \tag{2}$$

In Eq. (2),

$$|L| = L_{11}L_{22} - L_{12}L_{21} \tag{3}$$

Utilising Kedem and Katchalsky's theory [16, 17] for permeability of composite membranes, one can write the following relationship among the resistance coefficients, R_{ik}^* for the series composite membrane (the supporting membrane sandwiched between the two layers of the cholesterol liquid membranes) and the corresponding resistance coefficients for the constituent membrane elements,

$$R_{ik}^* = R_{ik}^s + 2\ R_{ik}^l \tag{4}$$

The superscripts s and l stand for the supporting membrane and the liquid membrane, respectively. Similarly, for the situation when one of the compartments of the transport cell (Fig. 2), the compartment C, was filled with cholesterol solution of concentration equal to its CMC, and the other compartment, the compartment D (Fig. 2), was filled with water, one can write,

$$R_{ik}^T = R_{ik}^s + R_{ik}^l \tag{5}$$

where the superscript T stands for the series composite membrane consisting of the supporting membrane and the cholesterol liquid membrane in series array. Using Eq.(5), Eq. (4) can be rewritten as

$$R^*_{ik} = 2R^T_{ik} - R^s_{ik} \quad (6)$$

The values of R^T_{ik} and R^S_{ik} can be computed from experimentally determined value of coefficients L_{ik} (Table 1). Values of the various resistance coefficients R^*_{ik} match with the experimental values (Table 2), lending support to the existence of the liquid membrane bilayers-one layer of the liquid membrane on either side of the supporting membrane.

Table 2. Values of the resistance coefficients for the case when both the compartments of the transport cell were filled with cholesterol solution of concentration equal to its CMC (Ref. 3).

	Computed values using Eq. (6)	Experimental values
$R^*_{11} \times 10^{-7} / m^{-3}\ N\ s$	7.10±0.16	7.47±0.23
$-R^*_{12} \times 10^{-2} m^{-1}\ A^{-1}\ J$	10.33±0.31	9.79±0.42
$-R^*_{21} \times 10^{-2} m^{-1}\ A^{-1}\ J$	10.23±0.40	9.62±0.36
$R^*_{22} / ohm\ m^{-2}$	8.41±0.26	7.92±0.25

5.2.2. Liquid membranes from lecithin and lecithin-cholesterol mixtures [4]

Similar experiments have been conducted [4] on lecithin and lecithin-cholesterol mixtures to demonstrate the formation of liquid membranes and bilayer of liquid membranes by them. In these studies the same transport cell, as used in experiments with cholesterol, has been used (Fig.2), and a Sartorius cellulose acetate microfiltration membrane/aqueous interface has been used as site for the formation of liquid membrane. The data on hydraulic permeability has been exploited to demonstrate the formation of liquid membranes and bilayers of liquid membranes. The CMC value for aqueous lecithin was found to be 12.951 ppm. For measurement of hydraulic permeabilities, the two compartments of the transport cell (Fig.2) were filled with the aqueous solutions of lecithin or lecithin and lecithin-cholesterol mixtures of desired composition. The aqueous solution of lecithin, cholesterol and their mixtures were prepared using the method described by Greshfeld and Pagano [6]. The concentration ranges chosen for hydraulic permeability data in the case of lecithin, were such that the data are obtained on both the lower and the higher side of the CMC of lecithin. During the hydraulic permeability measurements the solution in the compartment C was kept well stirred, and the electrodes E_1 and E_2 were short-circuited (Fig. 2).

The hydraulic permeability data [4] for various concentrations of lecithin and for lecithin-cholesterol mixtures of various compositions is reproduced in Fig.8. In all the cases, the proportional relationship, $Jv = L_{11}\ \Delta P$, where L_{11} is the hydraulic conductivity coefficient, is obeyed. The values of the coefficients L_{11} for various concentrations of lecithin and for lecithin-cholesterol mixtures of various compositions estimated from the slopes of the curves in Fig.8 are recorded in Table 3 and Table 4. The data in Table 4 are for the solutions of various concentrations of cholesterol prepared in 15.542 ppm solution of lecithin. The trend in the values L_{11} at various concentrations in the case of lecithin is similar to that observed in the case of cholesterol, i.e., in accordance with the liquid membrane hypothesis [9-11] and indicates complete formation of the liquid membrane in series with the supporting membrane when concentration of lecithin equals its CMC.

The values of L_{11} at several concentrations of lecithin below its CMC, computed using mosaic membrane model [12-14], i.e., using Eq. (11) of chapter 4 are in agreement with the corresponding experimentally determined valued (Table 3). This furnishes additional support in favors of liquid membrane formation.

The values L_{11} when both the compartments of the transport cell (Fig.2) were filled with the lecithin solutions of concentration equal to its CMC, were utilized to demonstrate the formation of bilayers of lecithin liquid membranes. Since at the CMC the interface is completely covered with the liquid membrane, the cellulose acetate microfiltration membrane in this case, would be sandwiched between the two layers of the liquid membrane generated on either side of it. Utilizing Kedem and Katchalsky's theory [15-17] for the permeability of composite membranes and following the arguments given in the case of cholesterol, Eqs. (4) to (6), one can write for such case,

$$(1/L^{*}_{11}) = (2/L^{T}_{11}) - (1/L^{s}_{11}) \qquad (7)$$

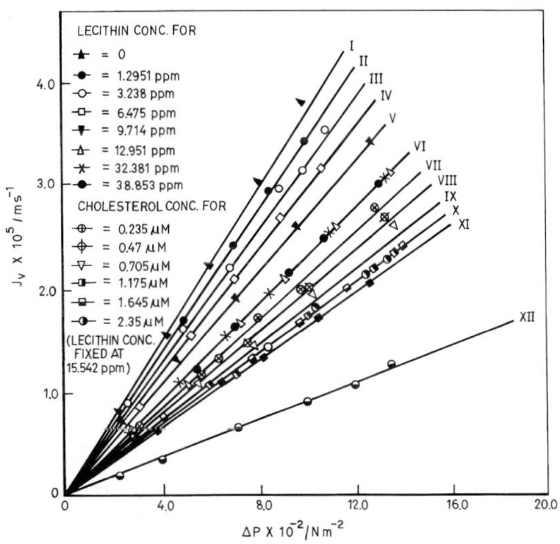

Fig.8. The hydraulic permeability data. Curves I to VI are for the case when compartment C was filled with lecithin solutions and compartment D with water. Curves VII to X are for the cases when compartment D was filled with water and compartment C was filled with lecithin cholesterol mixture of varying concentration of cholesterol keeping lecithin concentration constant at 15.542 ppm (>CMC) Curve XI is for the case when both compartments were filled with lecithin solution of concentration equal to its CMC. Curve XII is for the case when both compartments were filled with lecithin-cholesterol mixture of concentration 15.542 ppm with respect to lecithin and 1.175 μM with respect to cholesterol (Ref. 4).

The superscripts in Eq.(7) have the same meaning as in Eq.(6). The value of L_{11}^{*} computed from Eq.(7), using the values of L_{11}^{T} and L_{11}^{S} (Table 3) is in agreement with experimentally determined value (Table 3) lending support to the formation of the liquid membrane bilayer, i.e., one layer of the liquid membrane on either side of the supporting membrane.

Table 3. Value of $L_{11}/m^3s^{-1}N^{-1}$, at various concentrations of lecithin (Ref. 4).

C_L^*/ppm	0.0	1.2951	3.238	6.475	9.714	12.951[a]	32.381	38.853	C[b]
$L_{11}^{c}\times 10^{8}$	3.721 ±0.077	3.457 ±0.094	3.222 ±0.089	2.958 ±0.061	2.631 ±0.072	2.304 ±0.077	2.281 ±0.040	2.327 ±0.018	1.681 ±0.036
$L_{11}^{d}\times 10^{8}$	-	3.580 ±0.076	3.367 ±0.077	3.012 ±0.077	2.658 ±0.078	-	-	-	1.669 ±0.077[c]

[a] CMC. [b] Values for the system when both the compartments, C and D were filled with lecithin solution of concentration equal to its CMC. [c] Experimental Values. [d] Calculated values using the mosaic model. [c] Calculated value suing Eq.(7). *C_L, Lecithin concentration.

The value of L_{11} decreases regularly with increase in concentration of lecithin, and becomes constant when the concentration of lecithin equals or exceeds its CMC (Table 3). An examination of the values recorded in Table 4 reveals that when cholesterol is added to a solution of lecithin of concentration equal to or greater than its CMC, 15.542 ppm, in the present case, the value of L_{11} decreases further and goes on decreasing with the increasing concentration of cholesterol holding the concentration of lecithin constant at 15.542 ppm. The decreasing trend in the values of L_{11} continues upto the cholesterol concentration equal to 1.175×10^{-6} M and, thereafter, it again becomes constant. An obvious implication of this observation is that the increase in the resistance to water flow is due to incorporation of added cholesterol in the lecithin liquid membrane which already exists at the interface. At cholesterol concentration equal to 1.175×10^{-6} M, the lecithin liquid membrane is saturated with cholesterol. The decreasing trend in the values of L_{11} with increasing concentration of cholesterol is consistent with the results obtained on phospholipid-cholesterol BLM [18, 19]. There also water permeability has been found to decrease with the increase in concentration of cholesterol, and has been attributed to the fact that cholesterol strengthens the hydrophobic core and increases its viscosity. In order to assess whether the added cholesterol reaches straight upto the interface or not, surface tensions of solutions of various concentration of cholesterol prepared in 15.152 ppm aqueous solution of lecithin were measured. Surface tensions of all such solutions were found to be equal to the surface tension of 12.951 ppm solution of lecithin. This indicates that the added cholesterol probably does not reach deep upto the interface. This is in keeping with the literature reports that in mixed phospholipid-cholesterol films, cholesterol molecules occupy the cavities in lecithin monolayers caused by the thermal motions [18].

The value of L_{11} for the case when both the compartments, C and D of the transport cell, were filled with a solution of lecithin-cholesterol mixture, which were 15.152 ppm with respect to lecithin and 1.175×10^{-6} M with respect to cholesterol, was utilized to demonstrate the existence of bilayers of the liquid membrane generated by the lecithin-cholesterol mixture. By the same arguments as given in the case of lecithin, the supporting membrane in this case also will be sandwiched between the two liquid membranes generated on either side of it by the lecithin-cholesterol mixtures. The values of hydraulic permeability coefficient L_{11}^{*} in this case also were computed using Eq. (7). Of course, in this case L_{11} represented the value of L_{11} for the case when compartment C of the transport cell was filled with the solution which was 15.542 ppm with respect to lecithin and 1.175×10^{-6}M with respect to cholesterol, i.e., the concentration at which the supporting membrane is completely covered with the mixed liquid membrane and the compartment D was filled with water. The value of L_{11}^{*} thus computed, comes out to be equal to $(0.871 \pm 0.09) \times 10^{-8}\ m^3\ N^{-1}\ s^{-1}$ which agrees with the experimentally determined value (Table 4) lending support to the formation of bilayers of liquid membranes generated by the lecithin-cholesterol mixture.

Table 4. Values of L_{11} at various concentrations of cholesterol lecithin cholesterol mixtures; lecithin concentration kept constant at 15.542 ppm (>CMC) (Ref. 4).

$C_c^{*} \times 10^6$ M	0.0	0.235	0.470	0.705	1.175	1.645	2.350	C[a]
$L_{11} \times 10^8$	2.304	2.152	2.033	1.885	1.741	1.755	1.723	0.925
($m^3 s^{-1} N^{-1}$)	±0.077	±0.032	±0.044	±0.068	±0.092	±0.087	±0.054	±0.101

[a]Values for the systems when both the compartments C and D were filled with the lecithin-cholesterol mixture of the composition, 15.542 ppm with respect to lecithin and 1.1750×10^{-6} M with respect to cholesterol. *C_c cholesterol concentration.

Because of the surface active nature of lecithin and cholesterol, it is natural to expect that in the liquid membrane bilayers generated by lecithin, cholesterol and the lecithin-cholesterol mixture, the hydrophobic ends of the lipid molecules will be preferentially oriented towards the hydrophobic supporting membrane, the cellulose acetate microfiltration membrane in these experiment [3, 4] and their hydrophilic ends will be drawn outwards away from the supporting membrane. The values of electrical resistance of freely formed [2] BLMs, in general, are very high [20-23], much higher than those reported for biomembrane [24]. Also the rate of ionic diffusion through BLMs is much slower [25, 26] than through biomembrane [27, 28]. This is ascribed [29] to a tight molecular arrangement in BLMs in contrast to biomembranes where the lipid bilayer is in a fluid state [1]. The lipid bilayers generated in these experiments are expected to have more fluidity and, hence, in this respect are expected to be closer to the state of lipid bilayers in biomembranes. One can, therefore, look for at least a qualitative agreement between the transport data on the liquid membrane bilayers generated in these experiments and the data on biomembranes. Values of cation permeabilities, cationic transport numbers and electrical resistance for the liquid membrane bilayers generated by lecithin, Cholesterol and the lecithin-cholesterol mixtures have been measured [4] with a view to comparing them with the values for biological membranes.

For solute permeability (ω) measurements of sodium, potassium and calcium ions, the compartment C of the transport cell was filled with a solution of known concentration of sodium, potassium or calcium chlorides prepared in aqueous solutions of known concentrations of the liquid membrane generating substances, viz., lecithin, cholesterol or the lecithin-cholesterol mixture and the compartment D was filled only with the solutions of the liquid membrane generating substances. The concentrations of liquid membrane generating solutions were such at which a complete liquid membrane is expected to be formed in series with the supporting membrane. The concentrations of lecithin and cholesterol chosen were 15.542 ppm and 1.175×10^{-6} M, respectively, and that of the lecithin-cholesterol mixture was 15.542 ppm with respect to lecithin and 1.175×10^{-6} M with respect to cholesterol. In control experiments no liquid membrane generating substances were used. To estimate the value of the solute flux J_s, the condition, $J_v=0$, i.e., no net volume flux was imposed on the system and the contents of the compartments C and D were analysed for the cation concentration after a known period of time which was of the order of 6 to 8 hours. The amount of the permeant gained by the compartment D divided by the time and the area of the membrane, gave the value of the solute flux (J_s). Knowing the values of J_s, the values of ω were estimated using the Eq. (18) of chapter 4. The value of osmotic pressure difference $(\Delta\Pi)$ used in the estimation of ω in Eq. (18) of chapter 4 was the average of the values of $\Delta\Pi$ at the beginning of the experiment $(t = 0)$ and at the end of the experiment. During the measurements of ω, the solutions in the compartment C were kept well stirred.

For the measurement of transport numbers, the electrodes E_1 and E_2 of the transport cell (Fig. 2) were converted into Ag-AgCl electrodes using the method described by Carmody [30, 31]. The compartment C of the transport was filled with the electrolyte (chlorides of sodium or potassium) solution of known concentration prepared in aqueous solution of the known concentrations of the liquid membrane generating substances, and the compartment D was filled only with the solutions of the liquid membrane generating substances. The concentrations of the liquid membrane generating substances were the same as those used in ω measurements. For control experiments, however, solutions of electrolytes in water alone were used. The condition, $J_v=0$ was imposed on the system and the open circuit voltage; $(\Delta\phi)_{I=0,\, Jv=0}$ across the electrodes E_1 and E_2 was measured. The asymmetry potential of the electrodes was taken into account while measuring the potential difference, $\Delta\phi$. The transport numbers t_1 of the cations were calculated using the equation [15, 16]

$$(\Delta\phi)_{I=0,\, Jv=0} = (-t_1 / v_1 Z_1 F)\,(\Delta\Pi / \bar{C}_s) \tag{8}$$

where $\bar{C}_s$ is the average of the electrolyte concentration in the two compartments, v_1 and Z_1 are the number and the valency of the cation and F is the Faraday. The values of solute permeability of cations through the lecithin bilayers, cholesterol bilayers and the bilayers of liquid membranes generated by the lecithin-cholesterol mixture $(\omega^{bilayer})$ were estimated using the equation [15-17],

$$(1/\omega^*) = (1/\omega^s) + (1/\omega^{bilayer}) \tag{9}$$

obtained from the non-equation thermodynamics theory for the permeability of composite membranes. Since the values of cation permeabilities for living cells and for BLMs reported in literature, are in cm s^{-1}, these, in fact, are the values of ωRT; the values of cation permeabilities obtain using Eq. (9) were changed into s^{-1} for the sake of comparison. These are recorded in (Table 5). Perusals of (Table 5) reveals that the values of cationic permeabilities of the liquid membrane bilayers generated in these experiments [4] are several orders of magnitudes larger than the values reported for BLM [25]. It is noteworthy, however, that the values of Na^+ and K^+ permeabilities through the liquid membrane bilayers generated in these experiments [4] by the lecithin-cholesterol mixture are close to the values for the living membrane [27, 28] at least in their order of magnitudes (Table 5).

Transport numbers of, Na^+ for the liquid membrane bilayers were computed using the relationship [16],

$$(t_1^* / \omega_1^*) = (t_1^s / \omega_1^s) + (t_1^{bilayer} / \omega_1^{bilayer}) \quad (10)$$

the values of the cationic transport numbers of Na^+ and K^+ ions for lecithin liquid membrane bilayer (Table 5) are in agreement with the values reported for the, BLM prepared from egg phosphatidyl choline [32].

The electrical resistance of the liquid membrane bilayers was also computed from the experimentally determined values of R_m and R^* using the equation

$$R^* = R^s + R^{bilayer} \quad (11)$$

and were converted to normalized trans-membrane resistance R_m by multiplying with the area of the membrane. The values of R_m (Table 5), for liquid membrane bilayers generated from lecithin, cholesterol and the lecithin-cholesterol mixture are within the range reported for biomembranes, in general [24].

The agreement, though qualitative, of the transport data for the lecithin-cholesterol liquid membrane bilayer with those for living membranes/BLM is encouraging. It is indicative of the possibility that the liquid membrane bilayers generated using Kesting's hypothesis, from the constituents of biomembranes could be used as mimetic systems for biomembranes.

5.3 Mimicking light-induced transport [33-36]

5.3.1 Experiments with chloroplast extract

If the liquid membrane bilayers generated by the constituents of biomembranes have to act, as model systems for biomembranes, the agreements between the passive transport data on the liquid membrane bilayers with the data on living membranes, as discussed in the previous section, is not enough. It should be possible to mimic some of the biologically relevant transport processes on such liquid membrane bilayers. Light induced transport process, through thylakoid membrane of the chloroplast, affords an example where such a simulation can be attempted conveniently. BLMs in which light absorbing materials like chloroplast extract are incorporated have been shown [37] to act as mimetic system for thylakoid membrane of the chloroplast. Light gradients across chloroplast BLMs have been shown to induce volume flux. The light-induced volume flux, which can be termed as photo-osmosis, is considered to be a consequence of the electrical potentials developed across the BLMs due to action of light, the photo-electric effect [37]. The role of ultra thin lipid barrier in pigmented BLMs is mainly to provide a two dimensional support for orienting light-absorbing molecules which perform the task of energy transduction.

Table 5. Values of action permeabilities, normalized resistance and cationic transport numbers for various liquid membrane bilayers (Ref. 4).

	Permeability $s^{-1} \times 10^6$					Resistance (R_m)/ohm cm^2		Transport No. (t_l)			
	For liquid membrane bilayers (Ref.4)			Reported in Lit.		For liquid membrane bilayers (Ref.4)	Reported in Lit.	For liquid membrane bilayers (Ref. 4)		Reported in Lit.	
	Na^+	K^+	Ca^{+2}	Na^+	K^+			Na^+	K^+	Na^+	K^+
1. Lecithin bilayers	24.93	8.26	50.83	-	3.4×10^{-6}[b]	0.89×10^5	10^8[c]	0.54	0.52	0.55[d]	0.52[d]
2. Cholesterol bilayers	31.03	15.88	57.58	-	-	0.51×10^5	-	0.57	0.56	-	-
3. Lecithin-cholesterol bilayers	1.33	1.04	29.86	0.015[e]	0.56[e]	1.79×10^5	10^3	0.49	0.47	-	-
				0.14[f]	0.16[f]		10^5[g]				

[a]Undetectable: for phosphatidyl choline BLM (Ref.26). [b]For phosphatidyl choline BLM (Ref.25). [c]For phosphatidyl choline BLM (Ref. 22). [d]For phosphatidyl choline (egg) (Ref.32). [e]For squid axon (resting) (Ref.27). [f]For frog sartorious (resting) (Ref. 28) [g]For biomembranes (Ref.24).

Chloroplast extract is known to be surface active [38] in nature and hence, according to Kesting's hypothesis, [9-11] can generate a surfactant layer liquid membrane, which would cover the interface completely at a concentration equal to or greater than its critical micelle concentration CMC. Therefore, it should be possible, using the procedure adopted in the experiments with lecithin and cholesterol [3,4], to generate liquid membranes from chloroplast extract on either side of a hydrophobic supporting membrane. The chloroplast liquid membrane bilayers thus generated, should also show the phenomenon of photo-osmosis. Experiments have been designed to demonstrate that the liquid membrane bilayers generated from chloroplast extract do show the phenomenon of photo-osmosis [33, 34]. Variation of photo-osmotic velocity with parameters like wavelength and intensity of the exciting light, concentration and strength of the electron acceptors in the illuminated compartments, etc., has been studied [33, 34].

Trends observed were found to be consistent with the trends reported in chloroplast BLMs indicating workability of the liquid membrane bilayers as mimetic system for biomembranes. The liquid membrane bilayers generated from haemoglobin and from cytochrome-C were also shown [33-35] to exhibit the phenomenon of photo-osmosis. Surface conductivity [39] and photoconductivity [40, 41] of haemoglobin are documented in literature. Both haemoglobin and cytochrome-C have structural similarity with chlorophyll, the main light absorbing material in chloroplast extract; both have porphyrin ring in their structure. This observation coupled with the data on the variation of photo-osmotic velocity with wavelength of the exciting light, suggested that absorption of light by porphyrins which are present in chloroplast extract, haemoglobin and cytochrome-C, might be responsible for the observed phenomenon of photo-osmosis. Should it be so, protoporphyrin alone should not only show the phenomenon of photo-osmosis but should also reproduce the trends observed in case of liquid membrane bilayers generated from chloroplast extract, haemoglobin and cytochrome-C which, in turn, should be consistent with the trends reported in chloroplast BLMs. The experiments conducted on protoporphyrin did confirm this expectation.

Cyanocobalamin, whose central structure, the "Corrine" ring system, is very similar to that of porphyrin, was also experimented with [35]. In this case also the phenomenon of photo-osmosis was observed with similar trends in the data as those in case of chloroplast extract, protoporphyrin, haemoglobin or cytochrome- C. An account of these investigations [33-35] is summarized below.

Chloroplast extract for use in these experiments [33-34] was obtained from spinach leaves using the method described by Tien and Howard [42]. The final fraction was evaporated to dryness and the residue was dissolved in ethanol to make stock solution of known concentration. Aqueous solutions of chloroplast extract of desired concentrations, were prepared by adding known volume of ethanolic stock solution to aqueous phase with constant stirring. The amount of ethanol in the final solution was not allowed to exceed 0.1% by volume because it was shown by a control experiment that 0.1% solution of ethanol does not lower the surface tension of water to any measurable extent.

The CMCs of aqueous chloroplast extract, aqueous haemoglobin, aqueous protoporphyrin, aqueous cytochrome-C and aqueous cyanocobalamin, as determined from the variation of surface tension with concentration, were found to be 23.184 ppm, 11.997 ppm, 0.828.ppm, 0.214 ppm and 2.000 ppm, respectively.

A slightly modified version of the all-glass transport cell described in Fig.2 was used for hydraulic permeability measurements and photo-osmotic velocity measurements. It is depicted in Fig.9, which has been well labeled to make it self-explanatory. A Sartorius cellulose acetate microfiltration membrane of thickness 1.0 x 10^{-4} M and area 2.55 x 10^{-5} m^2, which acted as a support for the liquid membranes, separated the transport cell into two compartments C and D. The procedure for hydraulic permeability measurements has been described in the previous section. During hydraulic permeability measurements, the entire cell except the capillary (Fig.9), was covered with black paper to protect it from exposure to light and the electrodes E_1 and E_2 were short-circuited. For obtaining the hydraulic permeability data, which were exploited to demonstrate the existence of liquid membrane, the compartment C of the transport cell was filled with aqueous solutions of varying concentrations of the photoactive materials-chloroplast extract or haemoglobin or protoporphyrin or cytochrome-C or cyanocobalamin and the compartment D was filled with water.

For measurements of photo-osmotic velocity the experimental setup is described in Fig.9. The compartment C (Fig.9) was filled with aqueous solutions containing desired concentration of the photoactive materials and electron acceptors and the compartment D was filled with the aqueous solutions containing the same concentration of the photoactive materials as in compartment C and desired concentrations of electron donors. The concentration of photoactive materials was higher than their CMC. The pH of the solutions in the two compartments was maintained by using 0.1 M acetate buffer. The condition of no net pressure difference, $\Delta P = 0$ was imposed on the system by adjusting the pressure head. The light was then switched on, and the consequent movement of liquid meniscus in the capillary, L_1L_2 was noted with time. During the measurements of photo-osmotic velocity, a constant and stabilized voltage at 220 volts from A.C. mains was fed to the bulb B (Fig.9), and the distance between the transport cell and the bulb was kept fixed. In order to study the variation of photo-osmotic velocity with intensity of the incident light various voltages were fed to the bulb to alter the Intensity of the light and the consequent volume fluxes were noted in the capillary, L_1L_2. To measure photo-osmotic velocity in the presence of externally applied electric field, the electrodes E_1 and E_2 were connected to an electronically operated stabilized dc power supply. The electrode E_2 (Fig. 9) was connected to the positive terminal of the power supply to make the lower compartment C positive with respect to the dark compartment. The pressure head was suitably adjusted to balance the volume flux induced by externally applied voltage in the capillary, L_1L_2.

When the liquid meniscus in the capillary became stationary, the light was allowed to fall on the membrane in the lower compartment, and the consequent volume flux in the capillary, L_1L_2 was noted. This was repeated at several values of externally applied voltages. All experiments were done at constant temperature using a thermostat set at 40±0.1°C.

Table 6. Values of $L_P/m^3s^{-1}N^{-1}$, at various concentrations of chloroplast extract (ref. 34).

	Concentration/ppm					
	0.0	11.592	17.988	23.184[c]	46.368	69.552
$L_p^a \times 10^8$	2.611	2.185	1.697	1.513	1.480	1.450
	±0.114	±0.133	±0.083	±0.046	±0.036	±0.053
$L_p^b \times 10^8$	-	2.062	1.787	-	-	-
		±0.80	±0.063			

[a]Experimental values. [b]Calculated values using the mosaic model, Eq. (13). [c]CMC.

Table7. Values of $L_p/m^3s^{-1}N^{-1}$, at various concentrations of cytochrome-C (Ref. 35).

	Concentration/ppm							
	0.0	0.054	0.107	0.161	0.214[a]	0.428	0.642	C[b]
$L_p^c \times 10^8$	1.250	1.120	0.972	0.840	0.724	0.721	0.721	0.517
	±0.069	±0.042	±0.028	±0.055	±0.031	±0.014	±0.011	±0.006
$L_p^d \times 10^8$		1.118	0.987	0.885				0.509
		±0.059	±0.050	±0.040				±0.020[c]

[a]CMC, [b]Values for the system when both the compartments, C and D were filled with cytochrome-C Solution of concentration, 0.220 ppm., [c]Experimental values., [d]Calculated values using the mosaic model, Eq. (13), [c]Calculated values using Eq. (7).

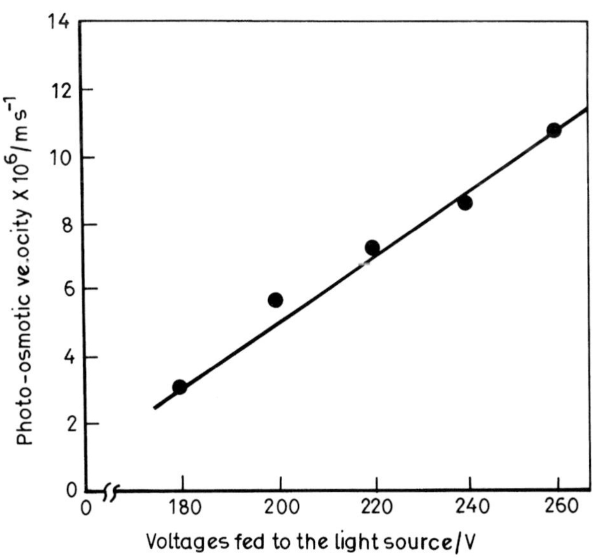

Fig.11. Variation of photo-osmotic velocity with intensity of light in case of cytochrome-C (Ref. 34).

Table 8. Values of photo-osmotic velocity using different election acceptors in the illuminated compartment (Ref. 33-35).

	Electron acceptor in the illuminated compartment	Photo-osmotic velocity $J_v \times 10^5$/m, s^{-1}
Chloroplast extract	$FeCl_3$ (1×10^{-3} M)	0.358 ± 0.027
	Na_2S (1×10^{-3} M)	0.419 ± 0.028
Haemoglobin	$FeCl_3$ (1×10^{-3} M)	0.248 ± 0.009
	Na_2S (1×10^{-3} M)	0.427 ± 0.004
Cytochrome-C	$FeCl_3$ (1×10^{-3} M)	0.726 ± 0.106
	Na_2S (1×10^{-3} M)	0.990 ± 0.163
Protoporphyrin	$FeCl_3$ (1×10^{-3} M)	0.408 ± 0.055
	Na_2S (1×10^{-3} M)	0.945 ± 0.101
Cyanocobalamin	$FeCl_3$ (1×10^{-3} M)	0.817 ± 0.012
	Na_2S (1×10^{-3} M)	0.852 ± 0.018

The dark compartment in al the cases contained *Fe^{2+} ions (1×10^{-3} M)*

Table 9. Values of Photo-osmotic Velocity using different electron donors in the dark compartment (Ref. 33-35).

	Electron donors in the dark compartment	Photo-osmotic velocity $J_v \times 10^5$/ms^{-1}
Chloroplast extract	Nal (1×10^{-3} M)	0.615 ± 0.001
	$K_4Fe(CN)_6$ (1×10^{-3} M)	0.497 ± 0.010
	$Na_2S_2O_3$ (1×10^{-3} M)	0.450 ± 0.003
	$FeSO_4(NH_4)_2$ SO_4 (1×10^{-3} M)	0.358 ± 0.027
Haemoglobin	Nal (1×10^{-3} M)	0.381 ± 0.019
	$K_4Fe(CN)_6$ (1×10^{-3} M)	0.330 ± 0.001
	$Na_2S_2O_3$ (1×10^{-3} M)	0.293 ± 0.013
	$FeSO_4(NH_4)_2$ SO_4 (1×10^{-3} M)	0.248 ± 0.009
Cytochrome-C	Nal (1×10^{-3} M)	1.797 ± 0.677
	$K_4Fe(CN)_6$ (1×10^{-3} M)	0.114 ± 0.313
	$Na_2S_2O_3$ (1×10^{-3} M)	0.835 ± 0.132
	$FeSO_4(NH_4)_2$ SO_4 (1×10^{-3} M)	0.726 ± 0.106
Protoporphyrin	Nal (1×10^{-3} M)	2.699 ± 0.068
	$K_4Fe(CN)_6$ (1×10^{-3} M)	2.293 ± 0.084
	$Na_2S_2O_3$ (1×10^{-3} M)	1.982 ± 0.151
	$FeSO_4(NH_4)_2$ SO_4 (1×10^{-3} M)	1.408 ± 0.065
Cyanocobalamin	Nal (1×10^{-3} M)	1.256 ± 0.030
	$K_4Fe(CN)_6$ (1×10^{-3} M)	0.964 ± 0.030
	$Na_2S_2O_3$ (1×10^{-3} M)	0.913 ± 0.020
	$FeSO_4(NH_4)_2$ SO_4 (1×10^{-3} M)	0.817 ± 0.012

The illuminated compartment in all the cases contained Fe^{3+} ions (1×10^{-3} M).

Table 10. Values of photo-osmotic velocity *($J_v \times 10^5/ms^{-1}$)* at various concentrations of the electron acceptor *(Fe^{3+} ions)* in the illuminated compartment (Ref. 33-35).

	$[Fe^{3+}]/M$	Photo-osmotic velocity
Chloroplast extract	1×10^{-4}	0.096 ± 0.020
	5×10^{-4}	0.186 ± 0.056
	1×10^{-3}	0.358 ± 0.027
	5×10^{-3}	0.598 ± 0.041
	1×10^{-2}	0.756 ± 0.013
Haemoglobin	1×10^{-4}	0.071 ± 0.003
	5×10^{-4}	0.172 ± 0.004
	1×10^{-3}	0.248 ± 0.009
	5×10^{-3}	0.406 ± 0.017
	1×10^{-2}	1.308 ± 0.016
Cytochrome-C	1×10^{-4}	0.398 ± 0.241
	5×10^{-4}	0.631 ± 0.181
	1×10^{-3}	0.726 ± 0.106
	5×10^{-3}	0.925 ± 0.079
	1×10^{2}	1.247 ± 0.063
Protoporphrin	1×10^{-4}	0.731 ± 0.047
	5×10^{-4}	1.301 ± 0.110
	1×10^{-3}	1.408 ± 0.055
	5×10^{-3}	2.101 ± 0.188
	1×10^{-2}	2.422 ± 0.128
Cyanocobalamin	1×10^{-4}	0.496 ± 0.038
	5×10^{-4}	0.567 ± 0.016
	1×10^{-3}	0.567 ± 0.016
	5×10^{-3}	0.714 ± 0.089
	1×10^{-2}	1.055 ± 0.094

The dark compartment in all the cases contained *Fe^{2+} ions ($1 \times 10^{-3}M$).*

Table 11. Values of photo-osmotic velocity *(J_v×10^5/m s^{-1})* through chloroplast liquid membrane bilayers at different externally applied voltages (V) (Ref. 34).

	Applied voltage	Photo-osmotic velocity
(a) When the dark compartment contained Fe^{2+} ions (1×10^{-3}M) and the illuminated compartment contained Fe^{3+} ions (1×10^{-3} M)	0.1	0.121 ± 0.003
	0.2	0.236 ± 0.022
	0.3	0.317 ± 0.011
	0.4	0.403 ± 0.013
	0.5	0.485 ± 0.002
(b) In the absence of Fe^{2+} or Fe^{2+} ions in either compartment	1.0	0.661 ± 0.003
	1.1	0.829 ± 0.005
	1.2	1.044 ± 0.010
	1.3	1.246 ± 0.015
	1.4	1.358 ± 0.001
	1.5	1.509 ± 0.002

Table 12. Values of the photo-osmotic velocity *(J_v×10^6/ms^{-1})* at different wavelength ranges (Ref. 34, 35).

Wavelength range/nm Filter No.	White light	365-445 (N-Hg-2)*	465-565 (B-.5052)*	560-660 (B-610)*	600-660 (N-630)*
(a) Chloroplast extract	3.580	2.382	1.315	1.599	2.031
	± 0.027	±0.003	±0.001	±0.001	±0.002
(b) Haemoglobin	2.480	1.712	1.383	1.064	2.495
	± 0.009	±0.074	±0.027	±0.030	±0.028
(c) Protoporphyrin	14.080	7.400	2.560	3.080	4.580
	± 0.550	±0.290	±0.060	±0.050	±0.190
(d) Cytochrome-C	7.261	6.006	5.344	3.608	-
	± 0.106	±0.189	±0.251	±0.052	
(e) Cyanocobalamin	8.174	6.732	4.654	4.116	4.515
	± 0.124	±0.086	±0.049	±0.054	±0.045

* Obtained from Photo-volt Corporation, New York. The dark compartment contained Fe^{2+} ions and the illuminated compartment contained Fe^{3+} ions, both 1×10^{-3} *M*.

Although the experiments on photo-osmosis were carried out under constant temperature conditions, the possibility of the thermal gradients produced by light absorption, causing the observed flow, has to be ruled out. It was observed in these experiments that as soon as the light was switched on, movement of the liquid in the capillary L_2L_2 (Fig.9) was noticed and also as soon as light was switched off the flow stopped. This strongly suggests that the observed volume flux cannot be on account of thermal gradients because establishment and abolition of thermal gradients cannot be an instantaneous process. It was also observed that in all cases, viz., chloroplast extract, haemoglobin, cytochrome-C, protoporphyrin and cyanocobalamin, on short-circuiting the electrodes E_1 and E_2 (Fig.9), the light induced volume flux stopped completely. This observation not only rules out the possibility of thermal gradient being a cause for the observed flow but also confirms that the light induced voltage across the membrane is the primary cause for the observed photo-osmosis as suggested by Tien in the case of chloroplast BLMs [43].

Magnitudes of electrical potentials developed across the pigmented BLMs, when it is illuminated from one side is known [44] to be enhanced manifold in asymmetrical systems, e.g., when different redox chemicals are present in the two bathing solutions separated by the pigmented BLM. Since the light induced voltage difference is the primary driving force for the photo-osmotic flux, the magnitude of photo-osmotic velocity should vary with the choice of the redox chemicals in the two compartments of the transport cell. To study this, two sets of experiments were performed [34,35]. In the first set of experiments, Fe^{2+} ions of concentration 1×10^{-3}M were kept in the dark compartment, and two different electron acceptors of concentration 1×10^{-3} M, namely, ferric chloride ($FeCl_3$) and sodium sulphide (Na_2S) were taken in the illuminated compartment. The data in Table 8 reveal that in all cases, viz., chloroplast extract, haemoglobin, cytochrome-C, protoporphyrin and cyanocobalamin, the magnitude of the photo-osmotic velocity when Na_2S was present in the illuminated compartment is greater than the magnitude when ferric chloride was taken instead. This is consistent with the fact that Na_2S is stronger electron acceptor than ferric chloride [44]. Relative electron accepting and donating strength of a variety of compounds tested [44] on chloroplast – BLM using e.m.f measurement is depicted in Fig.12. Similarly in the second set of experiments, ferric chloride of concentration was kept in the illuminated compartment and different electron donors of concentration 1×10^{-3} M were taken in the dark compartment. The data (Table 9) on photo osmotic velocity in all cases, for various electron donors, are in order of their electron donating strengths. The electron donating strengths of the various electron donors are reported to be in the following order [44,46]:

$NaI > K_4Fe(CN)_6 > Na_2S_2O_3 > FeSO_4(NH_4)_2SO_4$.

Values of the photo e.m.f of chloroplast-BLMs using ferric chloride as electron acceptor and each of the above listed compounds as electron donors are also reported to be in the same order [46], i.e., the value of photo e.m.f when NaI was used is greater than when $K_4Fe(CN)_6$ was used and so on. These observations indicate that in these experiments the observed photo-osmotic flow is primarily due to light induced voltage across the liquid membrane bilayers.

If electrons generated by light and their capture by the electron acceptors present in the illuminated compartment are responsible for the observed photo- osmotic effect, the values of photo-osmotic velocity should increase with the increasing concentrations of electron acceptors. The data on photo-osmotic velocity in case of all the five substances-chloroplast extract, haemoglobin, cytochrome-C, protoporphyrin and cyanocobalamin, at various concentrations of Fe^{3+} ions in the illuminated compartment keeping the concentration of Fe^{2+} ions in the dark compartment constant (1×10^{-3} M), confirm such a trend (Table 10). As a corollary of this, it should be expected that if an electric field is applied across the liquid membrane bilayer marking the illuminated compartment positive with respect to the dark compartment, the magnitude of the photo-osmotic velocity should increase with increase in applied voltage across the membrane. The data on the variation of photo-osmotic velocity with externally applied voltage were obtained in two cases-one in which Fe^{3+} ions (1×10^{-3} M) were present in the illuminated compartment and Fe^{2+} ions (1×10^{-3}M) in the dark compartment and the other in the absence of electron donor and acceptor species i.e., Fe^{2-} and Fe^{3+} ions in either compartment. The data in both the cases (Table 11) for chloroplast extract liquid membrane bilayer confirm this. The data on haemoglobin, protoporphyrin, cytochrome-C and cyanocobalamin liquid membrane bilayers could not be obtained, because the application of even small voltage caused electrolysis.

The open circuit photo-voltages (E_{op}) in the case of chloroplast -BLMs are known to be dependent on the intensity exciting light; the dependence has been found to be given by the following equation [44],

$$E_{op} = l\ log(1+I/L) \qquad (14)$$

where l and L are constants for a given chloroplast -BLM at a particular temperature. Under conditions of low light intensities, E_{op} becomes directly proportional to I as has, indeed, been found to be the case. As an implication of this, it follows that the photo-osmotic velocity through the liquid membrane bilayers should also show a similar dependence on the intensity of exciting light. The data in Figs.10 and 11, indeed, show such a dependence on the intensity of exciting light.

The values of photo-osmotic velocity, for chloroplast extract, haemoglobin, cytochrome-C, protoporphyrin and cyanocobalamin, induced by the light of different wavelengths obtained using different optical filters are recorded in Table 12. In the system containing chloroplast extract, chlorophylls are the main photoactive materials, whose major absorption peaks are at 400 nm and 660 nm [47]. The absorption peak at 400 nm is more intense than the peak at 660 nm. The magnitude of photo-osmotic velocity at various wavelength ranges (Table 12) shows the same gradation indicating that photo-osmotic flow is due to the absorption of light by the pigments. A perusal of Table 12 further reveals that in case of chloroplast extract, haemoglobin, cytochrome-C, an protoporphyrin, magnitude of the photo-osmotic velocity is maximum, amongst all the filters used, for the one corresponding to the wavelength range 365 nm -445 nm. This observation, which is common to the four substances, can be rationalized by the fact that porphyrins, which are present in all the four have most intense absorption band in the region of 400 nm, the Soret band [47]. The reported absorption maximum for protoporphyrin is at 408 nm [47]. The reported absorption maxima for cyanocobalamin are at 278, 361 and 550 nm, the band at 361 nm being [48] more intense than the band at 550 nm. The magnitudes of photo-osmotic velocity for cyanocobalamin, at various wavelength ranges, show the same gradation.

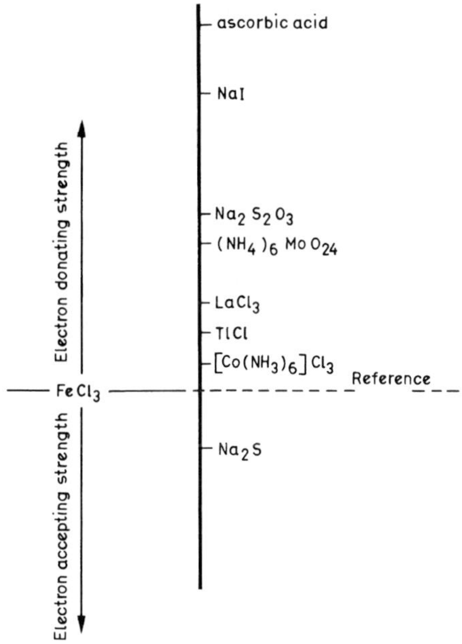

Fig.12 Relative electron accepting and donating strength of a variety of compounds tested on the chl-BLM. The cell arrangement: 10^{-3} M $FeCl_3$ in 0.1 M Na acetate pH 5 (Reference side) Chl-BLM/ Test compound in 0.1 M Na acetate pH5 (Ref. 44, 45).

Since photo-osmosis observed in these experiments was [33-35] shown actually to be photo-electro osmosis, the light induced electrical potential difference across the liquid membrane bilayers should also show the same trends as observed in the data on photo-osmosis and should also be consistent with the trends reported on BLMs. Experiments have also been conducted [36], with this object in view, on the liquid membrane bilayers generated, on a cellulose acetate microfiltration supporting membrane (average pore, size 0.2 μm), by chloroplast extract, haemoglobin and protoporphyrin. The data obtained from these studies are recorded in Tables 13 to 16 and in Fig.13, which have the same trends as observed in the data on photo-osmosis.

Photo-osmotic studies have also been conducted on liquid membrane bilayers generated by bacteriorhodopsin [49]. The trends observed were quite similar to those observed for the chloroplast extract liquid membrane bilayers [34]. The explanation offered for the origin of the effect in the case of bacteriorhodopsin was quite similar to that offered for chloroplast extract liquid membrane bilayers. One difference was that, when the bacteriorhodopsin membranes were asymmetrically illuminated, protons were pumped into the illuminated compartment, whereas the chloroplast extract/chlorophyll membranes pumped electrons into the illuminated compartment.

Table.13 Values of light-induced potential difference using different electron acceptors in the illuminated compartment (Ref. 36).

	Electron acceptor in the illuminated compartment*	Light potential difference/mV
Chloroplast extract	$FeCl_3$	28.36
	Na_2S	34.60
Haemoglobin	$FeCl_3$	19.20
	Na_2S	26.13
Protoporphyrin	$FeCl_3$	39.66
	Na_2S	42.40

*The concentration in all the cases was 1 mM. The dark compartment in all the cases contained 1 mM Fe^{2+} ions.

Table 14. Values of light-induced potential difference using different electron donors in the dark compartment (Ref. 36).

	Electron donors in the dark compartment*	Light induced potential difference / mV
Chloroplast extract	NaI	54.26
	$K_4Fe(CN)_6$	45.03
	$Na_2S_2O_3$	36.00
	$FeSO_4(NH_4)_2SO_4$	28.86
Haemoglobin	NaI	45.13
	$K_4Fe(CN)_6$	36.50
	$Na_2S_2O_3$	27.56
	$FeSO_4(NH_4)_2SO_4$	19.20
Protoporphyrin	NaI	63.43
	$K_4Fe(CN)_6$	47.03
	$Na_2S_2O_3$	44.60
	$FeSO_4(NH_4)_2SO_4$	39.66

*The concentration in all the cases was 1 mM. The illuminated compartment in all the cases contained 1 mM Fe^{3+} ions.

Table15. Values of light-induced potential difference at various concentration of electron acceptor (Fe^{3+} ions) in the illuminated compartment (Ref. 36).

	$[Fe^{3+}]$ in the illuminated compartment / mM	Light induced potential difference / mV
Chloroplast	0.1	19.73
	1.0	28.86
	10.0	40.96
	100.0	49.50
Haemoglobin	0.1	13.96
	1.0	19.20
	10.0	26.76
	100.0	30.53
Protoporphyrin	0.1	28.66
	1.0	39.66
	10.0	47.40
	100.0	55.0

The dark compartment in all the cases contained 1 mM Fe^{2+} ions.

Table 16. Values of the light-induced potential difference (*mV*) at different wavelength ranges (Ref. 36).

Wavelength range/nm	White light	365-445	465-565	560-660
Filter No.		(N-Hg-2)[a]	(B-505)[a]	(B-610)[a]
Chloroplast	28.86	23.23	14.60	18.23
Hemoglobin	19.20	16.46	13.40	10.43
Protoporphyrin	39.66	29.96	16.06	19.06

[a] Obtained from Photo-volt Corporation, New York. The dark compartment in all the cases contained 1 mM Fe^{2+} ions and the illuminated compartment contained 1mM Fe^{3+} ions.

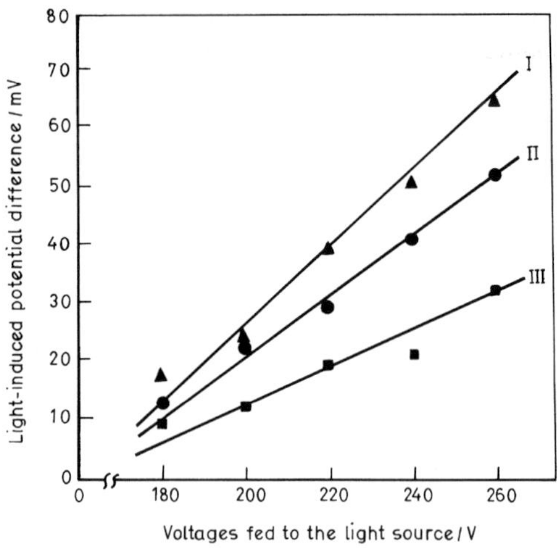

Fig.13. Variation of light induced potential difference with intensity of light. Feeding different voltages to the light source varied the intensity. Curves I, II and III are for protoporphyrin, chloroplast extract and haemoglobin respectively (Ref. 36).

5.3.2. Experiments with bacteriorhodopsin [49, 50]

Since bacteriorhodopsin acts as a photo-electric energy transducer [51-60] and generates electrical potential difference across the membrane under, the influence of light, by acting as light driven proton pump, the phenomenon of photo-osmosis should also be observable in the liquid membrane bilayers generated b bacteriorhodopsin. Experiments carried out [49] with a view to demonstrating the phenomenon of photo-osmosis through the liquid membrane bilayers generated by bacteriorhodopsin are described and discussed in this sub-section. Data on hydraulic permeability in the presence of varying concentrations of bacteriorhodopsin, have been obtained to demonstrate the formation of liquid membranes by bacteriorhodopsin on a supporting membrane. Data on photo-osmotic velocity through the liquid membrane bilayers, thus generated by bacteriorhodopsin, have also been obtained to gain information on the variation of the photo-osmotic velocity with the intensity and wavelength of the exciting light and with the concentrations of proton acceptors present in the system.

Bacteriorhodopsin (Sigma cat. No. B 3636), 2,4-dinitrophenol (DNP) (E. Merck) and doubly distilled water in all Pyrex glass still were used in these experiments. All solutions in photo-osmosis experiments were maintained at pH 2 using, a 0.1 M Tris -HCI buffer. The CMC of bacteriorhodopsin was found to be 9.5×10^{-2} ppm. For all transport studies the cell described in Fig.9 was used. The hydraulic permeability data and the photo-osmotic permeability data were obtained in the manner described in the previous sub-section. For more details the original paper on bacteriorhodopsin may be referred to [49]. All measurements were-made at 37±1°C.

The hydraulic permeability data at various concentration of bacteriorhodopsin were found to obey the proportional relationship (12). The value of L_p as estimated from the J_v versus ΔP plots, show a progressive decrease with Increase In bacteriorhodopsin concentration up to its CMC beyond which they become more or less constant (Table 17) This trend is in accordance with Kesting's liquid membrane hypothesis and demonstrates the formation of liquid membrane in series with the supporting membrane. The values of L_p computed, using mosaic model [12-14], at concentration below the CMC of bacteriorhodopsin, compare favorably with the experimentally determined values (Table 17) lending further support to the formation of liquid membrane in series with the supporting membrane. Since a complete liquid membrane is generated at concentration equal to the CMC, it follows that if both the compartments C and D of the transport cell were filled with solutions of bacteriorhodopsin of concentration equal to or greater than its CMC, the supporting membrane would be sandwiched between the two layers of liquid membranes generated on either side of it. Evidence in favors of this is obtained from the analysis of L_p values when both compartments were filled with an aqueous solution of bacteriorhodopsin of concentration 0.1 ppm, which is greater than its CMC. Following the analysis given earlier [4] it can be shown that

$$\frac{1}{L_P^*} = \frac{2}{L_P^C} - \frac{1}{L_P^\circ} \tag{15}$$

where L_P^* is the value of L_P when both compartments of the transport cell (Fig. 9) are filled with aqueous solution of bacteriorhodopsin of concentration higher than its CMC. The superscripts c and o stand, respectively, for the series composite membrane consisting of the supporting membrane and the bacteriorhodopsin membrane in series array and the bare supporting membrane. The value L_P^* computed using Eq.(15) agrees with the experimentally determined value (Table 17).

Table 17. Values of $L_P/m^3s^{-1}N^{-1}$, at various concentrations of bacteriorhodopsin (Ref. 49).

Conc. $\times 10^2$/ppm	0.0	2.375	4.750	7.125	9.500[a]	11.875	14.250	C[b]
$L_p^c \times 10^8$	0.826	0.772	0.706	0.660	0.632	0.627	0.634	0.524
	±0.023	±0.029	±0.023	±0.012	±0.008	±0.019	±0.009	±0.016
$L_p^d \times 10^8$		0.778	0.729	0.681				0.512
		±0.020	±0.016	±0.012				±0.002[e]

[a]CMC. [b]Values for the system when both the compartments C and D were filled with bacteriorhodopsin solution of concentration 0.1 ppm. [c]Experimental values. [d]Calculated values using the mosaic model. [e]Calculated value using (15).

The data on photo-osmosis are recorded in (Table 18 and 19). The induction time for photo-osmotic movement to commence was ca. 10s. Photo-osmotic volume flow continued as long as the light was on and stopped when the light was switched off. Since these experiments were carried out under constant temperature conditions, the possibility of temperature gradient produced by absorption of light inducing the observed volume flow was eliminated. The observed induction time of 10s is too short for the establishment or abolition of any measurable temperature gradient. Moreover, it was also observed that on short-circuiting electrodes E_1 and E_2 the light-induced volume flux stopped completely; when the short-circuiting was removed, the volume flux recommenced. This observation not only eliminates the possibility of the thermal gradients causing the observed flow, but also establishes that the light-induced electrical potential difference across the membrane is the primary driving force for the observed photo-osmosis.

In these experiments it was also observed that the direction of the light-induced volume flux was always from the illuminated compartment to the dark compartment. This observation can also be explained in same manner as in the case of chloroplast extract liquid membrane bilayers, i.e., on the basis of electrical double layer theory and electrokinetics. The proton pumping action of bacteriorhodopsin depends totally on the presence of the chromophore known as purple complex. The retinal in the purple complex is linked to the lysine residue of the polypeptide chain [61-63] through what is called a Schiff base (Fig.14). It is the Schiff base that loses and regains a proton in the photoreaction alternating between the protonated and deprotonated forms (Fig.14). In these experiments on excitation by light, protons are released in the illuminated compartment and are captured by the proton acceptors present there. Thus, an electrical double layer is generated which consists of a negatively charged membrane phase and a positively charged mobile phase. Since the illuminated compartment where protons are generated due to the action of light, is positive with respect to the dark compartment, the positively charged mobile phase of the double layer moves from the illuminated compartment, to the dark compartment under the influence of the light-induced electrical field.

The electrical potential difference developed across the bacteriorhodopsin liquid membrane bilayers, which is responsible for the observed photo-osmotic volume flux, is a consequence of the light driven proton pumping action of bacteriorhodopsin. This implies that the magnitude of photo-osmotic velocity should increase with an increase in concentration of proton acceptors present in the illuminated compartment. The data in Table 18 on the variation of photo-osmotic velocities with the increase in the concentration of DNP (proton acceptor) in the illuminated compartment (Fig. 9) confirm this trend.

Variation of photo-osmotic velocity with the intensity of exciting light showed a linear dependence. Such dependence implies that the light-induced electrical potential difference across the liquid membrane bilayers generated by bacteriorhodopsin varies linearly with the intensity of the exciting light. Similar trend was observed in the case of chloroplast extract liquid membrane bilayers.

The values of photo-osmotic velocity induced by light of different wavelengths obtained using different optical filters are recorded in Table 19. The magnitude of the volume flux among all the filters used is maximum for the filter corresponding to the wavelength range of 540-610 nm. This observation is consistent with the fact that absorption for bacteriorhodopsin is maximum at 560 nm [53], and indicates that absorption of light by bacteriorhodopsin is responsible for the development of the electrical potential difference across the liquid membrane bilayer causing the phenomenon of photo-osmosis.

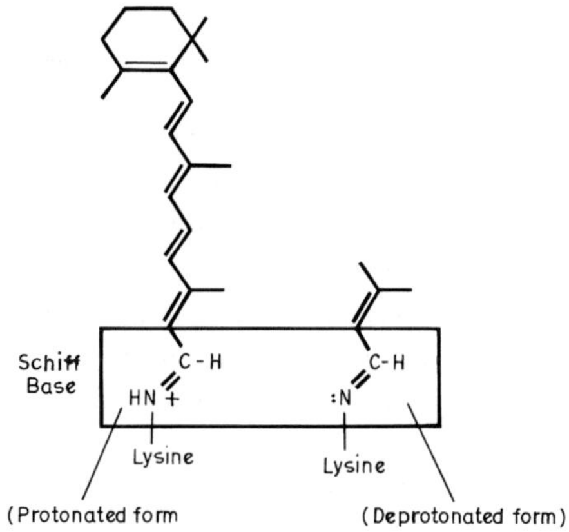

Fig. 14 The purple complex.

Table 18. Values of photo-osmotic velocity (J_v) at various concentration of proton acceptor (DNP) in the illuminated compartment (Ref. 49).

	Concentration of DNP in the illuminated compartment/M	*Photo-osmotic velocity* J_v×10^6/m s^{-1}
Bacteriorhodopsin[a]	1×10^{-4}	3.237±0.033
	5×10^{-4}	3.556±0.076
	1×10^{-3}	3.786±0.054
	5×10^{-3}	3.972±0.065
	1×10^{-2}	4.083±0.062

[a]Bacteriorhodopsin solution of concentration 0.1 ppm at pH=2 was taken in the compartments, C and D in all the cases.

Thus, the relevant conclusion from the studies on bacteriorhodopsin is that the bacteriorhodopsin liquid membrane bilayers when asyrnrnetrically illuminated extrude protons into the illuminated compartment.

Table 19. Values of photo-osmotic velocity *(J_v×10^6/ms^{-1})* a different wavelength ranges (Rcf. 49).

Wavelength range/nm	Photo-osmotic velocity
White light	3.786±0.054
(Filter No. 622[a], peak value of 440 ppm) 400-530	2.733±0.045
(Filter N, 624[a], peak value of 520 nm) 490-560	3.021±0.58
(Filter No. 626[a], peak value of 50 nm) 540-610	4.107±0.095
(Filter No. 608[a], peak value of 720 nm) 630-760	3.134±0.063

[a]Filters where obtained from Systronics India, A bacteriorhodopsin solution of concentration 0.1 ppm at pH=2 was used in the compartments, C and D in all the cases. The illuminated compartment in all the cases contained DNP (1×10^{-3}M)

5.4 Hydrophilic Pathways

5.4.1 Transport in presence of polyene antibiotics [64]

Another attempt has been made to demonstrate the workability of liquid membrane bilayers as mimetic systems for biomembranes by conducting transport studies in presence of polyene antibiotics. Studies on the permeability of liquid membrane bilayers generated by lecithin, cholesterol and lecithin-cholesterol mixtures in the presence of polyene antibiotics, namely, nystatin and amphotericin B, have been conducted [64]. The results obtained indicated the formation of aqueous pores in the liquid membrane bilayers and were consistent with those reported on BLMs. In this section an account of these studies is presented.

Both nystatin and amphotericin B were found to be surface active and their CMCs, as determined form the variation of surface tensions with concentrations, were found to be 7.0×10^{-11} *M* and 6.0×10^{-11} *M*, respectively. The transport studies were conducted using the all-glass cell diagrammed in Fig.2. In these studies, a Sartorius cellulose nitrate microfiltration membrane (average pore size, 0.2 μm) was used as a supporting membrane for the liquid membranes. Data on hydraulic permeability, transport numbers and solute permeability of ions in presence of polyene antibiotics were obtained to indicate the formation of aqueous pores by the antibiotics in the liquid membrane bilayers.

To obtain hydraulic permeability data, the two compartments of the transport (Fig.2) were filled with the mixtures of aqueous solutions of desired composition of lecithin, cholesterol and the polyene antibiotics. Known pressures were applied on compartment C and the consequent volume flux was measured by noting the rate of advancement of liquid meniscus in the capillary L_1L_2 (Fig.2). The hydraulic permeability data were obtained in the following sets of experiments:

1. The compartment, C was filled with the solutions of varying concentrations of the polyene antibiotics, viz., amphotericin B and nystatin, prepared in the aqueous solution of cholesterol of concentration 3.87×10^{-8} *M* and the compartment D contained only distilled water.

2. Both the compartments, C and D were filled with the solutions of varying concentrations of the polyene antibiotics prepared in (a) the aqueous solution of cholesterol of the same concentration as used in Set 1 above and (b) the aqueous solution of lecithin-cholesterol mixture of concentration, 15.542 ppm with respect to lecithin and 1.175×10^{-6} M with respect to cholesterol. In these experiments the same antibiotic was taken in the two compartments.

3. When in Set 2 of the experiments, amphotericin B and nystatin of fixed concentrations were taken in the compartments, C and D, respectively.

The concentrations of lecithin, cholesterol arid their mixture used in the above experiments were those derived from the earlier studies [3, 4]; these are the concentrations at which it was experimentally shown [3,4] that the liquid membrane generated completely covers the supporting membrane. Thus in Set I of the experiments, the supporting membrane would be completely covered by the liquid membrane generated by cholesterol and in Set 2 and 3 of the experiments respectively, the supporting membrane would be sandwiched between the two layers of the liquid membrane, generated on either side of the supporting membrane, by cholesterol, by lecithin and by the lecithin-cholesterol mixture. As has been indicated in the earlier studies [4], in the liquid membranes generated by lecithin, cholesterol or their mixture, hydrophobic ends of these surface-active substances will be preferentially oriented towards the hydrophobic supporting membrane the cellulose nitrate micro-filtration membrane in these experiments, and their hydrophilic moieties will be drawn outwards away from it.

For measurements of both, the transport number and the solute permeability (ω) of the ions, the compartment C of the transport cell (Fig.2) was filled with the solution of potassium chloride of known concentration prepared in the aqueous solution of desired composition of lecithin, cholesterol, or their mixture containing the desired concentration of one of the antibiotics. Compartment D contained only the aqueous solution of lecithin, cholesterol or their mixture of the same composition as contained in compartment C, along with the desired concentration of the antibiotics.

The compositions of the solution of lecithin, cholesterol, and their mixture were such that complete liquid membranes were formed at the interface and the chosen concentrations of the antibiotics were those at which the hydraulic penneability was found to be maximum indicating complete formation of aqueous pores in the liquid membrane bilayers. Transport number (t_2) of the anions were determined using the Eq.(8) and the well known relationship

$$t_1 + t_2 = 1 \qquad (16)$$

For determination of solute permeability (ω) for chloride ions through the liquid membrane bilayers generated on the supporting membrane in the presence of polyene antibiotics Eq. (18) of chapter 4

$$\omega = \left(\frac{J_s}{\Delta\pi} \right)_{J_v=0}$$

was utilized. The condition $J_{v=0}$ was imposed on the system and the solute flux J_s was estimated by measuring amount of chloride ions transported to the compartment D in a known period of time which was of the order of a few hours. The amounts of the chloride ion transported to the other compartment were measured by spectrophotometric determination of its reaction product with brucine and potassium persulfate at 540 nm [65]. The details of the procedure for the measurements of transport numbers and solute permeability have already been described in this chapter. Readers may also refer to the original publication [64].

All measurements of hydraulic permeability, solute permeability, transport numbers and critical micelle concentrations were carried out at constant temperature using a thermostat set at 40 ± 0.1°C. The hydraulic permeability data in all cases were found to be in accordance with the proportional relationship (12). The values of L_P for the various cases estimated from J_v versus ΔP plots are given in the Tables 20 and 21.

The values of L_P in the first set of experiments, when only the cholesterol solution was present in the compartment C, do not show any perceptible change due to the presence of the antibiotics in the same compartment but in the experiments where cholesterol solution was present in both the compartments, C and D along with the antibiotics, the values of L_p do show an increasing trend indicating the presence of aqueous pores in the liquid membrane bilayers. This is in conformity with the reported observation [66] on BLMs that the action of these antibiotics is strongly facilitated by their addition to the solutions on both sides of the BLM. The addition of the antibiotics in only one of the compartments, creates, what has been called a "half pore" and the complete aqueous pore is formed by the union of f two such "half pores" generated by the antibiotics present in the two compartments [67].

In the experiments where amphotericin B was added to the cholesterol solution in the compartment, C and nystatin to the cholesterol solution in the compartment, D (the concentrations of the antibiotics being those at which an increase in the value of L_P was maximum when they were separately added to the two compartments in the second set of experiments), an increase in the value of L_P was noticed (Table 20). This observation is in keeping with the indication available in the literature [67, 68], that permeability characteristics of nystatin and amphotericin B pores are almost Identical [68], and it is possible to form an aqueous pore composed of both molecules by adding amphotericin B to one side and nystatin to the other [69].

In the experiments where lecithin alone was taken in the two compartments, no increase in the value of L_p was observed due to the presence of the antibiotics. On the contrary, there was a decrease (Tables 20 and 21). However, in the experiments repeated with lecithin-cholesterol mixture, an increasing trend in the values of similar to those observed in case of cholesterol alone, was observed (Tables 20 and 21). In this case also addition of amphotericin B in compartment C and nystatin in compartment D showed an increase in the values of (Table 20) indicating the formation of aqueous pores composed of the pores formed by the two antibiotics.

An obvious conclusion from these studies is that the presence of cholesterol is necessary for the formation of aqueous pores. A decrease in the values of, L_p in case of lecithin alone, indicates that the antibiotics are incorporated into the liquid membrane bilayers strengthening their hydrophobic core and, thus, decreasing their permeability to water. The inference that presence of cholesterol is necessary for the formation of aqueous pores is consistent with literature reports. It is reported [70] that polyene antibiotics are active against fungi but not against bacteria. The membranes of the latter do not contain sterols. It has also been reported [71] that polyene antibiotics can cause mechanical rupture of BLM formed from a mixture of lecithin and cholesterol, but have little effect on BLM formed from lecithin alone. It has also been concluded [69,72,73] from conductance data that sterol, particularly cholesterol, is a necessary membrane constituent for polyene antibiotics to be effective.

A perusal of (Tables 20 and 21) reveals that in the case of both cholesterol and lecithin-cholesterol mixture an increase in the value of L_p is observed only after concentrations of the antibiotics exceed a certain minimum value and thereafter, it remains more or less constant. It is also noteworthy that in the case of lecithin-cholesterol mixture, the value of when amphotericin B is taken in one compartment and nystatin in the other, is greater than the values of L_p when the same antibiotic, amphotericin B or nystatin, is taken in both the compartments. This observation indicates that the channels formed, when amphotericin B is present in one compartment and nystatin in the other, are more hydrophilic than the channels formed when the same antibiotic is present in both the compartments.

The data on transport numbers in presence of the antibiotics, in the three cases i.e., when the compartments, C and D, (Fig.2) contained cholesterol alone, lecithin alone, and the lecithin-cholesterol mixtures, are recorded in (Table 22). The concentrations of lecithin, cholesterol, and the lecithin-cholesterol mixture were the same as those used in the hydraulic permeability experiments; these were the concentrations, at which it has been experimentally, demonstrated [3,4] that a complete liquid membrane is formed at the interface. The concentrations of amphotericin B and nystatin chosen in these experiments were those at which there was a distinct indication of pore formation in the hydraulic permeability experiments (Tables 20 and 21).

Table 20. Values of $L_p/m^3s^{-1}N^{-1}$, at various concentration of amphotericin B in the case of various liquid membrane bilayers generated on the supporting membrane (Ref. 64).

Concentrations $\times 10^{11}$/M	0.0	2.0	4.0	6.0	8.0	a
Cholesterol monolayers	2.166	2.155	2.264	2.126	2.114	-
$L_p \times 10^8$	±0.050	±0.043	±0.070	±0.094	±0.006	
Cholesterol Bilayer	1.717	1.818	2.077	2.021	1.906	1.951
$L_p \times 10^8$	±0.122	±0.012	±0.082	±0.082	±0.050	±0.024[b]
Lecithin bilayer	1.643	1.644	1.203	1.267	1.2116	1.156
$L_p \times 10^8$	±0.050	±0.029	±0.017	±0.026	±0.028	±0.032[c]
Lecithin-cholesterol bilayer	1.167	1.313	1.294	1.300	1.295	1.688
$L_p \times 10^8$	±0.038	±0.039	±0.072	±0.053	±0.047	±0.062[d]

[a] When amphotericin B was taken in compartment C and nystatin in compartment D., [b]Amphotericin B (4.0×10^{-11}M) in compartment C and nystatin (5.0×10^{-11}M) in compartment D., [c]Amphotericin B (2.0×10^{-11}M) in compartment C and nystatin (1.0×10^{-11}M) in compartment D., [d]Amphotericin B (2.0×10^{-11}M) in compartment C and nystatin (7.0×10^{-11}M) in compartment D.

A perusal of Table 22 reveals that the transport numbers of chloride ions show an increase due to the presence of the antibiotics except in the case of lecithin. Both in the case of cholesterol and lecithin-cholesterol mixture, the value of the transport number is maximum when amphotericin B was present in one compartment and nystatin in the other. In the case when the solution of lecithin alone was taken in the two compartments, the transport numbers showed a decrease (Table 22) in the presence of the antibiotics, These trends once again indicate that the presence of cholesterol is necessary for the pore formation, The decreasing trends, in case of lecithin, appear to be due to the incorporation of the antibiotics in the liquid membrane bilayers resulting in a strengthening of their hydrophobic core. In the case of cholesterol, the value of the transport number of chloride ion in the presence of the antibiotics is greater than 0.6 (Table 22) indicating that most of the current through aqueous pores is carried by chloride ions. This observation is consistent with the literature reports on BLMs. It is reported that nystatin or, amphotericin B treated BLMs are anion selective [66, 69, 72, 74, 75]. The membrane, however, does not completely discriminate between anions and cations. In the case of lecithin-cholesterol mixtures, the transport number of the chloride ion, although it increases in presence of the antibiotics, has a value less than 0.5.

Table 21. Values of $L_p/m^3s^{-1}N^{-1}$, at various concentration of nystatin for various liquid membrane bilayers generated on the supporting membrane (Ref. 64).

Concentrations $\times 10^{11}$/M	0.0	1.0	3.0	5.0	7.0	9.0
Cholesterol monolayers	2.166	1.926	1.897	1.914	1.961	1.613
$L_p \times 10^8$	±0.050	±0.048	±0.051	±0.049	±0.037	±0.064
Cholesterol Bilayer	1.717	1.650	1.638	1.880	1.859	1.808
$L_p \times 10^8$	±0.122	±0.051	±0.056	±0.021	±0.036	±0.020
Lecithin bilayer	1.643	1.470	1.089	1.185	1.110	1.060
$L_p \times 10^8$	±0.050	±0.027	±0.013	±0.041	±0.024	±0.009
Lecithin-cholesterol bilayer	1.167	1.157	1.172	1.136	1.260	1.241
$L_p \times 10^8$	±0.038	±0.019	±0.021	±0.029	±0.014	±0.009

Evidence in favors of the formation of the aqueous pores in the liquid membrane bilayers is also obtained from the solute permeability (ω) data for ions. Since the aqueous pores are known to be anion selective, the data on solute pemleablltty (ω) for chloride ions through the liquid membrane bilayers generated from lecithin, cholesterol, and their mixture, were obtained in the presence of the antibiotics. In these experiments also, the concentrations of the antibiotics and of lecithin, cholesterol and their mixture, which filled the two compartments of the transport cell, were the same as in the transport number experiments. A perusal of the values (Table 22) reveals that, except in the case of the lecithin bilayer where there was a decrease, in the case of both cholesterol bilayers and the lecithin-cholesterol bilayers the solute permeability is enhanced in the presence of the antibiotics.

This further confirms that the presence of cholesterol is necessary for the formation of aqueous channels. It can also be seen that the enhancement in the value of ω is maximum (Table 22) when amphotericin B was present in one compartment of the transport cell and nystatin in the other. This indicates that the aqueous channels formed by amphotericin B and nystatin in association with cholesterol are more permeable to chloride ions than the channel formed by either one of them alone. Similar conclusions can be drawn from the trends in the transport number data (Table 22). Thus, these studies give ample indication of the formation of aqueous pores by the polyene antibiotics in the liquid membrane bilayers generated from cholesterol and lecithin-cholesterol mixtures.

5.4.2 Explaining pharmacological action of hydrocortisone [76]

Studies have been conducted [76] on the transport through liquid membrane bilayers generated by prostaglandin E_1 in the presence of hydrocortisone, The data indicate the formation of hydrophilic pathways by hydrocortisone in the liquid membrane bilayers generated by prostaglandin E_1.

The all-glass cell described earlier (Fig.2) was used to obtain the data on hydraulic and solute permeability. To obtain hydraulic permeability data, the two compartments of the transport cell were filled with an aqueous solution of mixtures of prostaglandin E_1 and hydrocortisone acetate of desired composition. Known pressures were applied to the lower compartment C, and the consequent volume flow was measured with time in the capillary attached to the compartment D of the transport cell.

Table 22. Values of solute permeability (ω/mol^{-1} s^{-1} N^{-1}) and transport number (t_2) for chloride ions in the presence of the polyene antibiotics, for various liquid membrane bilayers generated on the supporting membrane (Ref. 64).

		Cholesterol bilayer		
	a	b	c	d
t_2	0.313	0.648	0.801	0.826
$\omega \times 10^{11}$	4.543	5.060	13.865	16.041
		Lecithin bilayer		
	a	b_1	c_1	d_1
t_2	0.299	0.272	0.111	0.183
$\omega \times 10^{11}$	22.101	15.034	18.808	19.853
		Lecithin-cholesterol bilayer		
	a	b_2	c_2	d_2
t_2	0.053	0.222	0.130	0.276
$\omega \times 10^{11}$	6.474	9.409	7.683	14.375

a: control without antibiotics; b, b_1, b_2 : when both the compartments C and D contained amphotericin B, $b(4.0\times10^{-11}M)$, b_1 and b_2 $(2.0\times10^{-11}M)$; c, c_1, c_2 : when both the compartments C and D contained Nystatin, c(5×10^{-11}M), c_1(1.0×10^{-11}), c_2(7×10^{-11}M); d, d_1, d_2 : when the compartment C contained Amphotericin B and the compartment D contained Nystatin d (Amphotericin B, $4\times10^{-11}M$; Nystatin $5\times10^{-11}M$), d_1(Amphotericin B, $2\times10^{-11}M$; Nystatin, $1\times10^{-11}M$), d_2 (Amphotericin B, $2\times10^{-11}M$; Nystatin, $7\times10^{-11}M$).

For solute permeability (ω) measurements, compartment C of the transport cell was filled with aqueous solution of known concentrations of the permeant along with aqueous solution of mixtures of prostaglandin E_1 and hydrocortisone of desired composition and the compartment D as filled only with aqueous solutions of the mixtures of prostaglandin E_1 and hydrocortisone of desired composition. Eq. (18) of chapter 4, i.e.,

$$\omega = (J_s/\Delta\Pi)_{Jv=0}$$

was used for estimating the values of ω. All measurements were made at 37°±0.1°C. For details the original paper should be consulted [76].

Hydraulic permeability data were obtained in the following sets of experiments:

1. The compartment, C of the transport cell was filled with solutions of varying concentrations of hydrocortisone prepared in an aqueous solution of fixed concentration of prostaglandin E_1, equal to $3\times10^{-8}M$, and the compartment, D was filled with distilled water.

2. Both the compartments, C and D were filled with aqueous solutions of a mixture of prostaglandin E_1 and hydrocortisone of the same composition as that in the compartment, C in Set 1.

The particular concentration of prostaglandin E_1, used in these experiments, 3×10^{-8} M, is higher than its critical micelle concentration (CMC) and was derived from our earlier studies, in which it was shown, using Kesting's hypothesis, that when surface-active prostaglandin E_1 is added to an aqueous phase, a surfactant-layer liquid membrane which completely covers the interface at concentrations equal to or greater than its CMC, is generated. The CMC of prostaglandin E_1 was found to be 1×10^{-8} M. It is obvious that in the surfactant layer liquid membrane, thus generated by prostaglandin E_1, the hydrophobic portions of the prostaglandin E_1 molecules would be preferentially oriented towards the hydrophobic supporting membrane and the hydrophilic moieties would be drawn outward away from it. In the experiments (Set 1) in which an aqueous solution of prostaglandin E_1 was added only to the compartment C, the liquid membrane would be generated only in compartment C in series with the supporting membrane, whereas in Set 2, the supporting membrane would be sandwiched between two layers of liquid membrane generated by prostaglandin E_1.

The hydraulic permeability data, in all the cases were found to be in agreement with the relationship (12),

$$J_v = L_p \Delta P \qquad (12)$$

The values of L_p for the different cases estimated from J_v versus ΔP plots are recorded in Table 23.

Table 23. Values of $L_p/m^3s^{-1}N^{-1}$, at various concentrations of Hydrocortisone it the mixture of prostaglandin E_1^a and hydrocortisone (Ref. 76).

	Concentration of Hydrocortisone $\times$ 10^6/M					
	0	1	2	3	4	5
$L_p^b \times 10^9$	13.394	6.937	3.993	4.055	4.082	4.062
	±0.098	±0.105	±0.095	±0.040	±0.120	±0.079
$L_p^c \times 10^9$	3.061	3.271	3.501	3.467	3.458	3.478
	±0.047	±0.029	±0.099	±0.133	±0.071	±0.039

[a]Prostaglandin E_1 concentration kept constant 3×10^{-8} M. [b]Values obtained when the mixtures of prostaglandin E_1 and hydrocortisone were added only to the lower compartment C of transport cell and the compartment D was filled with distilled water (Set.1). [c]Values obtained when 1he mixtures of prostaglandin E_1 and hydrocortisone were added to both the lower and the upper compartments of the transport cell (Set 2).

The values of L_p in the experiments in which solutions of varying concentrations of hydrocortisone prepared in a solution of the fixed concentration of prostaglandin E_l, were added only to the lower compartment C (Set 1), do not show any increase with the increase in concentration of hydrocortisone; on the contrary, there is decrease (Table 23). The decreasing trend in the values of L_p (Table 23) may be due to incorporation of hydrocortisone (also surface active in nature), into the already existing prostaglandin E_l liquid membrane at the interface. The values L_p in the other set of experiments (Set 2), however, show an increasing trend with increase in the concentrations of hydrocortisone (Table 23); the value of L_p increases up to a certain concentration of hydrocortisone beyond which it becomes more of less constant (Table 23). These trends in the values of L_p indicate the formation of aqueous pores in the prostaglandin E_1, liquid membrane bilayer only when hydrocortisone is present on both the sides of the membrane.

A perusal of Table 23 reveals that the values L_p increase up to a hydrocortisone concentration of $2\times10^{-6}M$, beyond which they become more or less constant. Thus, the concentration of $2\times10^{-6}M$ hydrocortisone appears to be the concentration at and beyond which complete aqueous pores are formed in the prostaglandin liquid membrane bilayer. Similar observations have been reported in the case of aqueous pore formation by polyene antibiotics [64, 67].

The solute permeability (ω) data for histamine and serotonin were also obtained in two sets of experiments. In the first set, a solution of known concentration of the permeant, histamine or serotonin, prepared in an aqueous solution of desired composition of the mixture of prostaglandin E_l and hydrocortisone, was added to the compartment C of the transport cell, and the compartment D was filled with distilled water. In the second set of experiments for ω measurements, compartment D of the transport cell instead of being filled with distilled water, was filled with an aqueous solution of the mixture of prostaglandin E_l and hydrocortisone of the same composition as that in the compartment C. In the control experiments no hydrocortisone was used. The composition of the prostaglandin E_1 and hydrocortisone mixture used in the solute permeability experiments was that at which the value of L_p showed maximum increase indicating the formation of complete aqueous pores in the prostaglandin E_1 liquid membrane bilayer (Table 23).

Table 24. Normalized values (r) of solute permeability for histamine and serotonin through the liquid membrane generated by the prostaglandin E_1 – hydrocortisone mixture[a] (Ref. 76).

Permeant	rb	r[c]
Histamine	0.678	1.903
Serotonin	1.00	1.470

[a]Concentration of prostaglandin E_l and hydrocortisone in the mixture were 3×10^{-8} and 2.5×10^{-6} M, respectively. [b]Value obtained when the mixtures of prostaglandin E_1 and hydrocortisone were added only to the lower compartment C of the transport cell and compartment D was filled with distilled water., [c]Values obtained when the mixtures of prostaglandin E_l and hydrocortisone were added to both lower and upper compartments of the transport cell.

The normalized values *(r)* of solute permeability *(r* = $\omega/\omega_{control}$*)* for histamine and serotonin are recorded in (Table 24). It is obvious that the solute permeability whereas in the of both histamine and serotonin is enhanced in the second set of experiments whereas in the first set of experiments the solute permeability of serotonin remains unaltered while that of histamine decreases (Table 24). These observations are consistent with the conclusion drawn from the hydraulic permeability experiments (Table 23) that aqueous pores in the prostaglandin liquid membranes are formed only when hydrocortisone is present on both the sides of the liquid membrane.

These trends in the hydraulic permeability and in the solute permeability appear relevant to therapeutic action of hydrocortisone in the treatment of inflammation. On a microscopic level inflammation is usually accompanied by the familiar clinical signs of erythema, edema, hyperalgesia, and pain [77]. Prostaglandins are always released when cells are damaged and have been detected in increased concentrations in inflammatory exudates. During inflammation chemical mediators like histamine and serotonin, which stimulate sensory nerve endings and cause pain [77,78] are also liberated locally. Hydrocortisone and its synthetic analogs are known to suppress the inflammatory manifestations. The exact mechanism of these therapeutic effects of hydrocortisone, however, remains unclear [76]. These data indicate that the enhanced solute permeability of histamine and serotonin leading to there reduced concentration at the inflammation site and the enhanced volume flow (increase in the values of L_p due to the formation of aqueous pores in the prostaglandin liquid membrane) could be a plausible explanation for the observed suppression of inflammatory manifestation by hydrocortisone.

5.4.3. Studies with prostaglandin's [80]

In spite of the presence of a hydrophobic core in the lipid bilayers, plasma membranes are quite permeable to water. Several viewpoints have been advanced to explain this apparent paradox. One generally agreed view is that passive transport through biomembranes and bilayer lipid membranes is in many cases, controlled by existing small holes or pores in them [2, 81]. Several authors on physical grounds have discussed the formation and character of these holes. Kashchiev and Exerow [82] have developed a unified analysis of the permeation of the bilayer lipid membrane and rupture, when these are due to fluctuation in formation of holes or pores in the membrane. Since prostaglandins are ubiquitously distributed in almost every tissue and body fluid, the earlier study [76] on hydrocortisone-prostaglandin combinations tempts us to suspect that prostaglandin in association with cholesterol, which is structurally similar to hydrocortisone, may be responsible for the formation of hydrophilic pores in the plasma membranes, leading to their unexpectedly high passive permeability to water and hydrophilic solutes in spite of the presence of a hydrophobic core in the lipid bilayer.

To obtain hydraulic permeability data, aqueous solutions of lecithin-cholesterol and prostaglandin mixtures of desired composition were used to fill the compartments C and D of the transport cell (Fig.2). Known pressures were applied to the compartment C, and the consequent movement of the liquid meniscus in the capillary L_1L_2 of known diameter attached to the compartment D was measured using a cathetometer reading to 0.001 cm and a stopwatch reading to 0.1 s. During the hydraulic permeability measurements, the electrodes E_1 and E_2 were short circuited so that the electro-osmotic back flow due to the streaming potentials developed across the membranes, did not interfere with the data on hydraulic permeability.

For solute permeability measurements, the compartments, C and D of the transport cell (Fig.2) were filled with aqueous solutions of desired composition of lecithin-cholesterol-prostaglandin mixtures, and a known concentration of the permeant was introduced in to the compartment, C. The condition of no net volume flux ($J_{v=0}$) was imposed on the system, and the solute transported to the other compartment in a known period of time was estimated. The values of solute permeability (ω) were estimated using Eq.(18) of chapter 4. For details of the procedures, the original paper should be consulted [80]. All measurements were made at constant temperature using a thermostat set at 37±0.1°C.

Hydraulic permeability data were obtained in the following sets of experiments:

1. (a) An aqueous solution of the lecithin-cholesterol-prostaglandin (E_1 *or* $F_{2\alpha}$) mixtures of composition 1.919×10^{-5} *M* with respect to lecithin, 1.175×10^{-6} *M* with respect to cholesterol and 3×10^{-8}*M* with respect to prostaglandin, was taken into the compartment C of the transport cell and the compartment, D was filled with distilled water.

   (b) Both the compartments, C and D were filled with an aqueous solution of the lecithin-cholesterol mixture of the same composition as in Set 1(a) and the prostaglandin (E_1 *or* $F_{2\alpha}$) solution of concentration 3×10^{-8}*M* was taken in the compartment, C only.

   (c) Both the compartments, C and D were filled with an aqueous solution of the lecithin-cholesterol-prostaglandin (E_1 or $F_{2\alpha}$) mixture of the same composition as in Set 1(a) above.

2. (a) An aqueous solution of cholesterol-prostaglandin (E_1 *or* $F_{2\alpha}$) mixture of composition 1.175×10^{-6} *M* with respect to cholesterol and 3×10^{-8} M with respect to prostaglandin, was taken into the compartment, C of the transport cell and the compartment, D was filled with distilled water.

   (b) Both the compartments, C and D were filled with cholesterol solution of concentration 1.175×10^{-6} *M* and prostaglandin (E_1 *or* $F_{2\alpha}$) solution of concentration 3×10^{-8} *M* was taken in the compartment C only.

   (c) Both the compartments, C and D were filled with a solution of the cholesterol-prostaglandin mixture of the same composition as in Set 2(a).

3. (a) An aqueous solution of lecithin-prostaglandin (E_1 *or* $F_{2\alpha}$) mixture of composition 1.919×10^{-5} *M* with respect to lecithin and 3×10^{-8} M with respect to prostaglandin, was taken in the compartment, C and the compartment, D was filled with distilled water.

   (b) Aqueous solution of lecithin of concentration 1.919×10^{-5} *M* were taken into both the compartments, C and D and prostaglandin solution of concentration 3×10^{-8} *M* was added only to the compartment, C.

(c) Both the compartments, C and D were filled with aqueous solutions of lecithin-prostaglandin mixture of the same composition as in Set 3(a).

For each of these experiments, separate control experiments were also performed in which all the conditions were the same except that no prostaglandins were used. The composition of the solution of lecithin, cholesterol prostaglandins and their mixtures used in the above experiments, were those derived from earlier studies [4, 83, 84]. The composition of the lecithin-cholesterol mixture, 1.919×10^{-5} *M* with respect to lecithin and $1.175 \times 10^{-6} M$ with respect to cholesterol, is the one at which, as was shown experimentally, the liquid membrane generated by lecithin completely covers the interface and is fully saturated with cholesterol. The concentration of prostaglandin used in the present study is the one at which, as has been shown in the earlier studies [83, 84], the lecithin-cholesterol liquid membrane generated at the interface is saturated with prostaglandin.

Formation of liquid membranes in these experiments is bases on Kesting's hypothesis [11], according to which, when a surfactant is added to an aqueous phase, the surfactant layer which forms spontaneously at the interface acts as a liquid membrane and modifies the mass transfer across the interface. As the concentration of the surfactant is increased the interface becomes progressively covered with the surfactant layer liquid membrane and at the CMC of the surfactant, it is completely covered. Thus, in the experiments where surface active materials are taken only in the compartment C, the surfactant layer liquid membrane is formed in series with the supporting membrane in the compartment C, while in the experiments where surface active materials are taken in both the compartments, C and D of the transport cell, the supporting membrane is sandwiched between the two layers of the liquid membranes generated on either side of it. Since lecithin, cholesterol and prostaglandin are all surface active in nature, it is obvious that in the liquid membrane generated in these experiments, the hydrophobic tails of these molecules will be preferentially oriented towards the hydrophobic supporting membrane and the hydrophilic moieties will be drawn outwards away from it. The gross picture of the liquid membrane bilayers formed in the void regions of the micro-porous supporting membrane when the solutions of lipids occupy both the compartments, C and D of transport cell, is shown in Fig. 15.

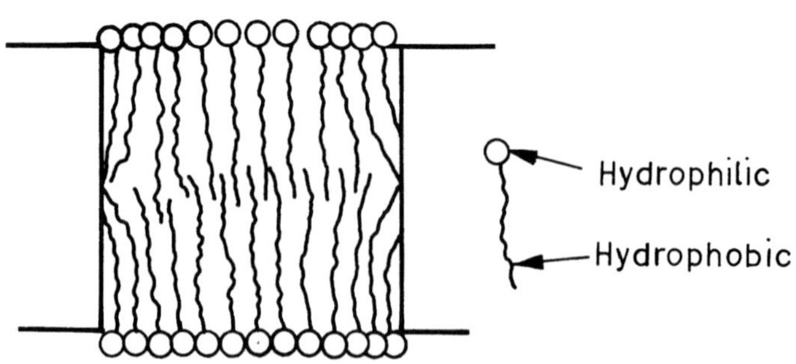

Fig.15 Gross picture of the liquid membrane bilayers formed in the void regions of the micro porous support.

In all the cases, the hydraulic permeability data were found to be in accordance with Eq.(12), i.e.,

$$J_v=L_p\Delta P \qquad (12)$$

where, J_v is the volume flux per unit area of the membrane, ΔP is the applied pressure difference across the membrane and L_p is the hydraulic conductivity coefficient. The normalized values of hydraulic conductivity coefficients $(L_p)/(L_p)_{control}$ estimated from the J_v versus ΔP plots, are recorded in Table 25.

Scrutiny of the data in Table 25 reveals the following trends: (1) when lecithin alone is taken either in one of compartments or in both the compartments, the presence of prostaglandin does not induce any increase in the values L_p. On the contrary, the value of L_p decreases; this naturally implies that prostaglandin gets incorporated in the lecithin liquid membrane at the interface and strengthens its hydrophobic region. (2) In the case, where cholesterol or lecithin-cholesterol mixture is taken in the compartment C only, the values of L_p again show a decrease. This trend is consistent with the earlier studies [76, 83, 84]. However, when cholesterol or the lecithin-cholesterol mixture is present in both the compartments, the values of L_p show a distinct increase due to the presence of the prostaglandin.

The normalized values of solute permeability $(\gamma=\omega/\omega_{Control})$ of several hydrophilic solutes are recorded in Table 26. The solute permeability data were obtained only in those cases where increase in hydraulic conductivity was observed. The trends observed in hydraulic conductivity data are corroborated by the trends in the solute permeability data (Table 26). The solute permeability is enhanced in the presence of prostaglandin when cholesterol or the lecithin-cholesterol mixture is present in both the compartments, C and D.

Thus, this study indicates that prostaglandin in association with cholesterol may be responsible for the formation of hydrophilic channels present in the lipid bilayers controlling passive transport through biomembranes. Further, the presence of cholesterol in each of the two constituent monolayers of the lipid bilayer appears essential for the pore formation in the presence of prostaglandins.

To explain the observed trends in the permeability data, one may invoke the formation of inverted micelles by prostaglandin within the hydrophobic core of the liquid membrane bilayers, where some water can always 'sneak in' from the aqueous exterior, i.e., the compartments C and D of the transport cell. Formation of inverted micelle is facilitated by the hydrophobicity of the medium i.e., the more hydrophobic the medium; the greater is the ease of formation of inverted micelles. Since lecithin is less hydrophobic than cholesterol, it appears that the hydrophobic core of the liquid membrane bilayers generated by lecithin alone is not able to provide enough hydrophobicity for the formation of the inverted micelles.

However, when the solution of the lecithin-cholesterol mixture is taken in the two compartments, the hydrophobic core of the liquid membrane bilayer appears to provide enough space of adequate hydrophobicity required for the formation-inverted micelles. The inverted micelles envisaged here are hollow cylinders, the inner surface of which are made up of the hydrophilic heads of the surfactant molecules.

Table 25. Normalized values $(K)^a$ of hydraulic conductivity coefficients $[K=(L_p)/(L_p)_{control}]$ of liquid membrane generated by lecithin, cholesterol and the lecithin-cholesterol mixture in the presence of prostaglandin.

	Lecithin-cholesterol[b]			Cholesterol[c]			Lecithin[d]		
	K^g	K^h	K^l	K^j	K^k	K^l	K^m	K^n	K^o
Prostaglandin	0.582	1.590	2.028	0.801	11.57	12.11	0.319	0.510	0.345
E_l^e	±0.031	±0.279	±0.347	±0.001	±0.12	±0.03	±0.006	±0.003	±0.009
Prostaglandin	0.794	1.133	1.263	0.830	8.23	8.85	0.399	0.599	0.435
$F_{2\alpha}^f$	±0.022	±0.077	±0.011	±0.010	±0.35	±0.29	±0.013	±0.012	±0.006

[a] Values reported as the arithmetic mean of 15 repeats *±S.D.* [b] Composition of lecithin-cholesterol mixture $=1.919\times10^{-5}M$ with respect to lecithin and $1.175\times10^{-6}M$ with respect to cholesterol. [c] Concentration of cholesterol-$1.175\times10^{-6}M$, [d] Concentration of lecithin=$1.919\times10^{-5}M$, [e] Concentration of prostaglandin $E_l=3\times10^{-8}M$, [f] Concentration of prostaglandin $F_{2\alpha}=3\times10^{-8}$ M, [g] Solution of lecithin-cholesterol mixture in compartment C along with the solution of the prostaglandin and water alone in compartment D, [h] Solution of lecithin-cholesterol mixture both in compartment C and D and the prostaglandin solution in compartment C only, [I] Solution of lecithin-cholesterol both in compartment C and D, prostaglandin solution also present in both compartments, [j] The cholesterol solution present in the compartment C along with the solution of prostaglandin, water in compartment D. [k] The cholesterol solution in both compartments C and D and prostaglandin, in solution in compartment C only. [l] The cholesterol solution and also the solution of prostaglandin present in compartment C and D. [m] The lecithin solution present in the compartment C along the solution of the prostaglandin, water along in compartment D. [n] The solution of lecithin in both compartments, C and D, and the prostaglandin solution in the compartment C only. [o] The solution lecithin and also the solution of prostaglandin present in compartments C and D.

5.4.4 Studies with hormones-Insulin and vasopressin [85]

Diphtheria toxin [86] botulinum neurotoxin [87] and tetanus toxin [88] are proteins that are similar in origin and macrostructure. It has been shown that all the three toxins form channels in bilayer lipid membranes (BLM), and that the amino terminus of the heavy chain in the structure of these toxins possesses a channel-forming domain. It was discovered that channels could be formed only in the presence of a pH gradient across the BLM, the toxin fragment being present on the acidic side of the membrane, and that reversing the pH gradient effectively blocked channel formation.

Because of the similarity of gross structure of both insulin and vasopressin with the clostridial toxins (Fig.16), the possibility of channel formation by vasopressin and insulin in the lipid bilayers cannot be ruled out and, hence, merits investigation. Transport studies on liquid membrane bilayers generated by a lecithin-cholesterol mixture in the presence of insulin / vasopressin have been carried out [85] with this object in view. Experiments on hydraulic permeability and on solute permeability of relevant permeants through lecithin-cholesterol liquid membrane bilayers in the presence of insulin / vasopressin on a supporting membrane have also been conducted. The data show trends comparable to those reported in demonstrating formation of channels in bilayer lipid membranes by toxin [89-92]. These studies can be viewed as further evidence in favors of workability of the liquid membrane bilayers as mimetic system of biomembranes.

Table 26. Normalised values $(\gamma)^a$ of the solute permeability of various solutes in the presence of the lecithin-cholesterol-prostaglandin mixture and the cholesterol-prostaglandin mixture (Ref. 80).

Permeant	Lecithin-cholesterol mixture[b]		Cholesterol[c]	
	$(\gamma)^f$	$(\gamma)^g$	$(\gamma)^h$	$(\gamma)^i$
	Prostaglandin E_1^d			
Calcium (Chloride)	1.477±0.003	1.673±0.134	1.339±0.118	1.528±0.070
Potassium (Chloride)	1.121±0.027	1.200±0.062	1.07±0.038	1.148±0.013
Sodium (Chloride)	1.1454±0.028	1.579±0.061	1.285±0.009	1.574±0.041
Aspartic acid	1.818±0.003	1.853±0.003	1.646±0.222	1.853±0.030
Glycine	1.23±0.215	1.804±0.262	1.412±0.314	1.543±0.177
	Prostaglandin $F_{2\alpha}^e$			
Calcium (Chloride)	1.932±0.067	2.018±0.085	1.862±0.094	1.984±0.130
Potassium (Chloride)	1.609±0.014	1.873±0.040	1.225±0.001	1.472±0.010
Sodium (Chloride)	1.309±0.008	1.504±0.002	1.090±0.002	1.235±0.023
Aspartic acid	1.889±0.004	1.903±0.006	1.818±0.003	1.853±0.003
Glycine	1.220±0.081	1.391±0.119	1.097±0.009	1.203±0.22

[a]Value of γ in each case is reported as the arithmetic mean of 15 repeats ± S.D. [b]Composition of lecithin-cholesterol mixture=$1.919\times10^{-5}M$ w.r.t., lecithin and 1.175×10^{-6} M w.r.t., cholesterol. [c] Concentration of cholesterol = 1.175×10^{-6} M, [d]Concentration of prostaglandin E_1 $=3\times10^{-8}$ M. [e] Concentration of prostaglandin $F_{2\alpha}=3\times10^{-8}$ M. [f] Solution of lecithin-cholesterol mixture both in the compartments, C and D and the prostaglandin solution along with the permeant in the compartment, C. [g] Solution of the lecithin-cholesterol-prostaglandin mixture both in the compartments, C and D and the permeants in the compartment, C. [h]The cholesterol solution in both the compartments, C and D and the prostaglandin solution along with the permeants in the compartment, C. [i]Solution of the cholesterol-prostaglandin mixture in both the compartments, C and D and the permeants in the compartment, C.

Hydraulic permeability and solute permeability data were obtained using the all-glass transport cell depicted in Fig.2. An aqueous solution of lecithin-cholesterol mixture of fixed composition was added to the compartment, C of the transport cell along with the aqueous solution of known concentration of insulin or vasopressin. The compartment, D was filled with an aqueous solution of lecithin- cholesterol mixture of the same composition as that in the compartment, C. The pH of the solution in the compartment, D was maintained constant at 7.4, and that of the solution in the compartment, C was varied form 7.4 to 3.6 using appropriate buffers. For maintaining pH between 7.4 and 5.4, a phosphate buffer was used, and for pH below 5.4, an acetate buffer was used.

The composition of the lecithin-cholesterol mixtures used in these experiments was $1.919\times10^{-5}M$ with respect to lecithin and 1.175×10^{-6} M with respect to cholesterol, because it had been shown in an earlier study [4] at this composition, the liquid membrane generated by lecithin at the interface is completely saturated with cholesterol. The concentration of insulin and vasopressin used in these experiments were 79.5 micro units per milliliter and 40 pg per milliliter respectively. This concentration of insulin is in the range of its concentration in plasma [93] under an appropriate stimulus and that of vasopressin is in the range of its concentration in blood under the conditions of dehydration or volume depletion [94]

To obtain data on hydraulic permeability, known pressures were applied on the compartment C and the consequent volume flux was measured by noting the rate of advancement of liquid meniscus in the capillary attached to the compartment D (Fig.2). The details of the procedure were the same as those described in earlier studies [3,4,64]

For measurement of solute permeability of the relevant permeants, Eq.(18) of chapter 4 was used. The compartments, C and D of the transport cell (Fig.2) were filled with solutions of the same compositions as those used in the hydraulic permeability experiments, along with a solution of the permeant of known concentration in the compartment, C. The condition $J_v = 0$ was imposed on the system and the amount of the solute transported to the compartment, D in a known period of time was estimated. For details, the original paper should be consulted [85]. Experiments on hydraulic permeability and on solute permeability were also performed when the direction of the pH gradient across the membrane was reversed. That is, the pH of the solutions in the lower compartment, C, was maintained fixed at 7.4 and that of the solution in the upper compartment, D was varied from 7.4 to 3.6.

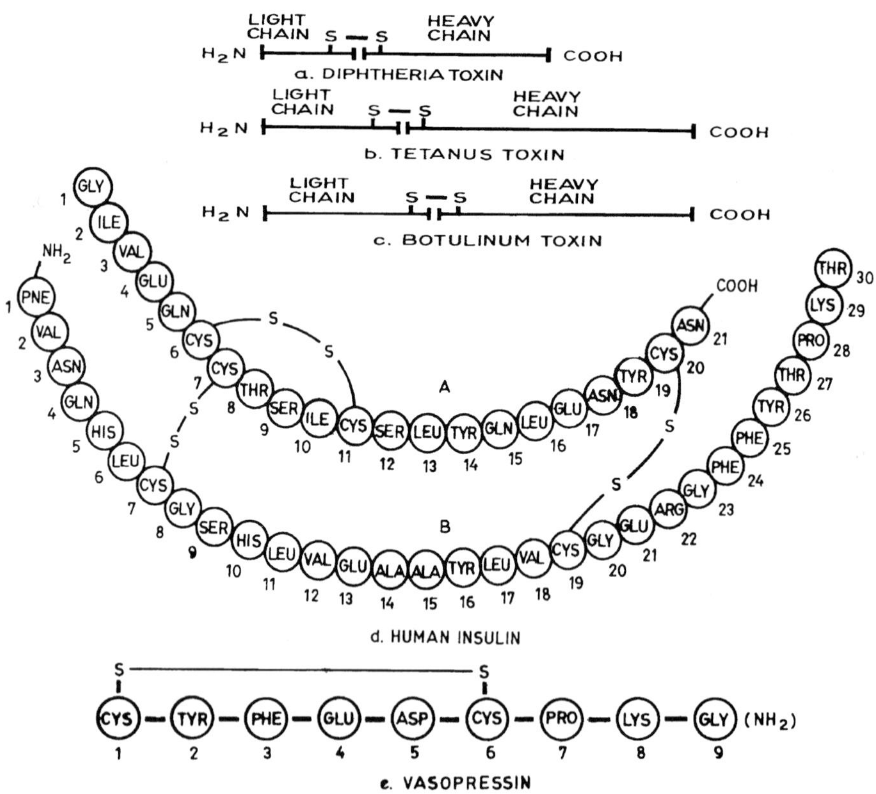

Fig 16. Structures of diphtheria toxin, tetanus toxin, botulinum toxin, insulin and vasopressin.

For the purpose of comparison, control experiments were performed for both hydraulic permeability and solute permeability in which everything was the same as in the actual experiments except that no insulin or vasopressin was used. All measurements were made at constant temperature using a thermostat set at 37±0.1°C. Because of the surface active nature of lecithin and cholesterol, it is natural to expect that in the liquid membrane bilayer generated in these experiments, the hydrophobic ends of the lipid molecules will be preferentially oriented toward the hydrophobic supporting membrane, the cellulose acetate microfiltration membrane in these experiments, and their hydrophilic ends will be drawn outward away from the supporting membrane.

The data obtained on the hydraulic permeability in the presence of insulin and also in the presence of vasopressin in the compartment C, in all cases, could described by the proportional relationship (12). The values L_p were estimated from the J_v versus ΔP plots and are recorded in Table 27. These values are normalized values, i.e., $L_p/(L_p)_{control}$ where $(L_p)_{control}$ is the values of L_p from the corresponding control experiment. As can be seen form (Table 27), hydraulic conductivity coefficient in the experiments where the pH of the solutions in both compartments was the same (pH 7.4), decreases in comparison to the value in the corresponding control experiment. This decrease may be due to incorporation of insulin in the lecithin-cholesterol liquid membrane, leading to a strengthening of the hydrophobic core of the bilayer. In the cases where the pH of the solutions in the compartment C, which contained insulin, was made acidic, maintaining the pH of the solution the compartment D constant at 7.4 an increase in the hydraulic conductivity coefficient in comparison to values from the corresponding control experiments was noted (Table 27). The increase was maximal when the pH in the compartment C was 4.6. The increase in the normalised values of L_p (Table 27) is indicative of an increased permeability in the liquid membrane bilayers.

These observations are consistent with the finding [91,92] in the case of botulinum, tetanus, and diphtheria toxins. All the three toxins have been shown to form aqueous channels in the bilayer lipid membranes (BLM) under pH gradient. It is also reported that channels are formed only when toxins are present on the acidic side. The pH, 4.6 of solutions in the compartment C, at which maximum increase in the values of the hydraulic conductivity coefficient was observed, also matches the pH at which the toxins have been reported to be most active in forming channels in the planar bilayer lipid membranes. In experiments where the pH of the solutions in the compartment C (containing insulin) was held constant at 7.4, and the pH of the compartment D was decreased, the hydraulic conductivity coefficient did not show any increase on the contrary, it decreased. This trend is also consistent with literature reports on clostridial toxins that reversing the pH gradient effectively blocked channel formation in the planar bilayer lipid membranes by the toxins [92]. Similar trends were observed in the data on hydraulic permeability in the case of vasopressin (Table 27).

Permeability changes induced by insulin and vasopressin in the lecithin- cholesterol liquid membrane bilayers, obtained from hydraulic permeability data, are corroborated by solute permeability data. The normalized values of the solute permeability, $\gamma=(\omega/\omega_{control})$, of several permeants relevant to the actions of insulin and vasopressin are recorded in Table 28. The pH gradient chosen in the solute permeability experiments was the one at which maximum increase n the value of hydraulic conductivity coefficient was observed (i.e., pH 7.4 to 4.6). It can be seen from Table 28 that the solute permeabilities, when the pH of the solution in the compartment D was 7.4 and that of the solution in the compartment C, which contains insulin or vasporessin, was 4.6, show an increase in comparison to the corresponding control values, and that when the pH gradient is reversed a decrease is observe. This agrees with the hydraulic permeability results.

Table 27. Normalized values of hydraulic conductivity coefficient $L_p = [L_p/(L_p)_{control}]$ in the presence of lecithin-cholesterol-insulin and lecithin-cholesterol-vasopressin mixtures[a] under various pH gradients (Ref. 85).

	7.4	6.8	6.0	5.4	5.0	4.6	4.0	3.6
				Lecithin-cholesterol-insulin mixture				
				pH of the solution in compartment C containing insulin				
L_p^b	0.83± 0.00	1.37± 0.00	1.36± 0.01	1.93± 0.01	2.09± 0.01	2.37± 0.01	1.17± 0.00	1.07± 0.00
				pH of the solution in compartment D				
L_p^c		0.74± 0.01	0.82± 0.01	0.86± 0.00	0.80± 0.00	0.83± 0.00	0.33± 0.00	0.35± 0.02
				Lecithin-cholesterol-insulin mixture				
				pH of the solution in compartment C containing insulin				
L_p^b	0.84± 0.01	1.43± 0.03	1.59± 0.00	1.63± 0.01	1.88± 0.01	2.04± 0.02	1.80± 0.01	1.78± 0.02
				pH of the solution in compartment D				
L_p^c		0.89± 0.01	0.91± 0.01	0.93± 0.01	0.88± 0.03	0.84± 0.00	0.89± 0.03	0.8± 0.00

Note: The value of L_p in each case is reported as the arithmetic mean of 10 repeats ±SD.
[a] Composition of lecithin-cholesterol-insulin mixture is 1.919×10^{-5} M w.r.t lecithin, 1.175×10^{-6}M w.r.t cholesterol and 79.5 micro units / ml w.r.t. lecithin, 1.175×10^{-6}M w.r.t. cholesterol, and vasopressin 40 pg per ml. [b] pH of the solution in the other compartment (D) fixed at 7.4. [c] pH of the solution in the other compartment (C) (containing insulin or vasopressin) fixed at 7.4.

Graziani and Livne [95] explained their observation on the increased water permeability of lipid membranes by vasopressin in terms of the permeation theory of Trauble [96], in which kinks are considered as carriers through lipid layers of the membranes. The kinks represent small mobile free volumes in the hydrocarbon phase of the membrane, which could harbor water molecules. These authors [95] did not observe any significant change in ion permeability due to vasopressin. Fettiplace et al. [97] observed that Iysine-vasopressin increased the electrical conductance of artificial bilayers, bet they did not describe the formation of typical pores. Schlieper [98] studied the interaction of the octapeptide hormone angiotensin II on BLMs, and observed the production of typical conductance fluctuations, characteristic of a pore-forming ionophore. These changes were accompanied by the translocation of water and ions across the membrane. In the experiments of Schilieper [98], the gating mechanism was found to be electrically driven, i.e., the pores are opened under an applied positive potential and are influenced by hydrostatic pressure.

Among the earlier reports [89-92, 95, 97, 98], studies on the clostridial toxins [89-92] because of their structural similarity to both insulin and vasopressin and the similarity in trends in the permeability data appear most relevant to the observations reported in this study. By analogy to the studies on clostridial toxins, formation of aqueous pores can be suggested to explain permeability changes in the case of insulin and vasopressin as observed in the these experiments. This suggestion, however, needs corroboration by additional experiments.

The study, besides demonstrating the workability of liquid membrane-bilayers as model systems for biomembranes, tempts us to suggest that the permeability changes likely to be induced by insulin and vasopressin in the plasma membranes should be incorporated in the mechanism of their relevant membranes during the course of action of these hormones (insulin of vasopressin) appears worthy of further investigation. The role of carbonic anhydrase enzyme in the generation of pH gradient in the action of vasopressin for example, deserves a thorough investigation.

Table 28. Normalized values (γ) of solute permeabilities $(\gamma = \omega/\omega_{control})$ in the presence of lecithin-cholesterol-insulin a lecithin-cholesterol-vasopressin mixtures under various pH gradients (Ref. 85).

Permeants	Initial conc. $(g\ L^{-1})$	Lecithin-cholesterol insulin mixture		Lecithin-cholesterol-vasopressin mixture	
Sodium (Chloride)	5.382	1.48±0.02	0.66±0.01	1.63±0.05	0.71±0.03
Potassium (Chloride)	10.43	1.77±0.04	0.52±0.01	1.38±0.01	0.44±0.03
Calcium (Chloride)	0.1776	1.21±0.01	0.88±0.02	1.56±0.04	0.74±0.09
Glycine	0.1	1.74±0.02	0.48±0.01	1.31±0.03	0.61±0.03
Glutamine	0.1	1.83±0.01	0.77±0.03	1.792±0.02	0.70±0.01
D(+) Glucose	0.8	1.36±0.08	0.60±0.03	1.45±0.02	0.53±0.05

Note: The values of γ are reported as the arithmetic mean of 10 repeats ±S.D

[a]Composition of lecithin-cholesterol-insulin mixture is $1.919\times10^{-5}M$ w.r.t., lecithin, $1.175\times10^{-6}\ M$ w.r.t. Cholesterol and insulin 79.5 microunits per ml. Composition of lecithin-cholesterol-vasopressin mixture is $1.919\times10^{-5}M$ w.r.t. Lecithin, $1.175\times10^{-5}\ M$ w.r.t. Cholesterol, and vasopressin 40 pg per ml. [b]pH in the compartment, C is 4.6 and that in the compartment, D is 7.4, [c]pH in compartment, C is 7.4 and that in the compartment, D is 4.6.

5.5. Mimicking electrical excitability of liquid membrane bilayers [99].

Several attempts to generate model/mimetic system stems of neuronal and other cell membranes are documented in literature [2, 102-104]. The early experiments of Teorell [100] and Meares and Page [103] continue to evoke interest even today from an electrophysiological point of view in addition to fundamental aspects of instabilities in the far from equilibrium regime. Mueller and Rudin [102] showed that lipid bilayers were not electrically excitable unless channel forming proteinaceous substances were incorporated in them. The significant conclusion from Mueller and Rudin's work was that excitability was associated with the membrane alone, and there was no need to invoke the detailed structure of the cell. This finding motivated several groups to understand investigations on artificial membrane systems in order to undertake the possible sources of excitability [105].

Efforts have been made to demonstrate the appearance of electrical oscillations in the absence of any channel-forming peptide or excitability inducing material (EIM). For example, Pant and Rosenber [101] experimented with lipids bilayer membranes separating KCl solutions that contained a redox couple at appropriate pH and demonstrated oscillations in the membrane potential. Several studies on oscillatory phenomena using filters doped with lipids or surfactants are documented in the literature [106-114]. Many of these studies were conducted on filters doped with di oleoyl phosphate (DOPH) wherein the oscillatory behavior was ascribed to phase transition / conformational changes of the lipid molecules coupled with salt transport. Recent detailed studies by Kim and Later [115] on a Millipore filter doped with a mixture of DOPH and oleoyl alcohol (OA), however, did not corroborate the mechanism implicating the phase transition. These researchers ascribed the oscillatory phenomena to transport processes in the macroemulsion gel-like phase formed on the low-pressure side of the membrane.

In an attempt to mimic the spike potential of neuronal membranes, Shashoua [116, 117] experimented with a polyelectrolyte bipolar membrane system prepared by layering a polycationic phase onto a polyanion phase, and separating aqueous solutions of sodium chloride. This membrane system under a dc electric field was shown to spontaneously generate spiking jumps in the transmembrane potential analogous to those observed in neuronal membranes. As for neuronal membranes, one of the primary requirements of their model system is that they should be bipolar in nature and, of course, their thickness should be as close as possible to the thickness of the plasma membranes. Shashoua's polyelectrolyte membranes, though very thick, were bipolar in nature.

Recently a new liquid membrane bilayer system has been generated using surfactants, which is bipolar in nature and shows electrical excitability in the absence of any channel former [99].

The set-up used in this study is shown schematically in Fig.17, which has been well labeled to make it self-explanatory. It essentially consists of two compartments, A and B, made from perspex glass, and separated by a Sartorius cellulose acetate micro filtration membrane, M, pore size 0.2μm. The compartments, A and B were filled respectively with aqueous solutions of sodium dodecyl sulfate (SDS) and cetylpyridinium chloride (CPC) of concentration equal to their respective CMCs (8.272 mM for SDS and 0.9 mM for CPC) along with the solutions of desired concentration / composition of NaCI and / or KCI. Known potential differences from an electronically operated electrophoresis power supply were applied across the platinum electrodes C and D and the transmembrane potentials were monitored with time using Ag / AgCI electrodes E and F connected to an x-t recorder. The

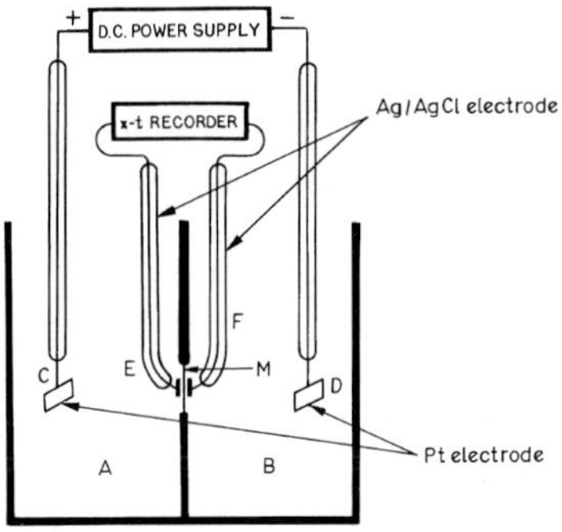

Fig.17. Schematic representation of the cell. M, supporting membrane (Ref. 99).

sensing Ag / AgCI electrodes E and F were placed as close to the membrane as possible (Fig. 17). Several repeats were conducted to check the reproducibility of the trends.

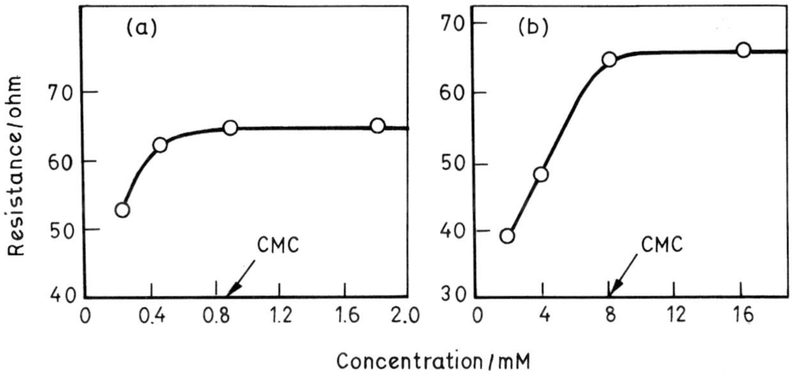

Fig.18. Variation of electrical resistance with the concentration of the surfactant. Curve, (a) is for CPC in the compartment B and curve (b) is for SDS in the compartment A. In each case the NaCl concentration in both the compartments is. 0.15 M. The area of the membrane is $5.024 \times 10^{-5}\ m^2$ (Ref. 99).

Evidence in favour of liquid membrane formation was obtained from electrical resistance data. The concentration of the surfactant (SDS) was varied up to its CMC and beyond in the compartment A, keeping the concentration sodium chloride in both the compartments A and B (Fig. 17) constant (0.5 M). Known current (I) was passed using Pt electrodes C and D, and the potential difference, $\Delta\varphi$ across the Ag / AgCI electrodes E and F were measured. The values of the electrical resistances were calculated from I-$\Delta\varphi$ plots at various concentrations the surfactant. The experiments were repeated with CPC in compartment B. For details the original paper [99] should be consulted.

The data on the variation of electrical resistance with the concentration of the surfactant (Fig. 18) reveal that electrical resistance increases with the increase in the concentration of the surfactant. The increasing trend continues up to the CMC of the surfactant beyond which it becomes more or less constant. The data on electrical resistance plotted in Fig. 18 are consistent with Kesting's hypothesis [9-11] indicating the formation of a complete surfactant layer liquid membrane at the interface at the CMC of the surfactant.

It appears therefore that when both the compartments A and B are filled with surfactant solutions of concentrations equal to their CMCs, SDS in A and CPC in B two surfactant layers would be formed, one on either side of the supporting membrane. However, formation of surfactant layer liquid membrane at the interface may be accompanied by the penetration of the surfactant into the pores as single molecules and also as micelles. Although the CMC value of the aqueous solution of the surfactant would not be the same as in sodium chloride solutions, the conclusion about the formation of the surfactant layer liquid membrane is not likely to change qualitatively.

The pores in the supporting membrane, Sartorious cellulose acetate microfiltration membrane in this case, though of uniform size and uniformly distributed, are tortuous pathways. One can, nevertheless, have a gross picture of the surfactant bilayers formed. The hydrophobic ends of the surfactant molecules will be preferentially oriented toward the hydrophobic supporting membrane and the hydrophilic ends will be drawn outward away from it. The surfactant bilayer formed in these experiments would be bipolar in nature; the face of the surfactant layer in the compartment containing SDS solution would be negatively charged, and that in the compartment containing CPC solution would be positively charged. The resting membrane potential of the system Ag/AgCI, NaCI (0.15 M), CPC (0.90mM)/ cellulose acetate membrane/SDS (8.27 mM) NaCI (0.15M), AgCI/Ag was found to be 4.98 *mV*, with the compartment containing CPC as positive. The potential difference was measured using a multimeter. On applying an electrical potential difference across the electrodes C and D, the transmembrane potential was found to oscillate with time. Details of the observations made and the trends discovered in the data are listed below:

i) When the two compartments (Fig. 17) contained only the aqueous solutions of the surfactants, i.e., SDS in compartment A and CPC in compartment B, no oscillations in the transmembrane potential were observed, but the moment a few drops of (aqueous solution of NaCI or KCI were added to the two compartments, immediate occurrence of electrical potential oscillations was noticed [Fig. 19 curve (a)]. This is because unless chloride ions are present in both the compartments, the Ag/AgCl electrodes will not detect the transmembrane potential differences.

ii) Experiments were conducted with NaCI of the same concentration on both sides of the membrane and also with KCI of the same concentration on both sides o.fthe membrane; the concentrations of NaCI and of KCl experimented with were 0.05, 0.1, 0.15, and 0.2 M. In each case, oscillations in the transmembrane potential were observed and it was found that the frequency of oscillations was initially high which showed a decreasing trend and finally, the oscillations ceased. But after sometime, the oscillations reappeared with the same qualitative trend in frequency, i.e., initially high, then low, and then no oscillations and so on. Typical traces in the two cases [(a) when 0.15 M NaCI was taken on both sides of membrane, and (b) when 0.15 M KCI was taken on both sides of the membrane] are shown in Fig. 19 (curves (b) and (c), respectively).

iii) Oscillations in transmembrane potentials were also observed when NaCI (0.15 M) was taken in the compartment A and KCI (0.15 M) in the compartment B (Fig.20, curve (a). Experiments with the mixture of NaCI and KCI on both sides of the membrane also showed oscillations in the transmembrane potential (Fig.20, curve (b). The composition of the mixture was so chosen that it more or less correspond to the concentrations of NaCI and KCI inside and outside the axons.

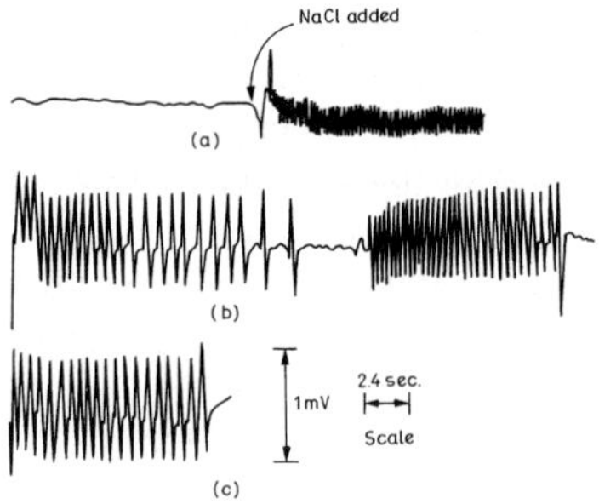

Fig.19. Traces of the transmembrane potential oscillations. In each case the compartment, A contained SDS (CMC) and the compartment, B contained CPC (CMC). The value of the imposed potential difference is 1.24. V, with the compartment containing SDS as positive. The scale shown is applicable to all traces a-c. Curve (a): for the situation when no electrolytes (NaCI or KCl) are added in the compartments, A and B and for the situation when they are added; the arrow indicates when a few drops of NaCI are added in both compartments. Curve (b): 0.15 M NaCI in both the compartments. Curve (c): 0.15 M KCI in both the compartments.

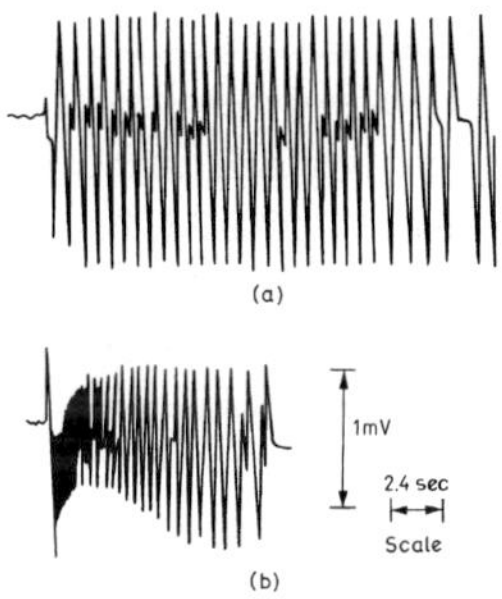

Fig.20. Traces of transmembrane potential oscillations. In each case, SDS (CMC) is in the compartment A and CPC (CMC) is in the compartment B. The value of the imposed potential difference is 1.24 V, with the compartment containing SDS as positive. The scale is applicable to both the traces (a) and (b). Curve (a) : 0.15 M NaCI in the compartment A and 0.15 M KCI in the compartment B. Curve (b) : Mixture of 0.015 M NaCI and 0.15 M KCI in the compartment A and mixture of 0.005 KCI and 0.15 M NaCI in the compartment B.

iv) In all cases studied, it was observed that the oscillations were observed only when the applied voltage across the electrodes C and D exceeded a certain minimum value. It was also observed that the oscillations ceased when the applied voltage exceeded a certain maximum value. For example, in the experiments with 0.15 M NaCl in compartment A and B, oscillations were observed only when the applied voltage was between 1.2 and 1.9 V. In the traces shown In Fig.19 and 20, the applied potential difference was 1.24 V, with the compartment containing SDS as positive. Thus there exists / a threshold value of the applied voltage for the occurrence of oscillations.

v) Another necessary requirement for the occurrence of oscillations is the proper polarity of the electrodes. The electrode C in the compartment A containing SDS should be positive and the electrode D in the compartment B containing CPC should be negative.

The mechanism of such oscillations has not yet been fully deciphered. The mechanism of membrane potential oscillations, postulated earlier [106-112] invoking phase transition in the case of surfactant / lipid-doped filters, was not corroborated by Kim and Later [115] In their experiments with filters doped with DOPH and OA. They suggested a mechanism based on the general mechanism proposed by Teorell [100, 118]. In Teorell's mechanism, oscillations arise due to superposition of the electro-osmotic flow and electro-osmotic pressure driven back flow. Kim and Larter replaced the electro-osmotic counter pressure driven flow in Teorell's experiments with the flow imposed externally with a syringe pump in the direction opposite that of the electro-osmotic flow.

There is not much similarity between this system and the one studied by Kim and Larter except that a proper electrode polarity was necessary for the occurrence of the membrane potential oscillations. In Kim and Larter's experiments, current was applied with the anode in the low-pressure compartment. This polarity was necessary for the formation of the gel like surface layer and for the occurrence of membrane potential oscillations. In these experiments, however, the oscillations were not observed unless current was passed with the anode in the compartment containing SDS. And, of course, unless the applied potential difference exceeded a certain threshold value, no oscillations were observed. A similar observation has also been made in many earlier works including the most recent one by Kim and Larter. The points of difference between the surfactant layer liquid membrane system and the one used by Kim and Larter are several; the two most important ones are (i) The membrane in these experiments was bipolar, whereas the gel like membrane in the experiments of Kim and Larter was negatively charged, and (ii) In Kim and Larter's experiments a pressure difference was created across the membrane using a syringe pump, whereas in these experiments there was no pressure difference applied across the membrane as such.

The membrane in this system being bipolar is closer to Shashoua's, which had three zones, a positively charged zone, a neutral zone, and a negatively charged zone, the neutral zone being the central zone. As already stated, in these experiments as well as in the experiments of Shashoua, the polarity of the electrode as indicated in Fig.17 was essential for the occurrence of the oscillations. Katchalsky [119] suggested the following explanation for the electrical potential oscillations in Shashoua's experiments. Consequent to the passage of current, NaCI is accumulated in the central neutral zone. The increase of the osmotic pressure leads to the flux of the solvent into the membrane and to an increase in the hydrostatic pressure in it. At the same time the increase of the salt concentration causes the polyelectrolyte molecule to compress which also increases the pressure. When this increase in pressure exceeds the osmotic pressure, the solvent flux changes its sign and the salt concentration inside the membrane increases even more. A concentration gradient arises and the salt leaves the membrane to flow out after the membrane has attained the maximum contraction. Then follows the relaxation and the membrane returns to its initial state to repeat the process.

Although the present system is analogous to Shashoua's, the explanation suggested above cannot be applicable to this system in total because it is not known whether the membrane in this system contracts in the presence of the salt or not. At present we are not in a position to give any definite mechanism of the phenomenon, but nonetheless, any mechanism should take into account the following facts / observations: (a) The bipolar nature of the membrane; (b) The electro-osmotic flow consequent to the applied electric field should be bi-directional, i.e., the electro-osmotic flow should occur from compartment A to B and also from B to A (Fig. 17) simultaneously. This bi-directional nature of the electro-osmotic flow may also contribute to the generation of instability in the system; (c) The oscillations are observed when the applied electrical potential difference exceeds a certain threshold value and cease when it exceeds a certain maximum value, and (d) In addition to the surfactant layer liquid membranes in series with the supporting membrane, existence of surfactants inside the pores as single molecules and also as micelles can not be ruled out and hence events related to these should also be considered in the mechanism in spite of the undeciphered mechanism, the qualitative resemblance of the trends observed in this study to certain aspects of neuronal membranes suggests that the bipolar surfactant layer liquid membranes can also be candidates for conducting membrane mimetic experiments related to excitability of neuronal membranes. The convenience with which such bipolar liquid membrane bilayers can be generated and their stability are added advantages of this new system. In addition, the observation reported on this new system appears interesting by itself, particularly from the point of view of electrokinetic phenomena in the far from equilibrium region, and merits deeper investigations.

5.5.1 Yagisawa's model of excitability

The mimetic studies on neuronal excitation can be put in three broad categories: (a) studies showing that no oscillations can be induced by the external stimulus of value greater than a certain critical values unless channel formers are incorporated in the membrane [102]; (b) studies showing that the external stimulus of value greater than a critical value, can induce oscillations even in the absence of channel formers [101, 109, 110, 115]; and (c) studies showing that oscillations can be obtained in the absence of an external stimulus and channel formers [106, 113, 116]. In the mechanism of oscillations, particularly in the categories (b) and (c) above, concept of dynamic channels, e.g., periodic gating and closing of channels, which has been linked to phase transitions, was invoked. Admittedly, all issues related to the hypothesized dynamic channels are not clear yet at the microscopic level. To clarify the mechanism of self-sustained oscillations of electrical potential between the two solutions divided by a lipid bilayer membrane, a microscopic model of the membrane system has been proposed by Yagisawa et al. [120]. The model assumes the existence of an electrical double layer at the interfaces and invokes phase transition between the gel phase and liquid crystalline phase as the driving force for the electrical potential oscillations. Although phase transition was invoked earlier also, e.g., by Antonov et al. [121], the proposal made by Yagisawa et al. [120] is different. Yagisawa et al. [120] proposed that the two lipid monolayers constituting the bilayer membrane undergo phase transition independently. Two types of repetitive phase transitions were postulated: (i) one half of the bilayer repeats the phase transition between the gel and the liquid crystalline state but the other half remains in the liquid crystalline state; and (ii) one half repeats the phase transitions, but other remains in the gel state. The repetitive phase transition of the lipid membrane was assumed to be driven by the concentration gradient of hydrogen ion (H^+) and metal ion (M^+) across the membrane. It was also invoked that the gel state of lipid layers is generally stabilized with the adsorption of H^+ ions on the polar heads of the lipid molecules while desorption of hydrogen ions stabilizes the liquid crystalline state. With these assumptions and a few more related to the existence of the electrical double layer at the Interfaces, Yagisawa et al. [120] were able to explain the occurrence of self-sustained oscillations. Using mathematics, which is a little too involved, they deduced the expression for the time dependence of membrane potentials from which they succeeded in producing the traces of potential oscillations using literature values of relevant parameters. By using the model, we see how and under what condition the repetitive phase transition may occur and induce oscillation of electrical potentials. Zukermann [122] while giving critical appraisal of the theoretical model has made one reservation, which is quite critical. To quote *"My one reservation about the biological significance of the work of Yagisawa et al. is that it is not clear at all that the main phase transition occurs in biological systems"*. By main phase transition is meant the transition between the gel phase and liquid crystalline phase of the bilayer. This in effect stresses the need for experimental substantiation.

Prompted by these, Srivastava et al. [123] have examined the experimental validity of the model, which apparently looks theoretically quite self-consistent. Experiments have been conducted on the liquid membrane bilayer system. [3,4]; the bilayers of liquid membranes are generated on a hydrophobic supporting membrane by a lecithin-cholesterol mixture using the liquid membrane hypothesis proposed by Kesting et al. [11]. The liquid membrane bilayer system is quite stable and its workability as a mimetic system of biomembranes has been well demonstrated through a good number of biomimetic transport experiments [4,33,36,49,50]. The experimental setup used which is very similar to the one shown in Fig. 17 is depicted in Fig. 21. The compartments, A and B (Fig. 21) were filled aqueous solution of lecithin-cholesterol mixture of composition 1.919×10^{-5} M lecithin and 1.175×10^{-6} M cholesterol and allowed to stand for several hours (5-6 h). This particular composition of lecithin-cholesterol mixture was chosen on the basis of an earlier study [4], wherein it was shown that at this composition a bilayer of liquid membranes are formed within the pores of the hydrophobic supporting membrane in such a way that the hydrophobic tails of the lipid molecules are anchored at the hydrophobic supporting membrane and the hydrophilic moieties are drawn outward away from it. The desired pH in the two compartments were maintained using histidine chloride buffer. The desired concentration of Na^+ ions was introduced in the two compartments by adding $NaCl$ solution of appropriate concentrations. The gradient of the concentration of H^+ ions was always kept opposite to that of M^+ ions; the compartment on the left-hand side of the membrane had a higher concentration of H^+ ions.

According to the model proposed by Yagisawa et al. [120], the self-sustained oscillations of membrane potential are induced by repetitive phase transition of the lipid membrane, which is driven by the concentration gradient of H^+ and M^+ (Na^+ or K^+) across the membrane. The essential conditions for the periodic reversal are: (i) at least one kind of cation, Na^+ or K^+, is included in the system besides protons and the variation in their permeability across the membrane due to phase transition is larger than that of proton permeability; and (ii) the phase transition has hysteresis. The model postulates that the gel state of lipid layers is stabilized with the H^+ adsorption on the ionized polar heads of the lipid molecules while the, desorption of H^+ ions stabilizes the liquid crystalline states. In the model proposed by Yagisawa et al., one repetitive phase transition cycle is visualized as follows: consider a lipid bilayer consisting of two mono layers juxtaposed in such a way that the hydrophobic tails of the lipid molecules in the two mono layers face each other. The bilayer separates two aqueous solutions; the concentration of H^+ ions on the left-hand side of the bilayer is greater than that on the right-hand side, whereas the concentration of M^+ ions on the right hand side is greater than that on the left hand side. When enough H^+ ions flow into the right half, being in the liquid crystalline state, it gets changed to the gel state. When the permeability of M^+ ions in the right of the bilayer decreases more (largely due to phase transition) than does that of H^+ ions, the direction of M^+ ion flux is reversed because the flow of M^+ ions from right solution into the right half of the bilayer decreases drastically. Then the direction of H^+ flux is also reversed because of the charge neutrality condition, the hydrogen ion adsorbed on the surface of the right half begin desorbing, which brings back the right half into the initial liquid crystalline state.

The model proposed by Yagisawa et al. [120] as summarized above implies that no external stimulus is necessary for the occurrence of oscillations. This implication, however, was not corroborated in all systems studied (Table 29). No oscillations were observed unless an external current of magnitude $\geq$ 2.4 mA, with the compartment having higher H^+ concentration as negative, was passed through the membrane. It was also observed that if the direction of the externally applied current is reversed, no oscillations are observed. Traces of electrical potential oscillations observed in all systems in (Table.29) are reproduced in Fig. 22. According to the model, if M^+(e.g,; Na^+ or K^+) are not included in the systems, the oscillations should not be observed. This implication was corroborated.

The basic premise of the model is the repetitive phase transition driven by repetitive adsorption and desorption of protons by the membrane surface; the repetitive phase transition-occurs only in one half of the bilayer and the other half stays in the gel state or the liquid crystalline state. The adsorption of protons stabilizes the gel state and desorption stabilizes the liquid crystalline state. Thus if the same acidic pH is maintained on the two sides of the membrane, both halves of the bilayer should stay in the gel state, whereas if the same alkaline pH is maintained on the two sides of the membrane, both halves of the lipid bilayer should stay in the liquid crystalline state. It is, therefore, obvious that if light is passed through the membrane, the value of absorbance of the light by the membrane should be different in the two cases, i.e., when it is in the gel state and when it is in the liquid crystalline state. Furthermore, if the solutions on the two sides of the membrane - are such that repetitive phase transition is induced, the value of light absorbance should oscillate with time. To test these implications an ingenious procedure was adopted.

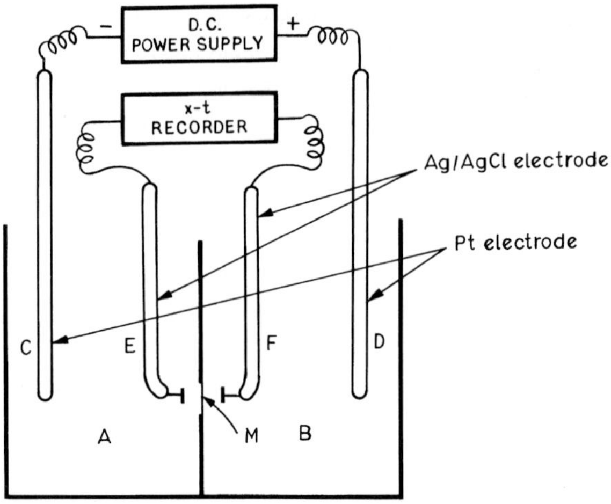

Fig.21. Schematic representation of the set-up used for monitoring electrical potential oscillations. N, supporting membrane (Sartorius cellulose acetate micro filtration membrane, cat no, 11107, pore size, 0.2 μm, thickness, 1×10^{-4} m). Compartments, A and B filled with aqueous solutions of the lecithin-cholesterol mixture and different concentrations of NaCl (Ref. 120).

A glass tube with a window which was covered with a Sartorius cellulose acetate microfiltration membrane of pore size, 0.2μm and thickness, 1×10^{-4} m (cat no. 11107) using an adhesive, was suspended in the curvette of the spectrophotometer. The dimensions of the tube and the window were such that it could easily be suspended in the cuvette and the micro filtration membrane on the window was in the path of the beam of light. The glass tube hung in the cuvette is schematically depicted in Fig. 23. The desired solution was put inside the tube d and cuvette, c (Fig. 23) and the cell was allowed to stand for several hours. The cuvette with the glass tube suspended in it was put in the spectrophotometer and the absorbance at λ=230 nm was measured. The systems on which measurements were made are given in Table 30 along with the values of light absorbance. The value of light absorbance was found to be almost the same irrespective of whether the solution on the two sides of the membrane were of the same acidic pH or the same alkaline pH or one side of the membrane contained acidic solution and the other side alkaline. Also in the case of system (d) of Table 30, where one should have observed oscillations of light absorbance with time because of repetitive phase transition as postulated in Yagisawa's model the same value of light absorbance was obtained as in other systems and it did not show any variation with time. This observation does not corroborate the basic premise of the model that adsorption of protons stabilizes the gel state and desorption of protons stabilizes the liquid crystalline state.

Thus the investigations carried out by Srivastava et al [123] did not corroborate the model proposed by Yagisawa et al. [120]. The observations made by Srivastava et al. [123] appear to be more in line with the model of Meares and Page [103]. As pointed out by Meares and Page [103], the basic ingredients which make oscillations possible are at least two independent transport processes driven by different forces and that the flows induced by these forces oppose each other. In the experiments of Srivastava et al. [123], the electrically driven (electro-osmosis) proton transport is opposed by osmotically driven M^+ flow. The following observations speak in favour of this hypothesis:

1. When polarity is reversed, no oscillations are observed.
2. In the absence of M^+, no oscillations are observed.

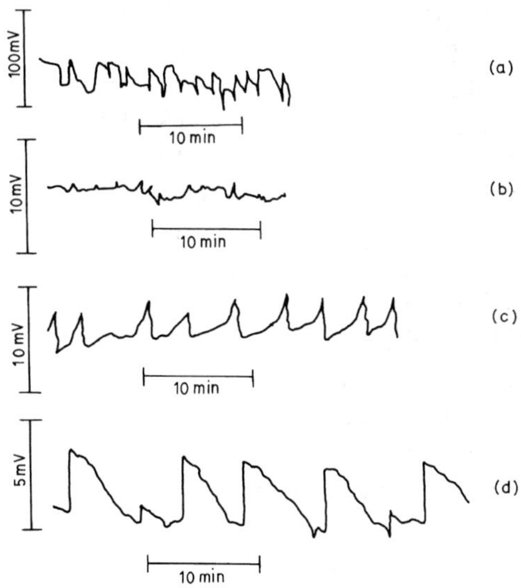

Fig. 22. Traces of transmembrane potential oscillations observed in the systems listed in Table 29 (Ref. 123).

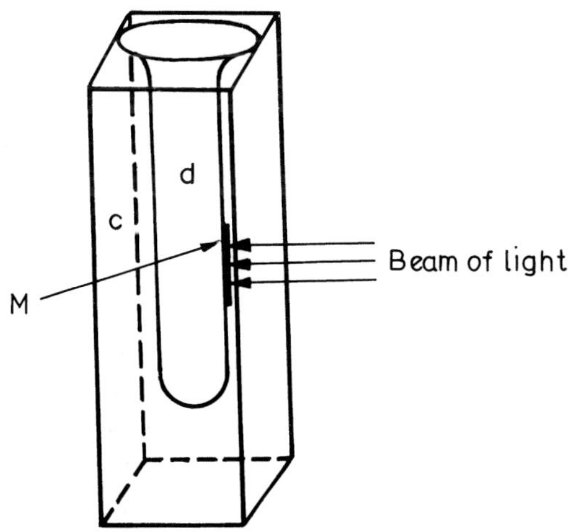

Fig. 23. Set-up used for measuring the absorbance of light: c and d glass tube with a window M, Sartorius microfiltration membrane fixed on the window (Ref. 123)

The transport studies described in this chapter give a strong indication of the workability of liquid membrane bilayers as mimetic system of biomembranes however, there is a need for carrying out many more biological mimicries to further establish the credentials of liquid membrane bilayers as mimetic system of biomembranes.

Table 29. Systems used for monitoring electrical potential oscillations [a] (Ref. 123).

	Systems		
	Negative		Positive
(a)	L.C.0.05 M NaCl (pH 6.2)	|	L.C.0.1 NaCl (pH 7.4)
(b)	L.C.0.05 M NaCl (pH 6.2)	|	L.C.0.1 NaCl (pH 7.4)
(c)	L.C.0.05 M NaCl (pH 6.2)	|	L.C.0.1 NaCl (pH 7.4)
(d)	L.C.0.05 M NaCl (pH 6.2)	|	L.C.0.1 NaCl (pH 7.4)

[a]*L* = Lecithin (1.919×10^{-5} M); C = cholesterol (1.175×10^{-6} M). The vertical lines symbolize membranes. In all the cases external current = 2.4 mA was passed

Table 30. Values of absorbance of light at λ=230 nm in different membranes sysetems (Ref. 123).

	System			Absorbance
(a)	L.C. (pH 6.2)	|	L.C. (pH 7.4)	0.35
(b)	L.C.(pH 6.2)0.5 M NaCl	|	L.C.(pH 6.2)0.5 M NaCl	0.34
(c)	L.C.(pH 7.4)0.1 M NaCl	|	L.C.(pH 7.4)0.1 M NaCl	0.35
(d)	L.C.(pH 6.2)0.5 M NaCl	|	L.C.(pH 7.4)0.1 M NaCl	0.37

[a]*L* = Lecithin (1.919×10^{-5} M); C = cholesterol (1.175×10^{-6} M). The vertical lines symbolize membranes. In all the cases histidine chloride buffer were used to maintain the pH.

REFERENCES

[1] S.J. Singer and J.L. Nicolson, Science, 175(1972) 720.

[2] H.T. Tien, "Bilayer Liquid Membranes-Theroy and Practice", Marcel Dekker, New York, 1974, see also(a) H.T. Tien and Angelica Ottova-Leitmanova, Membrane Biophysics as viewed from experimental Bilayer Lipid Membrane, Elsevier, Amsterdam 2000. (b)) H.T. Tien and A. Ottova-Leitmannova-Leitmanova, Planer Lipid Bilayers and their Applicatins, Elsevier, Amsterdam.

[3] R.C. Srivastava and R.P.S. Jakhar, J.Phys.Chem., 85(1981) 1457.

[4] R.C. Srivastava and R.P.S. Jakhar, J.Phys.Chem., 86(1982)1441.

[5] N.L. Gershfeld in E.D. Korn (Ed.), "Methods in Membrane Bioogy", Vol. 1, Plenum Press, New York, 1974, Chap.2.

[6] N.L Greshfeld and R.E. Pagano, J. Phys. Chem., 76(1972)1244.

[7] M.E. Haberland and J.A. Reynolds, Proc. Natl. Acad. Sci., U.S.A., 70(1973)2313.

[8] M.K. Jain, Curr. Top. Membr. Transp., 6(1975)1.

[9] R.E. Kesting, A. Vincent and J. Eberlin, OSW R&D Reprot No. 117, August, 1964.

[10] R.E. Kesting, Reverse Osmosis Process Using Surfactant, Feed Additives, OSW Patent Application SAL 830, Nov.2, 1965.

[11] R.E. Kesting, W.J. Subcasky and J.D. Paton, J. Colloid Interface Sci., 28(1968)156.

[12] K.S. Spiegler and O. Kedem, Desalination, 1(1966)311.

[13] T.K. Sherwood, P.L.T. Brain and R.E. Fischer, Ind. Engg. Chem. Fundam., 6(1967) 2.

[14] F.L. Harris, G.B. Humphreys and K.S. Spiegler, in P. Meares, (Ed.), "Membrane Separation Process Elsevier, Amsterdam, 1976, Chapter 4.

[15] A. Katchalsky and P.F. Curran, "Non-equilibrium Thermodynamics in Bio-Physics" Harvard University Press, Cambridge, MA, 1967.

[16] O. Kedem and A. Kachalsky, Trans. Faraday Soc., 59(1963)1941.

[17] A. Katchalsky and O. kedam, Biophys. J., 2(1962)53.

[18] N. Laxkshminarayanaiah, "Transport Phenomena in Membranes", Academic Press, New York, 1969, pp. 425, 449.

[19] M.K. Jain, "The bimolecular Lipid Membrane", Van Nostrand Reinhold, New York, 1972, pp. 121-122.

[20] P. Mueller, D.O. Rudin, H.T. Tien and W.C. Wescott, Nature (London) 194(1962) 979.

[21] T. Hanai, D.A. haydon and J. Taylor, J. Theor Biol., 9(1965) 278.

[22] R. MacDonald, Fed. Proc., 26(1967) 863.

[23] J.P. Reeves in R.M. Dowben, Ed., "Biological membrane", Little Brown and Company, Boston, 1969, pp. 238; see also Ref. 18 pp. 448 and Ref. 19, pp. 142.

[24] A.H. Maddy, C. Huang and T.E. Thompson, Fed. Proc., 25(1966)933; see also Ref. 2, p.8, and Ref. 19, p. 91.

[25] D. Papahadjopoulos and J.C. Watkins, Biochim. Biophys. Acta, 135(1967)135.

[26] H.J. Vreeman, Kon. Ned Akad. Wetensch Proc. Ser. B., 69(1966) 564.

[27] F.J. Brinley, J.r., and L.J. Mullins, J. Neurophysiol., 28(1965) 526.

[28] A.S. Troshin "Problems of Cell Permeability", Pergamon Press, 1966, see also Ref. 19, pp. 141-143 and Ref. 23 p. 231.

[29] Ref. 18, p. 248

[30] W.R. Carmody, J. Am. Chem. Soc., 51(1929) 2901; 54(1932) 3647.

[31] W.R. Carmody, Chem. Rev., 53(1953) 397.

[32] U. Hopper, A.L. Lehninger and W.J. Lennerz, J. Membrane Biol., 2, (1970)41; see also Ref. 19, p. 142.

[33] R.C. Srivastava, S.B. Bhise, A. Tandon, and R.K. Sharma, J. Colloid Interface Sci., 93(1983) 568.

[34] R.C.Srivastava, S.B. Bhise, A. Tondon and R.K. Sharma, J. Colloid Interface Sci., 99(1984) 71.

[35] R.C. Srivastava, A. Tandon, R.K. Sharma, S.B. Bhise and D.B. Madamwar, Indian J. Chem., 24A(1985) 918.

[36] R.C. Srivastava, A. Tandon, S. Kurup, S.B. Bhise and R.K. Sharma, J. Electroanal. Chem., 187 (1985) 325.

[37] See Ref. 2, Chapter 10.

[38] C.R. Stocking and V.R. Franceschi, Am. J. Bot., 68(1981)1008.

[39] D. Elbaurn, J. Harrington and R.L. Nagel., Prog. Clin, Biol., Res., 1(1975) 311.

[40] R.C. Nelson, J. Chem. Phys., 39(1963)112.

[41] Y. Okamotu and W. Brenner, Organic Semiconductors, Reinhold Publishing Corporation, Chapman and Hall Ltd., London, 1964, pp. 160.

[42] H.T. Tien and R.E. Howard, in R.J. Good, R.R. Stromberg and R.L. Patrick, (Eds.), "Techniques of Surface and Colloid Chemistry and Physics, Vol. 1, Marcel Dekker Ine., New York, 1972, pp. 205.

[43] Reference 2, pp. 361-365.

[44] H.T. Tien and B. Karvaly, in J.R. Bolton, (Ed.), Solar power and Fuels, Academic Press Inc., Now York, 1977, Chapter 7.

[45] H.T. Tien and V.K.H. Chen, in Progress in D.A. Candenhead, J.F. Danielli and M.D. Rosenberg, (Eds.), Surface and Membrane Science, Vol. 8, Academic Press Inc., New York, 1974. pp. 119-157.

[46] Reference 2, pp. 265.

[47] G.S. Marks Heme and Chlorophyll, D. Van Nostrand Company Ltd., London, 1969, pp. 94,97, 98; see also Remington, C., Biochem. J., 75(1980) 622.

[48] F.A. Robinson, The Vitamin Cofactors of Enzyme Systems, Pergamon Press, London, 1966, Chapter 10, pp 699-708.

[49] R.C. Srivastava and A. Tandon, Biotechnol. Bioeng., 31(1988) 511.

[50] R.P. Rastogi, R.C. Srivastava and S.N. Singh, Chem., Rev., 93(1993) 1945.

[51] W. Stoeckenius, R.H. Lozier and R.A. Bogomolni, Biochim. Biophys. Acta, 505(1979)215.

[52] W. Stoeckeniust and R.A. Bogouolni, Annu. Rev. Biochem., 52(1982)587.

[53] Th. Schreckenbach, in H.G. Shlegd and J. Barnea (Ed.), Microbial Energy Conversion, Pergamon Press, Oxford, 1977, p. 247.

[54] D. Oesterhelt and W. Stoeckenius, Proc. Nall. Acad. Sci., U.S.A. 70(1973) 2853.

[55] A. Danon and S.R. Caplan, Biochim. Biophys Acta, 423(1976)133.

[56] R.C. Srivastava, D.B. Madamwar, R.K. Sharma and A. Tandon, Experientia, 40(1984)773.

[57] R.C. Srivastava and R.P. Rastogi in : D. Mobious, R.M. Miller (Ed.), Transport Mediated by Electrified Interfaces, Vol 18, Elsevier, Amsterdam, 2003, pp 130.

[58] G. Stein, in J.R. Bolton (Ed.), Solar Power and Fuels, Academic Press, New York, 1977, p. 1-25.

[59] K. Singh and S.R. Caplan, Trends. Biochem. Sci., 5,(1980) 62.

[60] K. Singh, R.Korenstein, H. Lebdevaand S.R. Caplan, Biophys. J., 31(1980) 393.

[61] W. Stoeckenius, Sci., Am., 234(1976)38.

[62] D. Oesterhelt, Bioscience, 35(1985)18.

[63] R.C. Srivastava, R.K. Sharma, A. Tandon and S.B. Bhise, J. Colloid Interface Sci., 108(1985)249.

[64] R.C. Srivastava, in G.S. Singhal and T. Ramasarma (Eds), Trends in Bioenergetics and Biotechnological Process, Today and Tomorrow, Printers and Publishers, New Delhi, 1991, pp 211-240.

[65] F.E. Snell, and C.T. Snell, "Colorimetric Methods of Analysis", Van Nostrand, New York, 969, p. 610.

[66] H.T. Tien, Bilayer Lipid Membrane, Theory and Practice, Dekker, New York, 1974, Chapter 7, p. 205.

[67] A. Finkelstein and R. Holz, in G. Eisenman, (Ed.), Membranes, Vol. 2 Chapter 5, Dekker, New York, 1973.

[68] R. Holz and A. Finkelstein, J. Gen. Physiol., 56(1970)125.

[69] A. Cass, A Finkelstein and V. Krespi, J. Gen. Physiol., 56(1970)100.

[70] H.T. Tien Bilayer Lipid Membrane, Theory and Practice, Dekker, New York, 1974, p. 204.

[71] H. Van Zutphen, L.L.M. Van Deenen, and S.C. Kinsky, Biochem. Biophys, Res. Commun., 22(1966)393.

[72] T.E. Andreoli and M. Monahan, J. Gen. Physiol., 52(1968)300.

[73] A. Finkelstein and R. Holz, in G. Eisenman, (Ed.), Membranes, Vol.2, Dekker, New York, 1973, Chapter 5, p. 382.

[74] A. Finkelstein and R. Holz, in G. Eisenman, (Ed.), Membranes, Vol.7, Dekker, New York, 1973, Chapter 5, p. 387.

[75] A. Finkelstein and A. Cass, J. Gen. Physiol., 52(1980)145.

[76] R.C. Srivastava, G.S.K. Velu, A.N. Nagappa and D.B. Raju, J. Colloid Interface Sci., 131(1989)584.

[77] R.J. Flower, S. Moncada and J.R. Vane, in A.G. Goodman, L.S. Goodman and L.S. Gilman, (Eds.), The Pharmacological Basis, Therapeutics, Macmillan and Co., New York, 1980, 6th Edition, p. 683.

[78] W.C. Bowman and M.J. Rand, Text Book of Pharmacology, 2nd Edition, Blackwell Scientific, Oxford, 1980, p. 13.16.

[79] R.C. Haynes, Jr. and F. Murad, in A.G. Goodman, L.S. Goodman and A. Gilman, (Eds.), The Pharmacological Basis of Therapeutics, 6th Edition, Macmillan and Co., New York, 1980, p. 1479.

[80] R.C. Srivastava, D.B. Raju, V. Singh and S. Upadhyay, J. Membr. Sci., 58(1991)211.

[81] A. Kotyk and K. Janacek, Membrane Transport, Plenum Press, New York, 1977.

[82] D. Kashchiev and D. Exerowa, Biochim. Biophys. Acta, 732(1983)133.

[83] R.C. Srivastava, A.N. Nagappa and R.K. Sharma. J. Colloid Interface Sci., 117(1987)375.

[84] A.N. Nagappa, D.B. Raju and R.C. Srivastava, Indian J. Biochem. Biophys., 26(1989)172.

[85] V. Singh, D.B. Raju, C. Ravindran and R.C. Srivastava, J. Colloid Interface Sci., 139(1990)337.

[86] D.M. Gill, in J. T. Jelgaszewicz and T. Wadstrom, (Eds.), Bacterial Toxins and Cell Membrane, Academic Press, London, 1978, pp.291-332.

[87] L.L. Simpson, Pharmacol. Rev., 33(1981)155.

[88] J.P. Robinson and J.H. Nash, Mol. Cell Biochem., 48(1982)33.

[89] B.L. Kagan, A. Finkelstein and M. Colombini, Proc. Natl. Acad. Sci., USA, 78(1981)4950.

[90] J.J. Donovan, M Simon, R. Draper and M. Montal, Proc. Natl. Acad. Sci., USA, 78(1981)172.

[91] D.H. Hoch, M.R. Mira, B.E. Ehyilch, A. Finkelstein, B.R. Dasgupta and L.L. Simpson, Proc. Natl. Acad. Sci., USA, 82(1985)1692.

[92] L.L. Simpson, Annu. Rev. Pharmacol. Toxicol., 26(1986)427.

[93] W.C. Bowman and M.J. Rand, Text Book of Pharmacology, 2nd Edition, Blackwell Sclentlfic Publications, Oxford, 1980 p.19.45.

[94] M.H. Richard, in A.G. Goodman, L.S. Goodman, T.W. Rall and F. Murad, (Eds.), The Pharmacological Basis of Therapeutics, 7th Edition, Macmillan and Co., New York, 1985, p. 913.

[95] Y. Gyaziani and A. Livine, Biochim. Biophys. Acta, 291(1973)612.

[96] H. Trauble, J. Membrane Biol., 4(1970)193.

[97] R. Fettiplace, D.A. Haydon and C.D. Knowles, J. Physiol., 221(1971)18.

[98] P. Schlieper, Biochim. Biophys. Acta, 464(1977)448.

[99] R.C. Srivastava, A.K. Das, V. Agarwala and S. Upadhyay, Langmuir,9(1993)3313.

[100] T.Teorell, J. Gen. Physiol., 42(1959)831,847.

[101] H.C. Pant and B. Rosenberg, Biochim. Biophys., Acta, 225(1971)379.

[102] P. Mueller and D.O. Rudin, Nature, 213(1967)603; 217(1968)713.

[103] P. Meares and K.R. Page, Philos. Trans, Roy, Soc., A272(1972)1: and also Proc. Roy Soc. London, A339(1974)513.

[104] A. Mauro, P. Nanavati, E. Heyer, Proc. Natl. Acad. Sci., USA, 69(1972)3742.

[105] R. Larter, Chem. Rev., 90(1990)355, see also R.P. Rastogi, G. Srinivas, R.C. Srivastava, P.C. Pandey A.P. Misra and A.R. Singh, J. Colloid Interface Sci., 175(1995)262, R.P. Rastogi, G.P. Misra, P.C. Pandey, K.Kumar and Kanchan Bala, J. Colloid Interface Sci., 217(1999)275, R.P. Rastogi and R.C. Srivastava, Adv. Colloid Interface Sci., 93(2001)1, R.P. Rastogi, R.C. Srivastava and Prem Chand, J. Colloid Interface Sci., 263(2003)223, K. Bala, K. Kumar, S.K. Saha and R.C. Srivastava, J. Colloid Interface, Sci., 273(2004)320.

[106] T. Ishi, Y. Kurodu, K. Youshikawa, K. Sakabe, Y. Matubara and K. Iriyama, Biochem, Biophys. Res. Commun., 123(1984)792.

[107] Y. Kobatake, A. Irimajiri and N. Matsumoto, Biophys. J., 10(1970)728.

[108] M. Yoshida, N. Kamo, and Y. Kobatake, J. Membr. Biol., 8(1972)389.

[109] J. Arisawa, and T. Furukawa, J. Membr. Sci., 2(1977)303.

[110] J. Arisawa, and K. Misawa, J. Membr. Sci., 32(1987)223.

[111] K. Toko, M. Tusuka, S. Ezaki and K. Yamafuji, Biophys. Chem., 20(1984)39.

[112] K. Toko, MTusukuji, S. Liyama, and K. Yamafuji, Biophys. Chem., 23(1986)201.

[113] K. Yoshikawa, K. Sakabe, Y. Matsubara, and T. Ota, Biophys. Chem., 21(1985)33.

[114] K. Yoshikawa, K. Sakabe, Y. Matsubara, and T. Ota, Biophys. Chem., 20(1984) 107.

[115] J.T. Kim, and R. Larter, J. Phys. Chem., 95(1991)7948.

[116] V.E. Shashoua, Nature, 215(1967)846.

[117] V.E. Shashoua, in D.E. Tosteson, (Ed.), The Molecular Basis of Membrane Function; Prentice-Hall : Englewood Cliffs, NJ, 1968, p. 147.

[118] T. Teorell, Ark. Kemi, 18(1961)401.

[119] A. Katchalsky, Personal Discussion as cited in the text of Ref. 116, see also Volkenshtein, M.V., Biophysics, Mir Publishers, Moscow, 1983, pp. 554.

[120] K. Yagisawa, M. Naito, K. Gondaira and T. Karnbara, Biophys. J., 64(1993)1461.

[121] V. Antonov, V. V. Pertrov, A.A. Moluar, D.A. Predvoditelev and A.S. Ivanov, Nature 283(1980)585.

[122] M.J. Zukennann, Biophys. J., 64(1993)1369.

[123] R.C. Srivastava, S. Upadhyay, R. Sahney and R. L. Gupta, J Membr: Sci., 143(1998)75.

Chapter 6

Role of liquid membranes in drug action – Experimental studies

In this chapter are presented investigations carried out with a view to exploring the role of liquid membranes generated by surface active drugs in the mechanism of their action. For this the following drugs belonging to different pharmacological categories, many of them structurally dissimilar, have been experimented with:

(A) Antipsychotics

- (i) Haloperidol
- (ii) Chlorpromazine
- (iii) Reserpine

(B) Anticancer drugs

- (i) 5- Fluorouracil
- (ii) 1-Hexyl Carbamoyl-5 Fluorouracil
- (iii) 1-(2- tetrahydrofuryl-5-fluorouracil)

(C) Diuretics

- (i) Furosemide
- (ii) Triamterene

(D) Cardiac glycosides

- (i) Digitoxin
- (ii) Digoxin
- (iii) Oaubain

(E) Local aneasthetics

- (i) Procaine
- (ii) Tetracaine
- (iii) Lidocaine
- (iv) Dibucaine

(F) Antiarrhythmic drugs

- (i) Quinidine
- (ii) Disopyramide
- (iii) Procainamide
- (iv) Propranolol

(G) Barbiturates

(i) Sodium phenobarbital
(ii) Sodium pentobarbital

(H) Antehistamines-H_1antagonists

(i) Chlopheniramine maleate
(ii) Diphenhydramine hydrochloride
(iii) Tripelennamine hydrochloride

(I) H_2 – antagonists and histamine release blockers

(i) Cimetidine
(ii) Ranitidine
(iii) Famotidine
(iv) Disodium cromoglycate

(J) Steroids

(i) Testosterone propionate
(ii) Ethinyl estradiol
(iii) Progesterone
(iv) Hydrocortisone acetate

(K) Fat soluble vitamins

(i) Vitamin E
(ii) Vitamin A
(iii) Vitamin D

(L) Autacoids

(i) Prostaglandin E_1
(ii) Prostaglandin $F_{2\alpha}$

(M) Antidepressant drugs

(i) Imipramine hydrochloride
(ii) Clomipramine hydrochloride
(iii) Amitripryline hydrochloride

(N) Antiepileptic drugs

(i) Diphenylhydantoin
(ii) Carbamzepine
(iii) Valproate sodium

(O) Hypnotic and sedatives

(i) Diazepam
(ii) Nitrazepam
(iii) Chlordizepoxide

(P) β–Blockers

(i) Propranolol
(ii) Atenolol
(iii) Metoprolol

(Q) Antibacterials

(i) Ciprofloxacin
(ii) Norfloxacin

(R) ACE Inhibitors

(i) Captopril
(ii) Lisinopril

All drugs listed above were found to be surface active; the critical micelle concentrations (CMC), are recorded in are recorded in Table 1. The design of experiment conducted to unfold the role of liquid membrane in drug action and to throw light on the liquid membrane hypothesis of drug action was the following:

Table 1. Critical Micelle Concentrations (CMCs) of various Drugs (Ref. 1-31)

Drug	CMC/mol dm^{-3}
Haloperidol [1]	1.060×10^{-6}
Chlorpromazine hydrochloride [3]	4.500×10^{-5}
Reserpine [2]	1.600×10^{-6}
Imipramine hydrochloride [4]	1.480×10^{-4}
Clomipramine hydrochloride [5]	9.000×10^{-3}
Amitryptaline hydrochloride [5]	4.000×10^{-4}
Tetracaine hydrochloride [6]	1.210×10^{-5}
Dibucaine hydrochloride [6]	5.640×10^{-6}
Lidocaine hydrochloride [6]	1.160×10^{-3}
Procaine hydrochloride [6]	5.000×10^{-3}
Diazepam [7]	1.000×10^{-4}
Nitrazepam	8.000×10^{-6}
Chlordizepoxide [26]	2.000×10^{-5}
Chlorpheniramine maleate [8]	1.000×10^{-4}
Diphenhydramine hydrochloride [8]	1.000×10^{-3}
Tripelennamine hydrochloride [8]	1.000×10^{-3}

Table 1 contd.

Cimetidine [9]	5.102×10^{-6}
Ranitidine [9]	1.019×10^{-6}
Famotidine [10]	4.000×10^{-6}
Cromoglycate disodium [9]	1.593×10^{-6}
Furosemide [11,12]	8.300×10^{-5}
Triamterene [11]	1.000×10^{-5}
Quinidine hydrochloride [13]	3.960×10^{-7}
Disopyramide phosphate [13]	4.000×10^{-7}
Procainamide hydrochloride [13]	4.000×10^{-3}
Propranolol Hydrochloride [13,14]	4.750×10^{-5}
Atenolol [14]	4.000×10^{-3}
Metoprolol [14]	6.000×10^{-3}
Hydrocortisone acetate [15,16,17]	4.500×10^{-6}
Testosterone propionate [15,17]	3.870×10^{-6}
Ethinyl estradiol [15,17]	0.270×10^{-6}
Progesterone [15,17]	9.000×10^{-5}
5-Fluorouracil (5FU) [18]	8.000×10^{-10}
1-(2-tetrahydrofuryl) 5-flurouracil (FT) [18]	7.500×10^{-11}
1-Hexylcarbamoyl-5-fluorouracil (HCFU) [18]	6.100×10^{-11}
Vitamin E (α-tocopherol) [19]	5.000×10^{-8}
Vitamin D_3 (Cholecalciferol) [27]	8.000×10^{-9}
Vitamin A (Retinol acetate) [20]	6.000×10^{-9}
Prostaglandin E_1 (PGE_1) [16,21,22,23]	1.000×10^{-8}
Prostaglandin F_{2a} (PGF_{2a}) [16,21,22,23]	9.300×10^{-8}
Sodium phenobarbital [24]	7.500×10^{-5}
Sodium pentobarbital [24]	5.000×10^{-5}
Omeprazole [25]	3.000×10^{-6}
Lansoprazole [25]	1.000×10^{-6}
Ciprofloxin [28]	3.000×10^{-4}
Norfloxacin [28]	3.000×10^{-4}
Captopril [29]	6.000×10^{-4}
Lisnopril [29]	7.000×10^{-4}
Digitoxin [30]	5.600×10^{-9}
Digoxin [30]	9.800×10^{-8}
Oubain [30]	2.000×10^{-9}
Diphenylhydantoin [12, 31]	4.000×10^{-7}
Carbamzepine [31]	8.560×10^{-8}
Valproate Sodium [31]	7.970×10^{-5}

6.1 The design of experiments

The first step in these studies was to demonstrate the formation of liquid membrane in series with a hydrophobic supporting membrane. The hydraulic permeability data in the presence of various concentrations of the drugs below and above their respective CMCs were utilized to demonstrate the existence of a liquid membrane at the interface. The hydraulic permeability data at all concentration in case of all drugs studied, were found to obey the proportional relationship between volume flux J_v and the applied pressure difference ΔP $J_v = L_p \Delta P$. The values of hydraulic conductivity coefficient L_p or their normalized values i.e. L_p / L_p^0, L_p^0 being the value of L_p when drug concentration is zero, in case of all drugs were found to decrease progressively with increasing concentration of the surface active drugs, up to the CMC of the drug where after they become more or less constant. This trend is in keeping with Kestnig's hypothesis [32] and as argued in Chapter 5 is indicative of the complete coverage of the supporting membrane with the drug liquid membrane at CMC. The computed value of L_p at concentrates below the CMC, in case of all drugs, using mosaic model (Eq. 13 of chapter 5) were found to be in good agreement with the experimentally determined values, furnishing additional evidence in favour of the liquid membrane formation. The all glass transport cell used for obtaining the hydraulic permeability data and the experimental procedure has already been described in chapter 5. The all glass transport cell in diagrammed in Fig. 2 of chapter 5.

The next step in these studies was the measurement of solute permeability of relevant permanents in the presence of drug liquid membranes. For measurement of solute permeability also, the transport cell shown in Fig. 2 of chapter 5 was used. To acquire the data on solute permeability of relevant permeants in the presence of the liquid membrane, two sets of experiments were performed. In the first set, solutions of both drugs and the permeants were filled in the compartment C of the transport cell (Fig. 2 of chapter 5) and the compartment D was filled with water. In the second set of experiments the solution of the drug was taken in the upper compartment D and the solution of the permeant in the compartment C. After a known period of time that was of the order of several hours, the amount of permeant transported to the other compartment was estimated. The amount of permeant transported to the other compartment divided by the time and the area of the membrane gave the value of the solute flux J_s. The values of solute permeability (ω) were estimated using the equation,

$$\omega = \left(\frac{J_s}{\Delta \pi} \right)_{J_v = 0}$$

The value of the osmotic pressure difference $\Delta\pi$ used in the calculations ω was the average of the value of $\Delta\pi$ at beginning of the experiment ($t=0$) and at the end of the experiment. During the solute permeability measurement the solution in compartment C of the transport cell were kept well stirred and the condition volume flux $J_v=0$ was maintained by suitably adjusting pressure difference ΔP across the membrane. The details of the procedures for solute permeability measurements are described in Chapter 4 (Section 4.1.2) and 5 (Section 5.2). In all solute permeability experiments the concentration of the drugs were chosen always higher than their CMCs and the initial concentrations of the permeant were, as far as possible, comparable to the concentrations *in vivo*; concentrations of drugs higher than their respective CMC values were chosen to make sure that the supporting membrane was completely covered with the liquid membranes generated by the surface active drugs in accordance with Kesting's hypothesis [31]

The choice of cellulose membranes i.e. cellulose acetate or nitrate microfiltration membrane/aqueous interface, as site for liquid membrane formation was deliberate so that specific/active interaction of the drugs with the constituent of biomembranes as a cause for modification in the permeabilities of relevant permeants in the presence of drugs is totally ruled out and role of passive transport through the liquid membrane is highlighted.

The receptors in general are membrane proteins and hence should be surface active in nature. They should have both hydrophilic and hydrophobic domains in their structures. If for drug action the hydrophilic domains in the receptor are important then the transport of the permeant through the drug liquid membrane with its hydrophobic face facing the permeant would he relevant because in the formation of liquid membrane the hydrophilic moieties of the drug molecule will get attached with the hydrophilic domain of the receptor and the hydrophobic tails of the drug molecules would be drawn outward away from if facing the permeant. Similarly if the hydrophobic domain of the receptor were important for drug action the permeant in its transport would face the hydrophilic face of the drug liquid membrane. The two sets of experiments for the measurements of solute permeability were performed to be consistent with the two situations described above i.e. whether hydrophilic or hydrophobic domains of the receptor are important for drug action.

The orientation of the drug molecules in the liquid membrane generated in the two sets of experiments for solute permeability measurements would be different. Since hydrophobic tails of the surface active drug molecules will be preferentially oriented towards the hydrophobic supporting membrane, in the first set of experiments, the permeants would face the hydrophilic surface of the liquid membrane generated by the drugs, while in the second set of experiments, they would face the hydrophobic surface of the drug liquid membrane.

In cases where modification in the transport of permeants in the presence of drugs alone was not in keeping with trends reported on biological cells or, where interaction of the drug with membrane lipids was reported to be significant for the mechanism of its action, the solute permeability experiments were also carried out in the presence of a mixture of membrane lipids, namely lecithin and/or cholesterol and the drug. Here also two sets of experiments have been carried out- one in which the permeants would face the hydrophilic surface of the composite liquid membrane generated by the drug-lipid mixture and the other, in which the permeants would face the hydrophobic surface. The concentrations chosen for the lipids and drugs were such the liquid membranes generated by the lipids were saturated with the drugs.

These concentrations were derived from the hydraulic permeability data in the presence of lecithin-cholesterol-drug mixtures. To obtain the hydraulic permeability data in presence of lecithin-cholesterol-drug mixture, solutions of various concentrations of drugs prepared in an aqueous solution of lecithin-cholesterol mixture which was 15.542 ppm with respect to lecithin and 1.175×10^{-6} *M* with respect to cholesterol, were placed in the compartment C of the transport cell (Fig. 2 of chapter 5) and the compartment D was filled with water. This particular composition of lecithin-cholesterol mixture was chosen because, as has been shown in Chapter 5 at this composition the liquid membrane generated by lecithin is saturated with cholesterol and completely covers the supporting membrane.

The values of the coefficients, L_p were determined at various concentrations of the drugs, from the slopes of J_v versus ΔP plots. The values of L_p showed decrease with increasing concentration of the drug. The concentration of drug beyond which the value of L_p did not decrease further were taken to be the concentrations at which the liquid membrane generated by the lecithin-cholesterol mixture gets saturated with the drug. It was this particular composition of lecithin cholesterol-drug mixture, which was used in solute permeability (ω) experiments. In order to ascertain the location of the drugs in the lecithin-cholesterol liquid membranes, surface tensions of the solutions of various concentrations of drugs prepared in the aqueous solution of lecithin-cholesterol mixture of fixed composition (15.542 ppm with respect of lecithin, and 1.175×10^{-6}M with respect to cholesterol), were measured. If the surface tension of the aqueous solution of the lecithin-cholesterol mixture showed further decrease with increase in the concentration of the drug, it was inferred that the drug penetrates the lecithin-cholesterol liquid membrane and reaches the interface. On the other hand, if the surface tension of the lecithin-cholesterol mixture did not show any change with the concentration of the drug, it was inferred that the drug although gets incorporated in the lecithin-cholesterol liquid membrane, does not reach the interface.

6.2 Experimental studies.

In this section we give an account of the experimental studies conducted on a wide variety of drugs belonging to different pharmacological categories to throw light on the liquid membrane hypothesis of drug action. The data on solute permeabilities of relevant permeants in the presence of the liquid membranes has been used to gain information on the role of liquid membranes generated by the surface active drugs in the mechanism of their action.

6.2.1 Neuroleptics

6.2.1.1 Haloperidol [1] and chlorpromazine [3]

Most of the potent Neuroleptics are known to behave like powerful surface active agents [33]. Haloperidol is known to act by modifying the permeabilities of catecholamines and a few neurotransmitter amino acids in biological cells. Similar effects of chlorpromazine have been noted with membranes containing units like mitochondria [34], nerve ending particles [35] platelets [36] adrenomedullary particles [37] muscle fibers [38] and the influence of phenothiazines on the uptake and release of various neurotransmitter molecules [39,40] has been shown to be of significance to their action. With a view to investigating the role of accumulation of the drug in biomembranes in the mechanism of its action, studies on the interaction of the drug with synthetic monolayers were also undertaken by various authors [41,42]. To what extent permeabilities of biogenic amines and amino acids are modified as a result of this interaction has not been reported.

The hydraulic permeability data given in Tables 2 and 3 in case of both haloperidol and chlorpromazine clearly indicate the formation of liquid membrane by these drugs in series with the cellulose supporting membranes: in the case of haloperidol a cellulose acetate microfiltration membrane (Sartorius Cat no. 11107) and in the case of chlorpromazine a cellulose nitrate microfiltration membrane (Sertorius Cat. No. 11307) were used as supporting membrane. The values of hydraulic conductivity coefficients L_p show a progressive decrease with the concentration of the drugs upto their CMCs beyond which they become more or less constant. This trend is in accordance with Kesting's hypothesis [32] and as argued in chapters 4 and 5 is indicative of liquid membrane formation in series with the supporting membrane.

Table 2. Values of L_p at various haloperidol concentrations (Ref. 1)

	Haloperidol Concentration $\times 10^7$,M				
	0	1.064 (0.1CMC)	5.320 (0.5 CMC)	10.64 (CMC)	106.4
$L_p^a \times 10^8$ (m^3/s N)	2.804 ± 0.4368	2.095 ± 0.1273	1.603 ± 0.2015	0.7993 ± 0.0692	0.7662 ± 0.0216
$L_p^b \times 10^8$ (m^3/s N)	_	2.6035 ± 0.3996	1.8017 ±0.2510	_	_

[a] Experimental values. [b] Calculated values on the basis of mosaic model

Table 3. Values of L_p at various concentrations of chlorpromazine (Ref. 3)

	Chlorpromazine $\times 10^5$, M				
	0	2.25 (0.5CMC)	3.775 (0.75CMC)	4.5 (1CMC)	18.0 (4CMC)
L_p^a ($\times 10^9$) ($m^3 s^{-1}N^{-1}$)	3.960 ± 0.112	3.621 ± 0.168	3.341 ± 0.089	3.102 ± 0.286	3.305 ± 0.184
$L_p^b \times 10^8$ (m^3/s N)	_	3.632 ± 0.148	3.468 ±0.166	_	_

[a] Experimental values.

[b] Calculated values on the basis of the mosaic model.

Analysis of the flow data in the light of mosaic model [43-45] furnishes additional support in favour of liquid membrane formation in series with the supporting membrane. Following the arguments given in chapter 4 section 4.1.1 and chapter 5 section 5.3.1, it can be shown that if the concentration of surfactant is n times its CMC, n being less than or equal to 1, the value of L_p would be equal to $[(1-n)L_p^c + nL_p^s]$ where L_p^c and nL_p^s are respectively the values of L_p at 0 and CMC of the surfactant. The values of L_p thus computed at concentrations lower than the CMCs of the drugs are in good agreement with the experimentally determined values (Table 2 and 3).

The data on solute permeability ω of relevant permeants in case of haloperidol and chlorpromazine, in both orientations i.e. permeant facing hydrophilic surface and hydrophobic surface of the drug liquid membrane are recorded in Tables 4 and 5.

Table 4. Solute permeability ω of endogenous amines, amino acids, and cations in presence of 4.256×10^{-6} M haloperidol (Ref. 1).

	$\omega_1{}^{a}\times10^{12}$ moles/s N	$\omega_1{}^{b}\times10^{12}$ moles/s N	$\omega_1{}^{c}\times10^{12}$ moles/s N	$\omega_1{}^{d}\times10^{12}$ moles/s N
Dopamine	887.3	680.0	2607.0	274.4
Noradrenalin	75.8	65.9	294.3	
Adrenalin	50.7	undetectable	237.4	
Serotonin	193.7	94.5	348.1	
Histamine	48.8	109.1	318.8	
Glutamic acid	58.9	47.3	81.0	
γ-Aminobutyric acid	119.8	86.6	152.2	
Sodium (Chloride)	172.9	53.4	70.7	
Potassium (Chloride)	175.5	157.1	101.3	
Calcium (Chloride)	119.2	111.7	106.8	

[a] ω_1: control value – when no haloperidol was used. [b] ω_2: haloperidol in Compartment D of the transport cell. [c] ω_3: haloperidol in Compartment C of the transport cell. [d] ω_4: in the presence of γ-aminobutyric acid and haloperidol.

Table 5. Solute permeability (ω) of biogenic amines, cations, glucose and amino acids in the presence of 1.8×10^{-4} M chlorpromazine hydrochloride (Ref. 3).

	Solute permeability (ω) (mol s^{-1} N^{-1}) (x 10^{12})			
	ω_1	ω_2	ω_3	ω_4
Dopamine [a]	1015.0	344.9	531.5	70.98
Noradrenalin [a]	778.7	166.3	609.0	
Adrenalin [a]	2535.0	301.5	2000.0	
5-Hydroxytryptamine [a]	842.8	164.1	334.2	
Glutamic acid [b]	426.0	325.1	366.0	
γ-Aminobutyric acid [c]	784.7	608.3	624.1	
Sodium (Chloride) [d]	37.0	29.8	27.6	
Potassium (Chloride) [e]	62.1	36.0	51.8	
Glucose [f]	74.8	57.1	51.9	

ω_1 Control value –when no chlorpromazine was used; ω_2, chlorpromazine in compartment D of the transport cell and permeable substance in compartment C; ω_3, Chlorpromazine in compartment C of the transport cell and permeable substance in compartment C; ω_4, chlorpromazine and γ-aminobutyric acid in compartment D and permeable substance in compartment C. [a] Initial concentration 10μg/ml, [b] Initial concentration 500μg/ml (pH 3.2), [c] Initial concentration 200μg/ml (pH 7.0), [d] Initial concentration 5.382mg/ml, [e] Initial concentration 10.430mg/ml, [f] Initial concentration 20.00mg/ml.

Haloperidol being surface active has both hydrophobic and hydrophilic parts in its structure. The orientation of its molecules will, therefore, be significant when it forms a liquid membrane. The hydrophobic ends of the haloperidol molecules would be preferentially oriented towards the hydrophobic supporting membrane and their hydrophilic ends will face outwards, away from the supporting membrane. When haloperidol is in compartment C of the transport cell (first set of experiments) the haloperidol liquid membrane will present a polar surface to the permeant present in the same compartment. In the second set of experiments, however, where haloperidol is in Compartment D of the transport cell (Fig. 2 Chapter 5) and the aqueous solution of the permeant is in Compartment C, the haloperidol liquid membrane would present a hydrophobic surface to the permeant. Therefore, the orientation of haloperidol molecules with respect to approaching permeant would be different in the two sets of experiments.

The values of solute permeability ω given in Table 4 indicate that when the hydrophobic surface of the haloperidol liquid membrane faces the approaching permeant (second set of experiment,) a marked decrease in their permeability is observed. The haloperidol liquid membrane, thus, offers resistance to the transport of these permeants in this specific orientation. This reduction in the passive transport of biogenic amines, amino acids, and cations is likely to be accompanied by a reduction in their active transport. This occurs because the access of these permeants to the active carrier site of the biological membrane is likely to be effectively reduced due to the resistance of the haloperidol liquid membrane. The results also indicate that this specific orientation of haloperidol molecules with hydrophobic ends facing the catecholamines and amino acids would be necessary for the liquid membrane to resist the flow of these species. In the first set of experiments where haloperidol orients its hydrophilic ends towards catecholamines or amino acids the permeability of these substances in increased in the presence of haloperidol. This indicates that orientation of haloperidol with its hydrophobic ends facing the permeants would be necessary even in biological cells.

In cells, haloperidol reduces the permeability of catecholamines [33]. Despite the fact that these experiments were carried out using a cellulose acetate membrane, the results are similar to those observed in biological cells. This indicates that the liquid membrane generated by haloperidol contributes to the resistance of the flow of catecholamines.

The data on solute permeability (ω) recorded in Table 5 clearly indicate the ability of the chlorpromazine liquid membrane to reduce the permeability of biogenic amines and amino acids. The data further indicate that the reduction in permeability is maximum when the approaching permeable substances face the hydrophobic surface of the liquid membrane – the second set of experiments. Since chlorpromazine is also known to act by reducing the permeability of biogenic amines [39,40] and amino acids, it appears that the specific orientation of chlorpromazine with the hydrophobic ends of the molecule facing the permeable substances may be necessary even in biological cells. This implies that the receptor should have hydrophilic moieties projected outwards to which the hydrophilic ends of the drug become attached. Such an orientation can be rationalized if one examines the nature of receptors, in general, in relation to the lipid bilayer part of the biomembranes.

The receptors generally are membrane proteins and hence have to be surface active in nature. Thus, they will have both hydrophilic and hydrophobic moieties in their structure. Since the exterior environment of biological cells is aqueous in nature, it is logical to expect that the hydrophobic part of these membrane proteins will be associated with the hydrophobic core of the lipid bilayers and that only the hydrophilic part will face the exterior. Thus, the hydrophilic part of the drugs will interact preferentially with the hydrophilic part of the receptor protein, leaving the hydrophobic part to face the permeable substances. Predictions about similar orientations of receptor proteins in general have been made [46].

The effects of chlorpromazine have been noted with membrane-containing units like mitochondria [34], nerve –ending particles [35], platelets [36], adrenomedullary particles [37] and muscle fibers [38]. The influence of phenothiazines on the uptake and release of various neurotransmitters [39,40] seems to be of much significance to its action. In order to investigate the role of accumulation of the drug in biomembranes in the mechanism of its action, studies on the interaction of the drug with synthetic monolayers were undertaken by various authors [41,42]. To what extent the permeability of biogenic amines and amino acids is modified as result of this interaction has not been reported. These experiments [3] provide evidence that the liquid membrane generated by chlorpromazine itself offers resistance to the flow of biogenic amines and neurotransmitter amino acids. Although this resistance is passive in nature, it is likely to be accompanied by reduction in their active transport as well. This is because the liquid membrane generated by the drug is likely to reduce access of the permeable substances to the active site located on the biomembranes.

The data in Tables 4 and 5 show that the liquid membranes generated by both haloperidol and chlorpromazine impede the transport of γ-aminobutyric acid (GABA) and glutamic acid. The major factor responsible for the antipsychotic action of haloperidol and chlorpromazine is reduction in permeability to dopamine [33], which is under the influence of the GABA-glutamic acid system [47] in biological cells. It is interesting to note that the data in Tables 4 and 5 show that the permeability of dopamine through the drug liquid membranes (both haloperidol and chlorpromazine) is reduced further in the presence of GABA. This effect appears to be due to the strengthening of the hydrophobic core of the liquid membrane generated by the drugs-haloperidol of chlorpromazine-by GABA. This is evident from the structural similarity of the hydrophobic components of their structures, given in Fig.1:

The reduction in the permeability of serotonin (Tables 4 and 5) is in agreement with the observations reported [48] on biological cells. The extra-pyramidal effects of antipsychotic drugs are reported to be resistant to levodopa therapy [49]. Since reduced concentration of serotonin in cerebrospinal fluid has also been linked with a defect of extra-pyramidal function [50,51], the reduced permeability of serotonin in the presence of antipsychotic drugs offers a clue to the causation of extra pyramidal symptoms.

It is reported [52] that haloperidol is considerably more potent on a milligram basis than chlorpromazine *in vivo*. The liquid membrane phenomenon might explain this. Because haloperidol is more surface active [33] than chlorpromazine, as is obvious from the CMC values of 1.064×10^{-6} M and 4.5×10^{-5} M, respectively, the former will form a complete liquid membrane at a lesser concentration, making it pharmacologically effective even at a comparatively lower concentration.

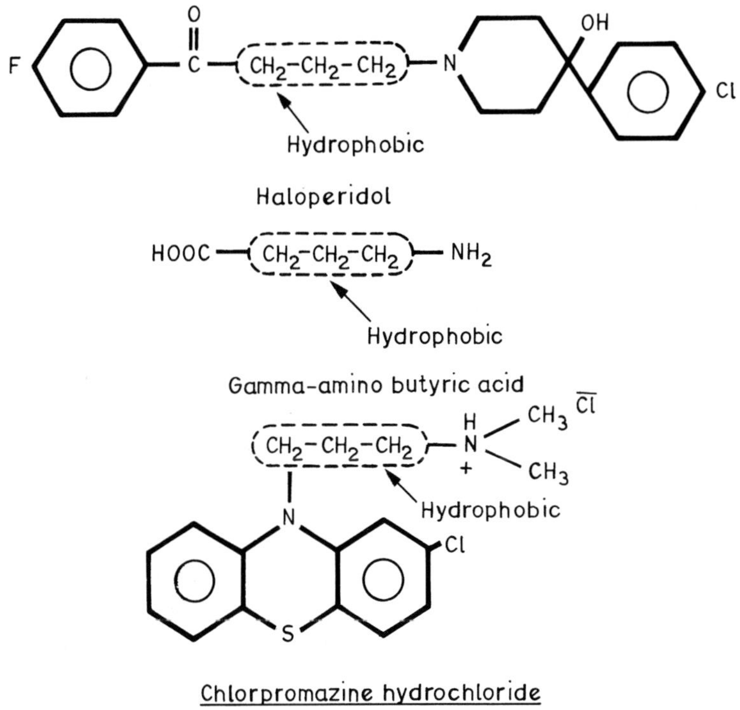

Fig 1. Structures of Haloperidol, γ-aminobutyric acid and chlorpromazine.

The observation of increased permeability of histamine in the presence of haloperidol, and its biological implication, if any remains to be explained. The resistance offered by haloperidol liquid membrane to the flow of sodium, potassium, and calcium cations is probably due to hydrophilicity of the ions. Unlike the observation in the case of endogenous amines and amino acids, even when the hydrophilic ends of haloperidol are facing the approaching cations, the permeability of these ions is reduced (Table 4). This observation may have some biological implications relative to nerve conduction.

The data in Tables 4 and 5 indicate that the resistance offered by the liquid membrane to the transport of cations and neutral molecules like glucose is much less in comparison to that offered to catecholamines. Thus, the increased resistance to the flow of dopamine in the presence of γ-aminobutyric acid, coupled with the resistance to the flow of glutamic acid offered by the liquid membrane generated by the drugs appear to make a significant contribution to their antipsychotic action.

The role of liquid membranes generated by these drugs in their action is further substantiated by the fact that haloperidol and chlorpromazine are structurally dissimilar. Of course, the specific orientation of the drug molecules in the liquid membranes with their hydrophobic ends facing the permeants appears crucial to their action.

6.2.1.2 Reserpine [2]

Reserpine, a drug structurally different from haloperidol and chlorpromazine has been experimented with. Existence of a liquid membrane generated by reserpine was demonstrated and data on the transport of biogenic amines and relevant neurotransmitter amino acids, through the liquid membrane generated by reserpine, were obtained [2].

Reserpine is a surface active drug and the CMC value of aqueous reserpine was found to be 1.6×10^{-6} M (Table 1). The data on hydraulic conductivity coefficient L_p at different concentration of reserpine, ranging from zero to 6.4×10^{-6} *M* are recorded in Table 6. The trend in the data in Table 6 is in accordance with the Kesting's hypothesis and is indicative of the formation of complete liquid membrane at the CMC of the drug in series with the supporting membrane: the value of L_p decreases progressively up to the CMC of the drug and also the values of L_p computed using mosaic model (Chapter 5) are in agreement with the experimentally determined values.

Table 6. Values of L_p at various concentrations of reserpine (Ref. 2)

Concentration of Reserpine $\times 10^6$,M	0	0.800 (0.5CMC)	1.200 (0.75 CMC)	1.600 (1CMC)	6.400 (4 CMC)
$L_p^a \times10^8$ (m^3s^{-1} N^{-1})	2.482 ±0.086	2.191 ±0.055	1.918 ±0.090	1.848 ±0.057	1.431 ±0.031
$L_p^b\times10^8$ (m^3 s^{-1} N^{-1})	---	2.165 ±0.071	2.006 ±0.064	--	--

[a] Experimental values. [b] Calculated values on the basis of mosaic model.

Data on the solute permeability (ω) of the biogenic amines and amino acids in the presence of the drug liquid membrane, in both orientations; permeants facing the hydrophilic surface of the liquid membrane and also the permeants facing the hydrophobic surface of the liquid membrane, have been obtained and are recorded in Table 7.

Table 7. Solute permeability (ω) of biogenic amines and amino acids in the presence of 6.4×10^{-6} M reserpine (Ref. 2).

	ω_1^a x 10^{12} moles s^{-1} N^{-1}	$\omega_2^b \times 10^{12}$ moles s^{-1} N^{-1}	$\omega_3^c \times 10^{12}$ mols s^{-1} N^{-1}
Dopamine[d]	1137.0	738.2	883.6
Noradrenalin[d]	1155.0	67.8	658.3
Adrenalin[d]	1165.0	567.3	880.2
5-Hydroxytryptamine[d]	1063.0	311.6	518.9
Glutamic acid[e]	403.6	217.5	491.7
γ-Aminobutyric acid[f]	695.1	407.1	1115.0

[a] Control value, when no reserpine was used. [b] reserpine in compartment D o the transport cell. [c] Reserpine in compartment C of the transport cell. [d] Initial concentration used, 10 μg/ml. [e] Initial concentration used, 500 μg/ml. [f] Initial concentration used, 200 μg/ml.

Data in Table 7 on the permeabilities of biogenic amines and amino acids reveal that the reduction in the permeabilities is maximum when the reserpine liquid membrane presents a hydrophobic surface to the permeants. Since reserpine is known to act by reduction in the uptake of biogenic amines [53], it appears that the particular orientation of the liquid membrane with its hydrophobic surface facing the permeants is relevant to reserpine's biological action. Reserpine is known to act by inhibiting the intraneuronal storage of catecholamines [53]. Although the ATP-Mg^{++} dependent uptake mechanism in isolated chromaffin granules has been considered to be a factor governing this mechanism [54], the effect on other subcellular particles is believed to be by a common unspecific mechanism [55]. The data in Table 7 indicate that the liquid membrane formation at very low concentrations (concentrations of the order of μ molar) can be one such common mechanism.

While some of the wide ranging actions of reserpine can be explained on the basis of blocking of uptake of catecholamines [53], it is difficult to find a common mechanism for other effects. Inhibition of experimentally provoked thrombus formation in rats [56] decreased oxygen utilization in brain [57] and liver [58] , the anti-tumor effect [59], extra-pyramidal symptoms [60], and reduction of thyroid secretion [61] are a few of them. Impairment of release of catecholamines by reserpine has also been reported [62] for which no explanation has been given at the molecular level. The liquid membrane phenomenon seems to offer a common mechanism for all such effects. Modification in the permeabilities of biologically relevant molecules by reserpine liquid membrane could be a plausible explanation.

Reserpine is also known to reduce permeability of biological cells to 5-hydroxytryptamine (serotonin) [62] which may have contributed to its sedative effect. The data in Table 7 also show a reduction in the permeability of 5-hydorxytryptamine because of the reserpine liquid membrane. Reserpine is known to lower the threshold to electro-shock in rats [63] which is related to depletion of γ-aminobutyric acid (GABA) in the brain. Since a reserpine liquid membrane reduces the permeability of GABA (Table 7), the above effect can at least partially be assigned to the formation of liquid membrane by reserpine *in situ*.

6.2.2 Anticancer drugs-5-flourouracil and its derivatives [18]

One of the important implications of the liquid membrane hypothesis of drug action [64] is that in a series of structurally-related drugs, which are congeners of a common chemical moicty and which act by altering the permeability of cell membranes, any structural variation which increases the hydrophobicity of the compound will increase the potency of the drug, while any alteration of the hydrophilic moieties of the drug may change the nature of its action qualitatively; a detailed discussion on this and other implications of the hypothesis will be presented in the next chapter (chapter 7) dealing with the assessment of the hypothesis.

It has been shown by Ligo [65] that the 1-hexylcarbamoyl-5fluorouracil (HCFU) synthesized by Ozak et.al [66] is more active against various tumors in mice and less toxic to host animals than its parent drug 5-fluorouracil (5FU). Ligo et al [65] have tested the activities of these drugs on Lewis lung carcinoma and B16 melanoma. It is evident from the structure of the two drugs (Fig. 2) that HCFU will be more hydrophobic and more surface active than its parent compound 5FU. Prompted by this clue, 5FU and two of its derivatives, HCFU and 1-(2- tetrahydrofuryl) 5-fluorouracil (FT), have been investigated [18] for the contribution of liquid membrane phenomenon to their action.

All the three drugs, 5FU, HCFU and FT, have been found to be surface active and shown to generate liquid membranes in series with a supporting membrane. Transport of relevant permeants through liquid membranes generated by these drugs in series with the supporting membrane has been studied. The data obtained from these model experiments indicate that the modification in the transport of relevant permeants, due to the drug liquid membrane likely to be generated at the sites of action, may also make a significant contribution to the biological actions of these drugs. In these studies also like all others, a non-specific non-living membrane has been chosen deliberately as the supporting membrane for the liquid membranes. Thus, the possibility of active and specific interactions of these drugs with the constituents of biomembranes as the cause for modification in the transport of relevant permeants is totally ruled out and the role of passive transport through the liquid membranes in the action of these drugs is highlighted.

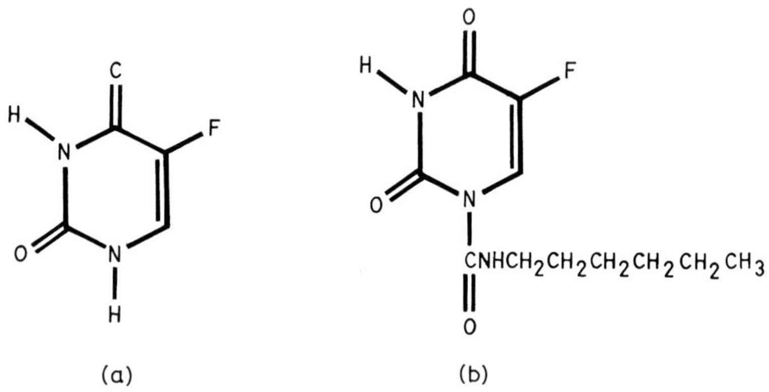

Fig 2. Chemical structures of (a) 5-fluorouracil and (b) 1-hexylcarbamoyl-5-fluorouracil.

CMCs of 5FU, HCFU and FT were estimated from the variation of surface tension with concentration and are recorded in Table 1. The hydraulic permeability data at various drug concentrations in the case of all the three drugs were found to be in accordance with the equation, $J_v = L_p \Delta P$. The values of L_p estimated from the slopes of J_v versus ΔP plots, in the case of all the three drugs, show a progressive decrease with increase in the concentrations of the drugs (Table 8) upto the respective CMCs of the drugs beyond which they become more or less constant. This trend in the values of L_p is in keeping with Kesting's liquid membrane hypothesis [32], and indicates the formation of drug liquid membranes in series with the supporting membrane. The values of L_p computed using the mosaic model, (Eq. 13, Chapter 5), at several concentrations of the drugs below their respective CMCs compare favorably with corresponding experimental values in the case of all three fluorouracil (Table 8). This fact further supports the formation of drug liquid membranes.

As explained in the design of experiments, for solute permeability (ω) measurements two sets of experiments were performed. In the first set of experiments, the compartment C of the transport cell was filled with an aqueous solution of the drug along with the permeant, and the compartment D was filled with distilled water (Fig. 2 Chapter 5) In the second set, the compartment D was filled with aqueous solution of the drug and the compartment C was filled with the aqueous solution of the permeant. The concentrations of the drugs used in the ω measurements were always higher than the respective CMCs. All measurements were made at constant temperature using a thermostat wet at 37±0.1°C.

Table 8. Values of L_p at various concentrations of 5FU, FT and HCFU (Ref. 18).

	Conc ($\times 10^{11}$M)	$L_p \times 10^8$ ($m^3 s^{-1} N^{-1}$) *	$L_p \times 10^8$ ($m^3 s^{-1} N^{-1}$)**
5FU	0.000	2.162 ± 0.064	-
	20.00(0.25 CMC)	1.930 ± 0.058	1.944 ± 0.056
	40.00(0.5 CMC)	1.720 ± 0.054	1.726 ± 0.059
	60.00(0.75 CMC)	1.573 ±0.074	1.508 ± 0.041
	80.00 (CMC)	1.290 ± 0.034	-
	160.00	1.260 ± 0.061	-
	240.00	1.266 ± 0.064	-
FT	0.000	2.162 ± 0.064	-
	1.875(0.25 CMC)	1.778 ± 0.086	1.805 ± 0.081
	3.750(0.5 CMC)	1.418 ± 0.049	1.406 ± 0.115
	5.625(0.75 CMC)	1.095 ± 0.059	1.106 ± 0.039
	7.500(CMC)	0.755 ± 0.025	-
	15.000	0.751 ± 0.031	-
	22.500	0.761 ± 0.031	-
HCFU	0.000	2.162 ± 0.064	-
	1.525(0.25 CMC)	1.795 ± 0.041	1.770 ± 0.049
	3.050(0.5 CMC)	1.422 ± 0.030	1.377 ± 0.036
	4.575(0.75 CMC)	0.999 ± 0.018	0.985 ± 0.023
	6.100(CMC)	0.592 ± 0.010	-
	12.200	0.594 ± 0.006	-
	18.300	0.582 ± 0.018	-

The values reported for L_p are arithmetic mean of 10 repeats ± S.D.

*Experimental values.

** Calculated values using mosaic model.

Since all three drugs, being surface-active in nature, have both hydrophilic and hydrophobic parts in their structure, it is expected that the hydrophobic ends of the drug molecules in the liquid membrane would be preferentially oriented towards the hydrophobic supporting membrane, in these experiments a Sartorius cellulose acetate membrane, Cat no. 11107, and hydrophilic moieties would be drawn outwards away from it. Thus, as explained in the design of experiments, section 6.1 of this chapter, in the first set of solute permeability experiments, the permeants would face the hydrophilic surface of the drug liquid membrane generated in series with the supporting membrane, while in the second set they would face the hydrophobic surface. The data on the solute permeability of relevant permeants in the two orientations of the drug molecules in the liquid membranes are recorded in Table 9 along with the corresponding values from control experiments where no drug was used.

Table 9. Solute permeability (ω) of various permeants in the presence of 5FU, FT and HCFU (Ref. 18).

Permeant	Initial concentration (mg/liter)	Control $\omega \times 10^{-9}$	5FU(1×10^{-9}M) $\omega\times10^{9}$ C	$\omega\times10^{9}$ D	FT (1×10^{-10}M) $\omega\times10^{9}$ C	$\omega\times10^{9}$ D	HCFU(1×10^{-10} M) $\omega\times10^{9}$ C	$\omega\times10^{9}$ D
Aspartic acid	150	0.856 ±0.011	0.628 ±0.004	0.688 ±0.006	0.475 ±0.020	0.715 ±0.062	0.398 ±0.008	0.568 ±0.042
Cyanocobalamin	30	0.488 ±0.018	0.281 ±0.024	0.365 ±0.021	0.316 ±0.015	0.379 ±0.018	0.282 ±0.026	0.347 ±0.037
Folic acid	0.05	8.715 ±0.266	6.013 ±0.557	7.541 ±0.316	6.631 ±0.496	7.590 ±0.010	4.406 ±0.220	5.743 ±0.334
Glutamine	500	0.474 ±0.031	0.759 ±0.065	0.694 ±0.052	0.160 ±0.011	0.363 ±0.014	0.417 ±0.013	0.399 ±0.008
Glycine	100	0.265 ±0.010	0.412 ±0.055	0.644 ±0.168	0.151 ±0.002	0.182 ±0.003	0.181 ±0.001	0.195 ±0.004

Values of ω are repoted as arithmetic mean of 10 repeats ±S.D. in mol. S^{-1} N^{-1}., C: drug in compartment C;D: drug in compartment D.

Antimetabolites, in general, are known to act by impairing the synthesis of purine and pyrimidine bases by interfering with folic acid metabolism or prevent the incorporation of the bases into nucleic acids [67]. The steps involved are known to be enzyme-catalysed. For example, 5FU is ultimately converted enzymatically into 5-fluorodeoxyuridine-5 phosphates, which inhibits the thymidylate synthetase enzyme system resulting in the blockade of DNA synthesis [68]. The data (Table 9), however, indicate that the passive transport through the liquid membranes, likely to be generated by the flurouracils (5FU, HCFU and FT) at the respective sites of action, may also contribute to their action.

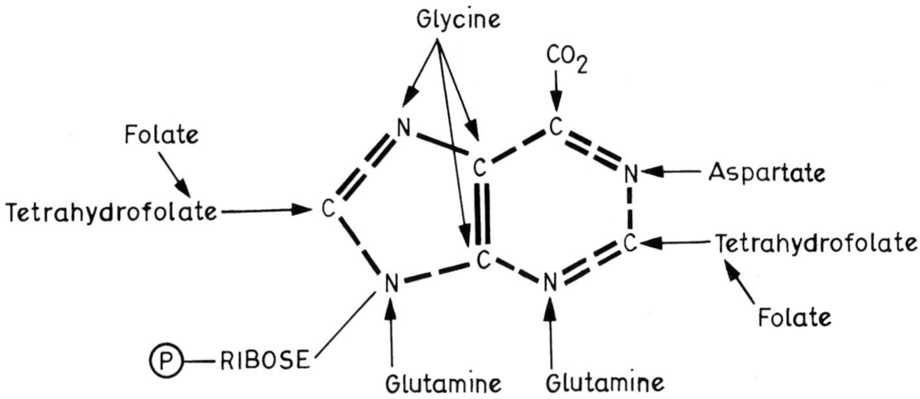

Fig 3. Compounds from which the atoms of the purine ring are derived in the biosynthetic pathway. The breaks in the bond separate the groups of atoms derived from each source (Ref. 73).

Vitamin B_{12} and folic acid, which are dietary essentials for man, are required for the synthesis of purine and pyrimidine bases and their incorporation into DNA. Their deficiency may result in defective synthesis of DNA in any cell that attempts chromosomal replication and division [69]. This impediment in the transport may contribute to the deficiency of vitamin B_{12} and folic acid inside the cells resulting in the defective synthesis of DNA. Thus, it appears that the phenomenon of liquid membrane formation may also contribute to the anticancer activities of 5FU and its derivatives. A perusal of Table 9 further reveals that inhibition in the transport of vitamin B_{12} and folic acid is more when the permeants face the hydrophilic surface of the liquid membranes than when they face the hydrophobic surface. This observation indicates that the specific orientation of the drug molecules in liquid membranes with their hydrophilic ends facing the permeants may be necessary on cancerous cells, while the drug molecules in the liquid membranes on the normal cells may have the other orientation – hydrophobic ends facing the permeants. This inference, in turn, implies that surface of the membranes of the cancerous cells should be less hydrophilic that those of the normal cells. Though there are some indications in literature [70-72] that he neoplastic state may also arise through an alteration in the surface properties of the cells, a thorough probe in terms of hydrophilicity of the cell surface is called for to substantiate this conjecture.

Amino acids like glycine, glutamine and aspartic acid are also required, in addition to folic acid, for the purine ring synthesis [73, 74]. Compounds from which the atoms of the purine ring are derived in the biosynthetic pathway are depicted in Fig.3. The data in Table 9 indicate that except in the case of 5FU, the transport of glycine, glutamine and aspartic acid is also impeded in addition to folic acid and vitamin B_{12}, by the liquid membranes generated by both FT and HCFU. In the case of 5FU, the transport of glycine and glutamine was enhanced. The impediment in the transport of the amino acids, also in the case of HCFU and FT, was more in the specific orientation of the drug molecules in the liquid membrane with their hydrophilic ends facing the permeants. This impediment in the transport observed in the case of FT and HCFU may also be a factor responsible for the impairment of the synthesis of purine bases contributing to the anticancer activity of these drugs.

It has been reported by Ligo et al. [65] that of the three drugs, HCFU, FT and 5FU, HCFU is most potent. This finding is consistent with the liquid membrane hypothesis of drug action [64]. The CMC of HCFU is the lowest. As CMC is the concentration at which a complete liquid membrane is generated at the interface, it would appear that of the three drugs, HCFU would require the lowest concentration for the development of a complete liquid membrane at the site of action. Since modification of the transport of the relevant permeants, which affects the biological effect, is maximum when a complete liquid membrane is generated, the concentration of HCFU required to produce the maximum biological effect would be the lowest amongst the three drugs, making HCFU the most potent drug.

Some of the adverse side effects of cytotoxic drugs include megaloblastic anaemia [75], neurological disorder relating to spinal column and cerebral cortex [76], ineffective haematopoiesis and pancytopenia [77]. These symptoms are also characteristic of deficiencies of vitamin B_{12} or folic acid or both [69,78-80]. The impediment in the transport of vitamin B_{12} and folic acid in the specific orientation of the drug molecules in the liquid membrane with their hydrophobic ends facing the permeants, which may be the orientation on the normal cells, could also be a plausible explanation fro the reported side-effects.

6.2.3. Diuretics [11]

Most of the high-ceiling diuretics [81] are known to act by altering the reabsorption of cations (e.g., Na^+) and anions (e.g. Cl^-) in the ascending limb of the loop of Henle [81]. Although diuretics act by modifying the membrane permeability, their surface activity was not documented in the literature, till Bhise et al [11] investigated furosemide and triamterene, which are structurally dissimilar and reported their CMC (Table 1). Bhise et al. demonstrated the formation of liquid membrane by them at the interface. Transport of relevant cations and anions in the presence of the liquid membranes generated by the drugs has been studied. The data indicate that the liquid membranes generated by the diuretic drugs contribute to the mechanism of their action. A cellulose nitrate microfiltration membrane (Sartorius Cat No. 11307)/ aqueous interface was chosen as a site for the formation of the liquid membranes to eliminate the possibility of active and specific interaction of the drugs with the constituents of the biological membranes and to highlight the role of passive transport through the liquid membrane. The hydraulic permeability data at various concentrations of the diuretic drugs, in the case of both furosemide and triamterene were shown to obey the linear relationship, $J_v = L_p \Delta P$ between the volume flux J_v per unit of the membrane and the applied pressure difference ΔP. The values of L_p at various concentrations of the diuretic drugs are shown in Table 10. The values of L_p (Table 10) show a progressive decrease with increase in drug concentration upto the CMC after which they become more or less constant. This gradation

(Table 10) is in keeping with the liquid membrane hypothesis [32] and indicates the progressive coverage of the supporting membrane with the liquid membrane with an increase in the concentration of the drug up to its CMC; at this concentration it is completely covered. Analysis of the flow data (Table 10) in the light of mosaic model [43-45] furnishes additional support for liquid membrane formation in series with the supporting membrane. The values of L_p (for both furosemide and triamterene), calculated using the mosaic model at concentrations below the CMC values of the drugs, match the experimentally determined values (Table 10) lending support to liquid membrane formation.

Table 10. Values of the hydraulic conductivity coefficient L_p at various concentrations of furosemide and triamterene (Ref. 11)

	Furosemide Concentration x 10^5 M					Triamterene Concentration x 10^6 M				
	0	2.08 (0.25CMC)	4.16 (0.5CMC)	8.3 (CMC)	24.9	0	2.0 (0.2 CMC)	5.0 (0.5 CMC)	10.0 (CMC)	30.0
$L_p^a \times 10^8$ ($M^3 . s^{-1} . N^{-1}$)	3.56 ±0.266	2.73 ±0.105	2.20 ±0.416	1.11 ±0.075	1.26 ±0.058	3.56 ±0.090	3.06 ±0.102	1.95 ±0.088	0.59 ±0.072	0.56 ±0.066
$L_p^c \times 10^8$ ($M^3 . s^{-1} . N^{-1}$)		2.94 ±0.105	2.33 ±0.105	-	-	-	2.96 ±0.102	2.08 ±0.088	-	-

[a] Expressed as mean ± SD. [b] Experimental values. [c] Calculated on the basis of the mosaic model.

The data on solute permeability (ω) of relevant permeants in the presence of liquid membranes generated by the diuretic drugs are recorded in Table 11. The primary action of furosemide is to reduce active absorption of chloride ions [81]. The results indicate that the liquid membrane formed by furosemide, even on an inert support, impedes the transport of chloride ions (Table 11). Similarly, the liquid membrane generated by triamterene offers resistance to the transport of Na^+ and K^+ ions (Tables 11). The significance of these observations is enhanced because the concentrations at which the complete liquid membranes are generated in series with the supporting membrane are low (of the order of μM) and comparers favourably with the concentrations of these drugs in renal tubules [82,83].

In the case of triamterene, the data indicate (Table 11) that the transport of potassium ions is impeded more than the transport of sodium ions. This agrees with the reported observations on biological cells that tramterene is a potassium-sparing diuretic [84]. In spite of the fact that in the this study an inert membrane like cellulose nitrate microfiltration membrane was used as support for the liquid membranes, the trend observed in the permeability of the cations is similar to that expected in biological cells. This strongly indicates that the liquid membranes generated by diuretic drugs, like triamterene, play significant role in the mechanism of its action.

An examination of Table 11 reveals that the resistance offered to the transport of chloride ions (in the case of furosemide) and that of potassium ions (in the case of triamterene) is maximal when the liquid membranes generated by these drugs presented a hydrophilic surface to the approaching permeating species (the first set of experiments: when drugs and permeating species were kept in compartment C of the transport; Fig. 2 Chapter 5).

Table 11. Solute permeability (ω)[a] of ions in the presence of furosemide or triamterene (Ref. 11)

	$\omega_1^b \times 10^{12}$ mol. s^{-1} .N^{-1}	$\omega_2^c \times 10^{12}$ mol. s^{-1} .N^{-1}	$\omega_3^d \times 10^{12}$ mol. s^{-1} .N^{-1}
	Furosemide [e]		
(Sodium) chloride	250.7 ± 35	189.0 ± 36	419.4 ± 79
	Triamterene [f]		
Potassium (chloride)	168.8 ± 12	91.6 ± 7	359.2 ± 9
Sodium (chloride)	111.2 ± 15	207.4 ± 15	232.5 ± 6

[a] Expressed as mean of fifteen repeats ± SD. [b] The drug in compartment D of the transport cell. [c] The drug in compartment C of the transport cell. [d] Control value: when no drug was used. [e] Concentration, 24.9×10^{-5} M. [f] Concentration, 3.0×10^{-5}M.

In the light of these observations, it appears likely that the action sites of diuretic drugs like furosemide and triamterene themselves may be hydrophobic so that the hydrophobic ends of these drugs get attached to them leaving the hydrophilic parts to face the permeating species. If the action sites are hydrophobic they should be located within the hydrophobic core of the lipid bilayer of the membranes. To substantiate these conjectures, which appear logical in the light of the trends observed in the these experiments, further investigations are needed.

The permeability of sodium ions is impeded most when the triamterene liquid membrane presents its hydrophobic surface to the cation (Table 11). The observation, however, is of limited biological significance because triamterene is known to be a potassium-sparing diuretic [84].

Diuretic drugs are also known to cause reduction in bile flow [85] and to alter ionic fluxes across isolated erythrocytes [86]. The phenomenon of liquid membrane formation may be a plausible explanation for these effects. The decrease in reabsorption of water, which results in diuresis, is considered mainly a consequence of modification in the permeability of ions [81]. This study, however, indicates that the liquid membrane generated by the diuretic drug itself offers resistance to volume flux of water.

Though the observed reduction in permeability of the ions (Table 11), due to the liquid membrane generated by the drugs, is passive in nature, it is likely to be accompanied by a consequent decrease in active transport. This would occur because access of the permeating species to the active sites on the biological membrane would be reduced due to the formation of the liquid membranes in series with the biological membrane. Thus, the liquid membranes generated by diuretic drugs may contribute significantly to the mechanism of drug action by impeding transport of ions as well as water.

There are a few reports [87,88] wherein it has been found that the response to diuretic drugs, such as furosemide is reduced in the presence of anticonvulsant drugs such as diphenylhydantoin (DHP). Since DPH is a membrane-stabilizing drug [89], it is likely to be surface active in nature and, hence, capable of generating a liquid membrane at the interface. It is, therefore, logical to assume that reduction in the response to furosemide in the presence of DPH may be due to the resistance offered to the transport of the former by the liquid membrane barrier generated by the latter (DPH). This point has been investigated by Srivastava and his group [12].

DPH, which was found to be surface active (CMC = 4.0×10^{-7} M),has been shown to generate liquid membrane at interfaces. Data on the transport of furosemide through the liquid membrane generated by DPH in series with a supporting membrane have been obtained. A non-living membrane, such as cellulose nitrate microfiltration membrane, was purposely chosen as the supporting membrane for the liquid membrane to highlight the role of passive transport through the liquid membrane in the reported reduction of furosemide response in the presence of DPH.

Hydraulic permeability data was obtained to demonstrate the formation of liquid membrane by DPH in series with a hydrophobic supporting membrane (Sartorius Cat. No.11307). The hydraulic permeability data at all concentration of DPH studied were found to obey the linear relationship, $J_v = L_p \Delta P$ between volume flux J_v and the pressure difference ΔP. The values of hydraulic conductivity coefficient L_p at various concentrations of DPH recorded in Table 12 show a progressive decrease with increase in concentration of DPH upto its CMC beyond which they become more or less constant. This trend is, as argued earlier, indicative of the fact that at CMC the liquid membrane generated by DPH completely, covers the supporting membrane.

Table 12. Values of L_p at various concentrations of diphenylhydantoin sodium (*DPH*) (Ref. 12).

	DPH concentrations x 10^7 M				
	0	1.0	2.0	4.0 (CMC)	8.0
$L_p \times 10^8$ (m^3 s^{-1} N^{-1})	0.808 ±0.029	0.486 ±0.018	0.368 ±0.004	0.265 ±0.008	0.242 ±0.008

Solute permeability (ω) for Furosemide was measured in the presence of DPH using the procedure already described. For ω measurements two sets of experiments were performed. In the first set, the compartment C of the transport cell (Fig. 2 Chapter 5) was filled with a solution of furosemide of known concentration, prepared in an aqueous solution of known concentration of *DPH*, and the compartment D was filled with distilled water. In the second set of experiments, the aqueous solution of *DPH* was placed in the compartment D, and the compartment C contained the aqueous solution of the permeant furosemide. In the control experiment no. *DPH* was used. Since the interface is completely covered with the liquid membrane at concentrations equal to or greater than the CMC, the concentration of *DPH* used in the experiment for ω measurements was 5.0×10^{-6} M, which is well above its CMC. Since DPH is surface active in nature, it should have both hydrophilic and hydrophobic moieties in its structure. The hydrophobic moieties would, therefore, be preferentially oriented towards the hydrophobic supporting membrane (the cellulosic microfiltration membrane in the present case), and the hydrophilic ends would be drawn outwards away from it. Therefore, in the first set of experiments for ω measurements, the permeant would face the hydrophilic surface of the liquid membrane generated by DPH. In the second set, however, where the permeant was present in the compartment C and DPH was present in the compartment D of the transport cell, the permeant would face the hydrophobic surface of the liquid membrane.

Table 13. Solute permeability $(\omega)^a$ of furosemide in the presence of 5×10^{-6}M diphenylhydantoin sodium (DPH) (Ref. 12)

	$\omega_1^b \times 10^{10}$ (mol. s^{-1} .N^{-1})	$\omega_2^c \times 10^{10}$ (mol. s^{-1} .N^{-1})	$\omega_3^d \times 10^{10}$ (mol. s^{-1} .N^{-1})
Furosemide[e]	15.99 ± 2.65	21.20 ± 1.93	8.40 ± 0.34

[a] The ω values given are arithmetic mean of 15 repeats ± mean deviation.
[b] ω_1Control values when no DPH was used.
[c] ω_2 Both DHP and furosemide present in compartment C and distilled water in compartment D.
[d] ω_3: DPH in compartment D and furosemide in compartment C.
[e] Initial furosemide concentration 10 μg ml^{-1}

The values or solute permeability, ω, for furosemide in presence of *DPH*, given in Table 13 indicate that in the first set of experiments where the permeant (furosemide) faces the hydrophilic surface of the *DPH* liquid membrane, the permeability is enchaned in comparison to that in the control experiments. In the second set, however, where the *DPH* liquid membrane presents its hydrophobic surface to the permeant, furosemide, the transport of furosemide is impeded. This observation on the impediment of furosemide transport by the liquid membrane in the specific orientation of the *DPH* molecules with hydrophobic ends facing the permeant, appears relevant to the observations reported on biological cells [87].

It has been reported [87] that in epileptic patients taking *DPH*, the mean diuretic effect of furosemide is reduced by about 50-68% of that of healthy subjects, and also the peak effect was observed to be delayed. It has also been reported [87] that diuretic response to furosemide was smaller in epileptic patients on anticonvulsant therapy including *DPH*. It has also been reported [88] that concurrent administration of *DPH* and furosemide results in malabsorption of furosemide.

This study indicates that the reduced permeability of furosemide may be a cause of its reduced response in the presence of anticonvulsant drugs such as *DPH*. It is likely that a liquid membrane may be generated by *DPH* at the site of action of furosemide in such an orientation that furosemide faces the hydrophobic surface of the liquid membrane, resulting in the impediment of furosemide transport to the relevant site. Consequently, this will lead to reduced and delayed response of furosemide in the presence of *DPH*.

6.2.4. Cardiac glycosides [30]

The liquid membrane phenomenon in the actions of digitalis glycoside (digitoxin, digoxin and ouabain) has been studied. Formation of liquid membranes, in series with a supporting membrane, by digitalis alone and by digitalis in association with lecithin and cholesterol has been demonstrated. The results obtained on the transport of relevant permeants, viz. sodium, potassium and calcium ions and dopamine, adrenaline, noradrenalin and serotonin, in the presence of the liquid membrane generated by digitalis in association with lecithin and cholesterol indicate that the liquid membrane barrier to transport may have a relevance with the biological actions of digitalis.

The hydraulic permeability data at varying concentrations of all the three digitalis drugs were found to be represented by the relationship, $J_v = L_p \, \Delta P$. The values of the hydraulic conductivity coefficients L_p recorded in Table 14 show a decreasing trend with increasing concentrations of the drugs upto their CMCs beyond which they become more or less constant. This trend in the values of L_p as argued earlier, is indicative of the formation of liquid membranes by the drugs in series with the supporting membrane, Sartorius cellulose acetate membrane Cat No. 11107 in this case.

Table 14. Values of L_p at varying concentrations of digitalis drugs (Ref. 30).

	Concentration ($\times 10^9$ M)	$L_p \times 10^9$ ($m^3.s^{-1}.N^{-1}$)*	$L_p \times 10^9$ ($m^3.s^{-1}.N^{-1}$)†
Digitoxin	0.00	7.023 ± 0.002	-
	32.666	6.541 ± 0.061	6.568 ± 0.008
	64.68	6.063 ± 0.053	6.112 ± 0.015
	98.00 (CMC)	5.643 ± 0.021	-
	130.34	5.593 ± 0.043	-
	196.00	5.633 ± 0.099	-
	266.68	5.602 ± 0.040	-
Digoxin	0.00	7.023 ± 0.002	-
	1.4	6.445 ± 0.105	6.422 ± 0.019
	2.8	5.913 ± 0.028	5.822 ± 0.038
	4.2	5.133 ± 0.118	5.221 ± 0.055
	5.6 (CMC)	4.621 ± 0.073	-
	11.2	4.639 ± 0.077	-
	16.8	4.625 ± 0.122	-
Ouabain	0.00	7.023 ± 0.002	-
	0.50	6.510 ± 0.009	6.529 ± 0.007
	1.00	6.127 ± 0.120	6.035 ± 0.012
	1.50	5.566 ± 0.144	5.540 ± 0.017
	2.00 (CMC)	5.046 ± 0.022	-
	4.00	5.056 ± 0.082	-
	6.00	5.043 ± 0.050	-

The values of L_p are arithmetic mean of 10 repeats ± SD

* Experimental values, † Calculated values using mosaic model.

The value of L_p computed using mosaic models at concentrations of the drug below their CMC compare favourably with the experimentally (Table 14) determined values. This fact gives additional support to the formation of liquid membrane in series with the supporting membrane.

Evidence in favour of incorporation of digitalis in the liquid membrane generated at the interface by the lecithin-cholesterol mixture is obtained from the data on hydraulic permeability at varying concentrations of these drugs in the lecithin-cholesterol mixture of fixed composition, 1.919×10^{-5}M with respect to lecithin and 1.175×10^{-6} M with respect to cholesterol. The hydraulic permeability data in this case too were found to be represented by the Eq. $J_v=L_p\Delta P$. The values of L_p decrease with increasing concentration of drugs up to certain concentration and then become constant (Table 15). The concentration of the drug beyond which the values of L_p become more of less constant can be taken to be the concentration at which the lecithin liquid membrane at the interface, which is already saturated with cholesterol, is also saturated with the drug (Table 15). Concentrations of the drugs in the lecithin-cholesterol mixture used in the solute permeability experiments were a little higher than the saturating concentrations obtained from these studies (Table 15). In these experiments pH was maintained at 7.4 using phosphate buffer and the temperature set at 37±0.1°C.

For solute permeability measurements, two sets of experiments were performed. In the first set of experiments aqueous solutions of mixtures of lecithin-cholesterol-digitalis of desired composition were filled in the lower compartment (C) of the transport cell (Fig. 2 Chapter 5) along with the solution of known concentration of the permeant and the upper compartment (D) was filled only with the phosphate buffer (pH 7.4) which was used to prepare aqueous solution filled in compartment C. In the second set of experiments an aqueous solution maintained at pH 7.4 using the phosphate buffer of the mixture of lecithin, cholesterol and the digitalis of desired composition was filled in the upper compartment (D) of the transport cell and the aqueous solution of the permeant of known concentration prepared in the phosphate buffer (pH 7.4) was filled in compartment C. Since lecithin, cholesterol and digitalis glycosides are all surface active in nature they have both hydrophilic and hydrophobic parts in their structure. The orientation of these molecules will therefore be significant when a liquid membrane is formed. The hydrophobic ends of the these molecules in the liquid membrane would be preferentially oriented towards the hydrophobic supporting membrane and their hydrophlic ends will be drawn outwards away from the supporting membrane. In the first set of experiments for the solute permeability experiments, therefore, the permeants would face the hydrophilic surface of the liquid membrane generated by the lecithin-cholesterol-digitalis mixture, whereas in the second set of experiments they would face the hydrophobic surface. The orientations in the first set and in the second set of solute permeability experiments will be referred to as orientation 1 and orientation 2, respectively, throughout the following discussion.

The composition of the aqueous solution of lecithin-cholesterol-digitalis mixture used in the solute permeability experiments was the one at which the liquid membrane generated by lecithin at the interface was completely saturated with both cholesterol and the digitalis.

This composition was derived from our earlier studies [90] and from the present data on hydraulic permeability in the presence of varying concentrations of digitalis in the mixture of lecithin and cholesterol of fixed composition, i.e. 1.919×10^{-5} M with respect to lecithin and 1.175×10^{-6} M with respect to cholesterol. For details of the methods of measurement of solute permeability the original paper should be consulted [30].

Table 15. Values of L_p at varying concentrations of digitalis drugs in the presence of lecithin-cholesterol mixture of fixed composition (1.919×10^{-5} M with respect to lecithin and 1.175×10^{-6} M with respect to cholesterol) (Ref. 30).

	Concentration ($\times10^9$ M)	$L_p\times10^{9*}$ ($m^3s^{-1}N^{-1}$)
Digitoxin	0.00	10.671 ± 0.039
	1.96	5.903 ± 0.051
	3.92	2.231 ± 0.023
	5.88	2.201 ± 0.027
	7.84	2.187 ± 0.032
Digoxin	0.00	10.671 ± 0.039
	1.12	6.363 ± 0.027
	2.24	3.773 ± 0.007
	3.36	3.727 ± 0.034
	4.48	3.755 ± 0.026
Ouabain	0.00	10.671 ± 0.039
	0.40	8.728 ± 0.024
	0.80	5.780 ± 0.082
	1.20	5.719 ± 0.022
	1.60	5.736 ± 0.018
	2.00	5.700 ± 0.062
	4.00	5.692 ± 0.043

* The values of L_p are arithmetic mean of 10 repeats ± SD.

The major component to the positive inotropic action of digitalis is the inhibition of the membrane-bound (Na^+, K^+) ATPase. Digitalis glycosides bind to (Na^+, K^+) ATPase from extracellular side of the plasma membrane, inhibit its enzymic activity and impair the active transport of intracellular calcium [91, 92]. Magnitude of the inotropic effect of digitalis is proportional to the degree of inhibition of the enzyme [93, 94].

The active grouping in the cardiac glycosides is thought to be a carbonyl group in conjugation with a C=C double bond located in the lactone ring. Since the carbonyl group is electronegative, it acts as a proton acceptor and can, therefore, build up a hydrogen bond with a hydroxyl group of the phosphoric acid residue in the phosphorylated enzyme intermediate. The single hydrogen bond permits free rotation of the cardiac glycoside molecule so that the correct face of the steroid nucleus comes into close relationship with the complementary enzyme surface. In view of this mode of interaction of cardiac glycosides with its pharmacological receptor [92], the (Na^+, K^+) ATPase enzyme [91], present data on the solute permeability of cations in the first set of experiments i.e., in orientation 1, appear relevant (Table 16).

The values of ω given in Table 16 indicate that when permeants face hydrophilic surface of the liquid membrane, the permeability of Na^+ and Ca^{2+} ions is enhanced while that of K^+ ions is reduced. Since the activity of (Na^+, K^+) ATPase enzyme is stimulated by Na^+ ions, at inner surface of the cell membrane and by K^+ ions, at the outer surface and inhibited by Ca^{2+} ions [95], reduced access of K^+ ions and enhanced access of Ca^{2+} ions to the enzyme sites due to the liquid membrane, likely to be formed by digitalis at the external side of the cell membrane, in association with membrane lipids at its site of action may be an important factor leading to the inhibition of the (Na^+, K^+) ATPase enzyme. The enhanced permeability of Na^+, ions (Table 16) in the specific orientation 1, in the presence of liquid membranes however, would not be affective in activating the enzyme because it is only the intracellular Na^+ ions, which can activate the enzyme.

It has also been suggested [96] that inotropic action of cardiac glycosides may also depend upon the level of catecholamines, particularly noradrenaline in the heart tissues. The present observation on the increased permeability of catecholamines, particularly noradrenaline in the presence of the liquid membranes (Table 16) appear relevant to this suggestion.

Of the three drugs presently studied, ouabain is found to be the most potent [91-97] and digitoxin is the least potent. This gradation in the potencies is consistent with the present observation of the concentrations of the digitalis required to saturate the lecithin-cholesterol-liquid membrane at the interface (Table 15): digitoxin > digoxin > ouabain. Since modification in the transport of relevant permeants, which affect the biological effect, is maximum when the lecithin-cholesterol liquid membrane is completely saturated with digitalis, it would appear that the concentration required to produce maximum biological effect would be the lowest for ouabain and the highest for digitoxin making ouabain most potent and digitoxin least.

The gradation observed in CMC values (Table 1) of the drugs (digitoxin > digoxin > ouabain) is the reverse of the gradation reported in their potencies [91-97]. This observation is consistent with the conclusion drawn from the liquid membrane hypothesis for drug action [64] that lower the CMC more potent is the drug.

Modification in the transport of dopamine, noradrenaline, adrenaline and serotonin in the presence of the liquid membrane generated by the lecithin-cholestarol-digitalis mixture in the specific orientation 1 also appear relevant to some of the other reported biological effects of digitalis.

Cardiac glycoside, particularly ouabain, have been used to produce experimental dysrhythmias [98]. It is also documented that β-adrenoreceptor blocking agents like propranolol are useful in the treatment of digitalis induced dysrhythmias [99]. Evidence from animal experiments indicates that serotonin containing systems in the hypothalamus; amygdala and colliculi may be sites of action of cardiac glycosides in increasing sympathetic discharge [99]. These observations appear consistent with the enhanced permeability of adrenaline, noradrenaline, and serotonin in the presence of the liquid membranes generated by digitalis in association with lecithin and cholesterol in the specific orientation 1 (Table 16).

It has been suggested that digitalis may block dopamine receptors in brain and that this blockade may also contribute to the increase in sympathetic outflow produced by digitalis [100]. Since for actions of dopamine, hydrophilic portions of dopamine receptors have been considered important [101, 102], the liquid membrane formed by digitalis at the dopamine receptors would present its hydrophobic surface to the permeant dopamine. The reduced permeability of dopamine in the specific orientation 2 (Table 16), therefore, appears consistent with this suggestion.

It has been reported [100] that administration of digitalis causes several behavioral changes in mice due to the blocking of CNS dopamine receptors. The reduced permeability of dopamine in the specific orientation 2 (Table 16) appears consistent with this observation.

Prolonged treatment with a cardiac glycoside may produce endocrine disorders like gynaecomastia [103]. These effects, which are ultimately linked with the reduced concentration of dopamine in hypothalamic region, may also be ascribed to the reduced permeability of dopamine in the presence of the liquid membrane in the specific orientation 2 (Table 16).

Cardiac glycosides are known [104] to produce nausea and vomiting as side effects by acting on chemoreceptor trigger zone (CTZ). This action is mediated by dopamine. The enhanced permeability of dopamine in the presence of the liquid membrane in the specific orientation 2 (Table16) as observed in this study could be plausible explanation for the causation of nausea and vomiting by digitalis.

Thus it appears from the above discussion that the liquid membrane phenomenon is also likely to make a significant contribution to the biological actions of digitalis.

6.2.5. Local anaesthetics [6]

Four local anaesthetic drugs namely procaine, tetracaine, lidocaine and dibucaine, as hydrochloride salts, have been investigated [6] to unfold the role of liquid membrane phenomenon in the mechanism of their action. Most of the useful local anaesthetics contain both a hydrophilic and hydrophobic part in their structure [105] and hence are surface active in nature. They act by modifying the permeabilities of nerve cell membranes to sodium and potassium ions. Ionic surfactants are reported to impede ion transfer across interface and this inhibition is ascribed to the formation of a lipid-like layer at the interface [106]. Existence of a liquid membrane generated by each of these drugs at the interface has been demonstrated. Data on the transport of sodium and potassium ions through the liquid membranes generated by these drugs in series with supporting membrane have been obtained to gain information on the contribution of the liquid membrane in the action of the drugs. Since local anesthetics are known to interact with membrane lipids [107] the studies have been extended to the liquid membranes generated by lecithin-cholesterol-local anaesthetic drug mixtures.

Table 16. Solute permeability (ω)[a] of various permeants in the presence of liquid membranes generated by digitoxin[b] (ω_1), digoxin[c] (ω_2) and ouabain[d] (ω_3) in lecithin-cholesterol mixture of fixed composition (1.919×10^{-5} M with respect to lecithin and 1.175×10^{-6} M with respect to cholesterol) along with the control values (ω_0) when no drug (digitalis) was used (Ref. 30).

Permeants	Initial conc.(mg lit^{-1})	$\omega_0\times10^9$ (mole s^{-1} N^{-1})	$\omega_1 \times 10^9$ (mole s^{-1} N^{-1})		$\omega_2 \times 10^9$ (mole s^{-1} N^{-1})		$\omega_3 \times 10^9$ (mole s^{-1} N^{-1})	
			Orientation 1	Orientation 2	Orientation 1	Orientation 2	Orientation 1	Orientation 2
Dopamine HC1	10	0.382	0.4567	0.313	0.554	0.336	0.594	0.324
		±0.062	±0.014	±0.029	±0.016	±0.031	±0.022	±0.018
Adrenaline	10	0.507	0.709	0.675	0.624	0.589	0.822	0.662
		±0.027	±0.041	±0.037	±0.023	±0.035	±0.015	±0.008
Noradrenaline	10	0.774	0.928	0.875	0.896	1.372	1.172	1.124
		±0.040	±0.071	±0.011	±0.027	±0.076	±0.038	±0.027
Serotonin creatinine sulphate	10	0.193	0.352	0.253	0.397	0.421	0.652	0.748
		±0.017	±0.024	±0.032	±0.011	±0.020	±0.013	±0.017
Sodium (chloride)	5.382	0.169	0.210	0.136	0.245	0.118	0.268	0.103
		±0.021	±0.019	±0.026	±0.016	±0.031	±0.029	±0.015
Potassium (chloride)	10.43	0.156	0.076	0.182	0.093	0.163	0.104	0.188
		±0.018	±0.002	±0.011	±0.005	±0.015	±0.009	±0.026
Calcium (chloride)	0.22	0.128	0.142	0.079	0.162	0.106	0.191	0.111
		±0.018	±0.031	±0.005	±0.026	±0.013	±0.007	±0.014

a, values of ω reported as arithmetic means of 15 repeats ± SD
b, digitoxin concentration 6×10^{-9}M
c, digoxin concentration 3.5×10^{-9}M
d, ouabain concentration 1.5×10^{-9}M

As explained in earlier sections; section 6.1, in these experiments also a cellulose nitrate microfiltration membrane (Sartorius Cat no. 11307)/ aqueous solution interface was chosen as site for the formation of liquid membranes so that the possibility of active interaction of the drugs with biomembranes [108] as a cause for the modification of permeability is totally ruled out and the contribution of passive transport through the liquid membranes is highlighted.

The data on hydraulic permeability, which were utilized to demonstrate the formation liquid membrane in series with the supporting membrane, were obtained at concentration ranges chosen were such that the data are obtained above and below the CMC of the drugs. The data at all concentration ranges studied were found to the obeyed by the relationship $J_v=L_p\Delta p$. The decreasing trend in the values of the hydraulic conductivity coefficient L_p with the increase in the concentration, upto the CMC of the drugs in the case of all four local anaesthetic drugs, was in accordance with Kesting's hypothesis [32] and demonstrated the formation of complete liquid membrane in series with the supporting membrane at the CMC of the drug. The values of L_p computed at concentrations below the CMC of the drug using mosaic model were also found to the in agreement with the experimentally determined values which lent further support to the phenomenon of liquid membrane formation. The hydraulic permeability data in the partucular case of tetracaine hydrochloride are shown in Table 17 as an example.

Table 17. Values of L_p ($m^3\ sec^{-1}N^{-1}$) at various concentration of tetracaine hydrochloride (Ref. 6).

	Concentration of Tetracaine hydrochloride ($\times 10^6$ M)				
	0	3.053	6.106	12.212	24.424
$L_p \times 10^8$	5.021 ±0.077	4.615 ±0.116	4.284 ±0.144	3.560 ±0.230	3.550 ±0.290
$L_p^a \times 10^8$	-	4.655 ±0.114	4.291 ±0.154	-	-

[a] Computed values using mosaic model.

Since local anaesthetics are known [107] to interact with membrane lipids, incorporation of local anaesthetics in the liquid membrane generated by the lecithin-cholesterol mixture was demonstrated. For this also hydraulic permeability measurement were carried out at various concentrations of local anesthetic drugs prepared in an aqueous solution of lecithin-cholesterol mixture which was 15.542 ppm with respect to lecithin and 1.175×10^{-6} M with respect to cholesterol. The solution of various concentrations of the drugs prepared in the aqueous solution of lecithin-cholesterol mixture of the fixed composition were placed in compartment C of the transport cell (Fig. 2 Chapter 5) and compartment D was filled with water. This particular composition of lecithin-cholesterol mixture was chosen because it has been shown [90] that at this composition the liquid membrane generated by lecithin is saturated with cholesterol and completely covers the supporting membrane.

The hydraulic permeability data at various concentrations of local anaesthetic drug in presence of lecithin-cholesterol mixtures was also found to be in accordance with the equation $J_v = L_p \Delta P$. The values of L_p at various concentrations of the drugs in the presence of lecithin-cholesterol mixture decrease with the increase in concentration of the drug which indicative of incorporation of the drugs in the liquid membrane generated by the lecithin-cholesterol mixture. The concentration of the drug beyond which the values of L_p become more or less constant is the concentration at which the liquid membrane generated by the lecithin-cholesterol mixture is saturated with the drug. The data in the particular case of tetracaine hydrochloride are recorded in Table 18. Similar trends were obtained in the case of other local anaesthetic drugs namely, procaine dibucaine and lidocaine.

In order to ascertain the location of the drugs in the lecithin-cholesterol liquid membrane, surface tensions of solutions of various concentrations of drugs prepared in the aqueous solution of lecithin-cholesterol mixture of fixed composition, 15.542 ppm with respect to lecithin and 1.175×10^{-6} M with respect to cholesterol, were measured. The surface tension of the aqueous solution of lecithin-cholesterol mixture showed further decrease with increase in concentration of the local anaesthetics. This suggests that the drugs penetrate the liquid membrane generated by the lecithin-cholesterol mixture and reach the interface.

Table 18. Values of L_p at various concentrations of tetracaine hydrocholoride in lecithin-cholesterol-tetracaine hydrochloride mixtures (Ref. 6).

	Tetracaine hydrochloride ($\times 10^6$ M)					
	0.0	3.053	6.106	9.059	12.212	15.265
$L_p \times 10^8$	0.9117	0.8789	0.8120	0.7556	0.7067	0.7125
(m^3 sec^{-1} N^{-1})	±0.0169	±0.0093	±0.0330	±0.0300	±0.0320	±0.0310

Note. Lecithin and cholesterol concentrations kept constant at 15.542 ppm and 1.175×10^{-6} M, respectively.

For the measurement of solute permeability (ω) of sodium and potassium ions, two sets of experiments were performed. In the first set of experiments the compartment C of the transport cell (Fig. 2 Chapter 5) was filled with a solution of the electrolyte (Sodium chloride or potassium chloride) of known concentration prepared in the aqueous solution of known concentration of the local anaesthetics and compartment D was filled with distilled water. In the second set of experiments aqueous solution of known concentration of the permeants (sodium chloride or potassium chloride) were placed in compartment C and aqueous solution of known concentrations of the drug in compartment D of the transport cell. In control experiments, however, no drug was used. Concentration of the drugs in these experiments was always higher than their CMC to ensure complete coverage of the supporting membrane with the liquid membrane generated by the drugs. Because of surface activity, the hydrophobic portion of the drug molecules will be preferentially oriented toward the hydrophobic supporting membrane and the hydrophilic portion of the drug molecules will face outward away from it. Thus, in the first set of experiments where both the permeants and the drug are present in the same compartment, the permeants would face the hydrophilic surface of the liquid membrane generated by the drug while in the second set of experiments the permeants would face the hydrophobic surface of the drug-liquid membrane. Initial concentrations of sodium and potassium ions in the solute permeability experiments were chosen so as to correspond to the concentrations of the ions in the neighborhood of nerve cells.

Values of ω for sodium and potassium ions were also estimated in presence of lecithin-cholesterol-local anaesthetic drug mixtures. Since the transport of cations was observed to be impeded more when the permeants face the hydrophobic surface of the liquid membrane, the values of ω in presence of lecithin-cholesterol-drug mixtures were also measured in the second set of experiments. For this the solution of local anaesthetic drugs prepared in the aqueous solution of lecithin-cholesterol mixtures of composition 15.542 ppm with respect to lecithin and 1.175×10^{-6} M with respect of cholesterol was filled in the compartment D of the transport cell and aqueous solution of the permeants (sodium or potassium ions) were placed in compartment C. The concentrations of the local anaesthetic drugs in these experiments were those at which the liquid membrane generated by lecithin-cholesterol mixture becomes saturated by the drugs. These concentrations were derived from the hydraulic permeability data in the presence of lecithin-cholesterol-drug mixtures. All measurements were made at constant temperature using a thermostat set at 40 ± 0.1°C.

First indication of the role of liquid membrane in local anaesthesic came from the values of critical micelle concentration (CMC) of the local anaesthetic drugs. The critical micelle concentrations of the local anaesthetic drugs as determined in the present experiments are more or less equal to their minimum blocking concentrations (MBC) (Table 19). Since, according to Kesting's hypothesis [32] the CMC is the concentration at which the interface is completely covered with the liquid membrane, it appears, *prima facie*, that the liquid membranes generated by the local anaesthetic drugs at the site of their action have a role to play in the local anaesthetic action. This is further supported by the gradation in the binding of local anaesthetics to nerves and other tissues [109]. The binding affinities of these drugs increase in the following order: procaine < lidocaine < tetracaine < dibucaine. The fact that the CMC values of these drugs are in the reverse order, i.e. dibucaine < tetracaine < lidocaine < procaine (Table 19) is also indicative of the role of liquid membrane phenomena in their action. The higher the value of CMC, the higher the concentration required to generate a complete liquid membrane at the site of action. As reported by Skou [107] relative anaesthetic potencies of procaine, lidocaine, and tetracaine are in the following order: tetracaine > lidocaine > procaine. The above gradation is in agreement with the descending order of the CMC value of these drugs (Table 19)- the lower the CMC value more potent is the drug. This again indicates the role of liquid membrane phenomena in local anaesthesia.

The local anaesthetic action is linked with the inhibition of the transport of the sodium and potassium ions. In order to assess the inhibition of solute permeability in presence of drugs, let us define a parameter 'r' given by

$$r = \frac{\omega_0 - \omega_1}{\omega_0}$$

where ω_0 is the value of solute (cation) permeability in control experiments where no drug was used and ω_1 is the value in the presence of the drugs. The value of r should lie between 0 and 1 – the former indicating no inhibition and the latter indicating complete inhibition. The values of r for sodium and potassium ions in presence of the local anaesthetic drugs and also in presence of lecithin-cholesterol-drug mixture are recorded in Table 20. A perusal of Table 20 reveals that the permeabilities of sodium and potassium ions are inhibited in both the sets of experiments. However, the inhibition is maximum in the second set of experiments where the permeating cations face the hydrophobic surface of the liquid membrane generated by the local anaesthetic drugs alone.

Table 19. Critical micelle concentration (CMC) and minimum blocking concentration (MBC) of local anaesthetics.

	Tetracaine hydrochloride (μM)	Dibucaine hydrochloride (μM)	Lidocaine hydrochloride (mM)	Procaine hydrochloride (mM)
CMC	12.212	5.640	1.160	5.000
MBC	10.000[a]	5.000[a]	1.170[b]	4.460[a]

[a] Ref.110

[b] Calculated on the basis of anaesthetic potency (Ref. 107)

Table 20. Values of r for sodium and potassium ions in presence of local anaesthetic drugs and lecithin-cholesterol-local anaesthetic drug mixtures (Ref. 6).

	Sodium			Potassium		
	r^a	r^b	r^c	r^a	r^b	r^c
Dibucaine hydrochloride	0.150	0.217	0.154	0.352	0.496	0.202
Tetracaine hydrochloride	0.212	0.395	0.081	0.319	0.367	0.235
Lidocaine hydrochloride	0.0395	0.062	0.008	0.300	0.553	0.150
Procaine hydrochloride	0.042	0.065	0.048	0.103	0.316	0.049

[a] Both permeants and the drugs in the compartment C and water in the compartment D (Fig. 2 Chapter 5). [b] Aqueous solution of the drugs in the compartment D and aqueous solution of permeants in the compartment C. [c] Lecithin-cholesterol-drug mixture in the compartment D and permeants in the compartment C.

Local anaesthetic drugs are known [111] to reduce the permeability of a resting nerve to potassium as well as to sodium ions. They also reduce permeability to sodium and potassium ions in the membranes of muscle both in the resting state and during the generation of an action potential [112]. The net effect depends, however, on the extent to which cationic and non-ionic forms of the local anaesthetics are available. In the case of procaine, it is shown [113] that while the uncharged form causes a decrease in resting sodium conductance, the charge form decreases the resting potassium conductance. Thus the reduced permeability of both sodium and potassium ions as observed in these experiments (Table 20) is in confirmance with the literature reports related to biological cells.

Ability of local anaesthetics to block nerve impulse conduction has been shown to correlate with their interaction with the lipid monolayer [110]. Both the ability to interact with the lipid films and the ability to block nerve conduction are reported to be pH-dependent [107]. It has also been observed, particularly in the case of procaine, that the drug penetrates only into the ionic region of the lipid monolayers [114]. The action of local anaesthetics is known to be short-lived and reversible in nature. This may be because the interaction of local anaesthetics with cell membranes is limited only to the ionic surface of the membranes [114,115].

In the model proposed by Lee [116] for action of local anaesthetics, sodium channels are postulated to be surrounded by an annulus of lipid which is in the crystalline or gel state. It is the rigidity of the surrounding lipid microenvironment that keeps the sodium channel open. Addition of local anaesthetics triggers a change in the surrounding lipids to the fluid-liquid crystalline phase, allowing the sodium channel to close with resulting local anaesthesia. The drop in phase transition temperature due to the addition of local anaesthetics is cited [117] as an evidence for fluidization. Since in the lowering of phase transition temperatures, head group interactions are known [118] to play an important role, it appears that polar head of local anaesthetic molecules interact from within the channel, with polar heads of lipids surrounding the channel and fluidize them. As a consequence of loosening of the lipid microenvironment the channel is filled up with a hydrophobic core consisting mainly of hydrophobic moieties of local anaesthetic molecules impeding the transport of sodium ions.

In these experiments, however, resistance offered to the transport of cations in presence of lecithin-cholesterol-local anaesthetic drug mixture is less than that in presence of the local anaesthetic drugs alone (Table 20) in the second set of experiments. This is because the local anaesthetic drug molecules are incorporated within the liquid membrane generated by lecithin-cholesterol mixtures and their hydrophobic moieties are less available to approaching cations to impede their transport. Since the planar configuration of lipid films is known to be more stable than the circular configuration [119] the expected fluidization leading to hindrance of the transport of sodium ions owing to the hydrophobic core of local anaesthetic drug molecules, as in the case of lipid microenvironment of sodium channels is less likely in these experiments [6].

Thus the liquid membranes generated by local anaesthetic drugs in the specific orientation of the drug molecules with hydrophobic ends facing the cations seem to make a significant contribution to the mechanism of their action.

6.2.6 Antiarrythmic Drugs [13]

Antiarrhythmic drugs are known to contain both hydrophobic and hydrophilic moieties in their structure [120] and hence are expected to be surface-active in nature. These drugs are known to cause increase in membrane surface pressure and stabilization of membranes.

The antiarrhythmic action is known to be [121] mainly on account of modification in the permeability of biomembranes to sodium ions. Since antiarrhythmic drugs are expected to be surface-active and hence capable of generating liquid membranes at the interface, it is logical to suspect that modification in the permeability of sodium ions may be on account of the liquid membranes generated by them at the respective sites of action. It is of interest to mention that many local anesthetics also show antiarrhythmic action. Since, in local anesthetics, the liquid membranes generated by them have been shown to contribute to the modification in cation permeability, it appears likely that the phenomenon of liquid membrane formation may also be important in the mode of action of antiarrhythmic drugs. This precisely was the point of investigation in this study [13].

Four structurally dissimilar drugs, viz. quinidine hydrochloride, disopyramide phosphate, procainamide hydrochloride and propranolol hydrochloride, were chosen for the study. The first three drugs belong to the class I and propranolol to the class II in Vaughan William classification [121]. Existence of liquid membranes generated by each of these drugs at the interface has been demonstrated. Data on the permeability of sodium ions in presence of the liquid membranes have been obtained to gain in formation on the role of the liquid membranes in antiarrhythmic action. As explained earlier in these experiments also, a sartorious cellulose nitrate/ aqueous solution interface was deliberately chosen as site for the formation of liquid membranes so that the role of passive transport through the drug liquid membranes is highlighted.

The critical micelle concentrations (CMC) of aqueous solutions of the drugs were determined from the variation of surface tension with concentrations. These are recorded in Table 1. The hydraulic permeability data at various concentrations of antiarrhythmic drugs, were utilized to demonstrate the existence of liquid membrane in series with the supporting membrane. The concentration ranges selected were such that data are obtained above and below the CMC of the drugs.

The hydraulic permeability data at various concentrations of the drugs, in case of all the four antiarrhythmic drugs, were found to obey the linear relationship, $J_v=L_p.\Delta P$. The normalized values of the hydraulic conductivity coefficient, the values of L_p/L_p^0 where L_p^0 is the value of L_p when no drug was used-estimated from the J_v versus ΔP plots – are plotted against the concentrations of the drugs in the values of L_p/L_p^0 in case of all the four drugs show a progressive decrease with increase in concentration of the drugs up to their CMC beyond which either they become more or less constant. This trend is in accordance with the liquid membrane hypothesis [32] according to which as concentration of the surfactant is increased, the supporting membrane gets progressively covered with the surfactant layer liquid membrane until it is completely covered at the CMC.

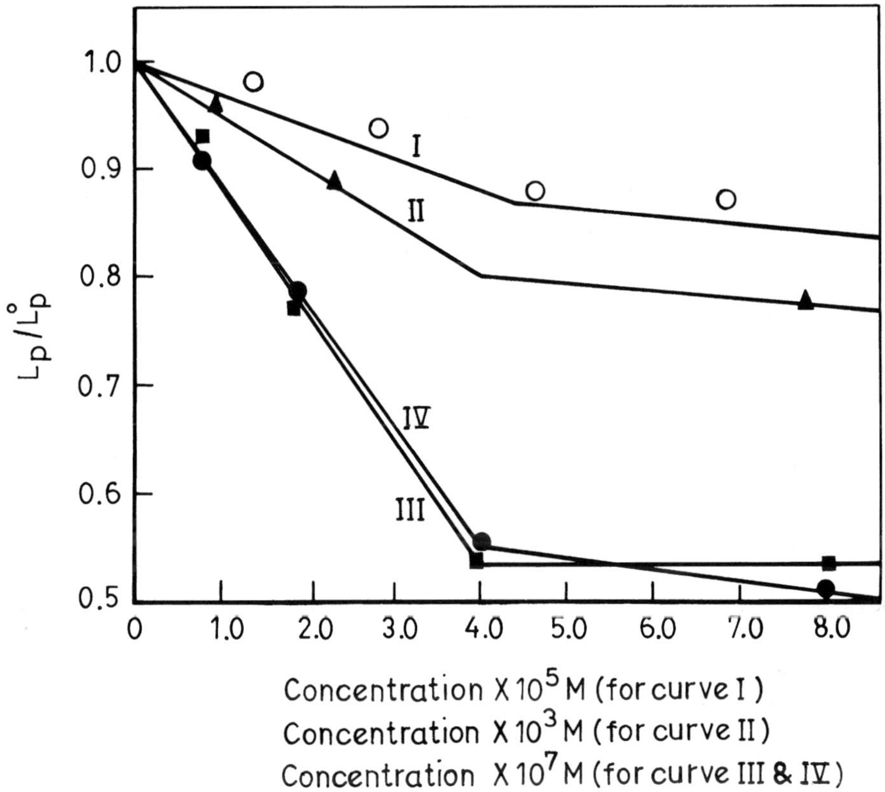

Fig. 4 Variation of L_p / L_p^0 with concentration of the drugs. Curves II, III, IV represent data for propranolol hydrochloride, procainamide hydrochloride, dysopyramide phosphate and quinidine hydrochloride and I respectively (Ref. 13).

Analysis of the flow data in the light of mosaic membrane model [43-45] further corroborated the existence of liquid membrane in series with the supporting membrane. For the measurement of solute permeability (ω) of sodium (Na^+), two sets of experiments were performed; one in which the drug liquid membrane presented its hydrophilic surface to the permeant and the other in which it presented its hydrophobic surface. For details of the experiment the original paper [13] and the section 6.1 of this chapter may be consulted. The values of ω for sodium (Na^+) ions in the presence of antiarrhythmic drugs obtained in the two sets of experiments are recorded in Table 21, along with the values in the control experiments where no drug was used.

Table 21. Permeability of sodium (Na^+)[a], (ω)[b]. in presence of antiarrhythmic drugs[c] (Ref.13).

Drug	$\omega_1 \times 10^{10}$ (mol. s^{-1}. N^{-1})	$\omega_2 \times 10^{10}$ (mol. s^{-1}. N^{-1})	$\omega_3 \times 10^{10}$ (mol. s^{-1}. N^{-1})
Quinidine hydrochloride	5.3255 ± 0.3521	2.8934 ± 0.1710	2.8757 ± 0.1405
Disopyramide phosphate	5.3255 ± 0.3521	3.0709 ± 0.1731	3.3823 ± 0.1540
Procainamide hydrochloride	5.3255 ± 0.3521	3.1132 ± 0.2108	3.1778 ± 0.2475
Propranolol hydrochloride	5.3255 ± 0.3521	2.5844 ± 0.1293	2.5885 ± 0.1542

ω_1 = control value-when no drug was used. ω_2 drug and sodium ion in compartment C and water in the compartment D: the first set of experiments

ω_3 = drug in compartment D and sodium ion in the compartment C.: the second set of experiments.

[a]Initial concentration of sodium ion 2117.2 ppm.

[b]Values of ω are reported as arithmetic mean of 10 repeats ± S.D.

[c]The concentrations of quinidine hydrochloride, disopyramide phosphate, procainamide hydrochloride and propranolol hydrochloride used were: 1.6×10^{-6} M, 1.6×10^{-6} M, 8.0×10^{-3} M and 7.315×10^{-5} M respectively.

A perusal of Table 21 indicates that the liquid membranes generated by the antiarrhythmic drugs, in both the orientations-hydrophilic ends facing the permeant and hydrophobic ends facing the permeant-impede the transport of sodium ions. Antiarrhythmic drugs are known [121] to stabilize cardiac membrane by a non-specific mechanism. This study indicates that the liquid membranes generated by antiarrhythmic drugs in series with the cardiac membrane impeding the transport of sodium ions may be such a mechanism. A persual of Table 21 further reveals that impediment in the transport of sodium ions is not significantly different in the two orientations of the liquid membranes generated by the drugs which implies that both hydrophilic and hydrophobic moieties in the structure of these drugs may be necessary for antiarrhythmic action. This conjecture is consistent with the literature report [120] that non-specific antiarrhythmic agents interact with hydrophilic and hydrophobic regions of the biomembrane. Propranolol which is primarily a β-blocker drug is also known [120] to exert a non-specific membrane-stabilizing action similar to that of quinidine at concentrations higher than those needed for β-blocking action. It is for this reason, that the transport of sodium ions in presence of propranolol was studied. The data on the inhibition of sodium ion transport in presence of propranolol (Table 21) is consistent with its reported antiarrhythmic action.

6.2.7 Barbiturates [24]

Barbiturates are known to be surface active [122-124] and hence should be capable of generating liquid membranes at the interface in accordance with Kesting's liquid membrane hypothesis [32]. There are several instances where the role of surface activity in the biological actions of barbiturates has been indicated [125-127]. In these studies the formation of liquid membranes in series with a supporting membrane, either by barbiturates alone or by barbiturates in association with membrane lipids (lecithin and cholesterol), has been demonstrated. For this, data on the hydraulic permeability in the presence of lecithin-cholesterol-barbiturate mixtures have been utilized. Data on modification in the transport of the relevant permeants, namely γ-aminobutyric acid (GABA), glycine, aspartic acid, serotonin and noradrenaline, in the presence of the liquid membranes generated by a lecithin-cholesterol-barbiturate mixture have been obtained and discussed in the light of the reported biological actions of barbiturates.

Sodium Phenobarbital and sodium pentobarbital were chosen for the study. The critical micelle concentrations (CMC) of aqueous sodium phenobarbital and sodium pentobarbital as determined by the variation of surface tension with concentration at 37°C were found to the 7.5×10^{-5} M and 5.0×10^{-5}M respectively.

The hydraulic permeability data at various concentrations in the case of both barbiturates (sodium phenobarbital and sodium pentobarbital) were found to be in accordance with the proportional relationship, $J_v = L_p\, \Delta P$. The values of L_p at various concentrations of the drugs, estimated from the J_v versus ΔP plots, are recorded in Table 22. The values of L_p decrease with increasing concentration of the drugs up to their CMC, beyond which they become more or less constant. This trend in the values of L_p is consistent with Kesting's liquid membrane hypothesis and is indicative of the formation of liquid membranes by the drugs, in series with the supporting membrane. Analysis of the values of L_p in the light of mosaic model [43-45] lends further support to the information of the drug liquid membrane in series with the supporting membrane. The values of L_p thus computed at several concentrations of the drugs below their CMC match with experimentally determined values (Table22), lending additional support to the formation of the liquid membranes.

Table 22. Values of the hydraulic conductivity coefficient $(L_p)^a$ at various concentrations of barbiturates (Ref. 24).

	Sodium phenobarbital concentration 10^5 (M)						
	0.0	1.875	3.75	5.625	7.5^d	15.0	30.0
$L^b \cdot 10^9$ ($m^3\ s^{-1}.\ N^{-1}$)	14.360 ±0.044	12.645 ±0.147	10.700 ±0.217	9.100 ±0.029	7.391 ±0.002	7.396 ±0.037	7.439 ±0.206
$L^c \cdot 10^9$ ($m^3\ s^{-1}.\ N^{-1}$)	-	12.618 ±0.034	10.876 ±0.023	9.133 ±0.013	-	-	-
	Sodium phenobarbital concentration 10^5 (M)						
	0.0	1.25	2.5	3.75	5.0^d	10.0	15.0
$L^b \cdot 10^9$ ($m^3\ s^{-1}.\ N^{-1}$)	14.360 ±0.044	12.791 ±0.147	10.929 ±0.179	9.457 ±0.127	7.767 ±0.063	7.746 ±0.015	7.810 ±0.012
$L^c \cdot 10^9$ ($m^3\ s^{-1}.\ N^{-1}$)	-	12.712 ±0.049	11.064 ±0.054	9.415 ±0.058	-	-	-

[a] The values of L_p reported as arithmetic means of 10 repeats ± S.D.

[b] Experiment values

[c] Computed values using mosaic model.

[d] Critical micelle concentration.

Evidence in favour of the incorporation of barbiturates in the liquid membrane generated by lecithin-cholesterol mixture can be obtained from the hydraulic permeability data at varying concentrations of these drugs in the lecithin-cholesterol mixtures of fixed composition. The hydraulic permeability data in this case were found to be represented by the equation $J_v = L_p \Delta P$. It was observed that as the concentration of drug is increased, the values of L_p first decrease and then become more or less constant (Table 23). The concentrations of drug beyond which the values of L_p become more or less constant can be taken to be the concentration at which the lecithin liquid membrane at the interface (which is already saturated with cholesterol) is also saturated with the drug Table 23.

Thus, the concentrations of sodium phenobarbital and sodium pentobarbital required to saturate the lecithin-cholesterol liquid membrane are 6.0×10^{-5}M and 2.0×10^{-5} respectively. The concentrations of sodium phenobarbital and sodium pentobarbital compare favourably with their reported [128, 129] plasma concentrations, at least in order of magnitude. The plasma concentration of sodium phenobarbital is in the range 0.2-0.5 m [128] and that of sodium pentobarbital ranges fro m 4.2 t o11 μM [129].

In view of these studies, the concentrations of barbiturates in the lecithin-cholesterol mixture of fixed composition used in the solute permeability experiments were either equal to or a little higher than the concentrations required to saturate the lecithin-cholesterol liquid membrane.

Table 23. Values of the hydraulic conductivity coefficient $(L_p)^a$ at various concentrations of sodium phenobarbital and sodium pentobarbital in the presence of lecithin-cholesterol mixture of fixed composition (Ref. 24).

	Sodium phenobarbital concentration $\times10^5$ (M)						
	0.0	1.5	3.0	4.5	6.0	7.5	9.0
$L_p \cdot 10^9$ ($m^3 s^{-1}. N^{-1}$)	12.226 ±0.075	10.036 ±0.022	8.195 ±0.207	6.913 ±0.142	5.155 ±0.111	5.194 ±0.085	5.201 ±0.108
	Sodium phenobarbital concentration $\times$ 10^5 (M)						
	0.0	1.0	2.0	3.0	4.0		
$L_p \cdot 10^9$ ($m^3 s^{-1}. N^{-1}$)	12.226 ±0.075	11.436 ±0.007	8.390 ±0.087	8.247 ±0.146	8.417 ±0.068		

[a] The values of L_p are reported as arithmetic means of 10 repeats ± S.D.

For solute permeability (ω) measurements, the composition of the lecithin-cholesterol-barbiturate mixture chosen was the one at which the liquid membrane generated by lecithin completely covered the supporting membrane and was saturated with both cholesterol and the barbiturate under investigation. This composition was derived from earlier studies [90, 130] and from the data on hydraulic permeability in the presence of the varying concentrations of barbiturates in the mixture of lecithin and cholesterol of fixed composition, i.e. 15.542 ppm with respect to lecithin and 1.175×10^{-6}M with respect to cholesterol. This particular composition of the lecithin-cholesterol mixture was chosen because it was shown in an earlier study [90] that at this composition the liquid membrane generated by lecithin at the interface is completely saturated with cholesterol. For the measurement of solute permeability, ω, one compartment of the transport cell was filled with the aqueous solution of the lecithin-cholesterol-barbiturate mixture (Fig. 2 Chapter 5), along with permeant, and the other compartment was filled with distilled water. The condition $J_v=0$ was imposed on the system and the amount of permeant transported to be compartment filled with distilled water in a known period of time was estimated. For details of the method of measurement of solute permeability, ω, original paper may be referred to [24]. All measurements were performed at constant temperature, using a thermostat setting of 37±0.1°C.

Table 24. Solute permeability (ω)[a] of various permeants in the presence of liquid membranes generated by sodium phenobarbital[b] (ω_1) and sodium pentobarbital[c] (ω_2) in lecithin-cholesterol mixture of fixed composition along with the control values (ω_0) when no barbiturates were used (Ref. 24).

Permeants	Initial Concentration (10^{-3} mol l^{-1})	$\omega_0 \cdot 10^9$ (mol s^{-1} N^{-1})	$\omega_1 \cdot 10^9$ (mol s^{-1} N^{-1})	$\omega_2 \cdot 10^9$ (mol $s^{-1}N^{-1}$)
γ-Aminobutyric acid (GABA)	1.940	0.284 ±0.014(6.6)	0.719 ±0.018(6.8)	0.855 ±0.006(6.8)
Glycine	1.333	1.077 ±0.021(7.05)	1.402 ±0.011(7.0)	1.674 ±0.017(6.9)
Serotonin creatinine Sulphate	0.0247	0.219 ±0.008(6.9)	0.652 ±0.013(6.5)	0.748 ±0.011(6.4)
Aspartic acid	1.127	0.269 ±0.014 (4.2)	0.136(4.0) ±0.015	0.192 ±0.017 (4.0)
Noradrenaline	0.059	0.752 ±0.020(7.1)	0.135 ±0.011(6.5)	0.595 ±0.003(6.2)

[a] Values of ω reported as arithmetic mean of 15 repeats ± S.D. The figures within parentheses indicate pH of the permeant solution in the lecithin-cholesterol-barbiturate mixture.

[b] Sodium Phenobarbital concentration 6.0 x 10^{-5}M

[c] Sodium pentobarbital concentration 2.0 x 10^{-5}M.

The solute permeability data recorded in Table 24 appear relevant to the biological actions of the barbiturates. Electrophysiological studies have indicated that sedative barbiturates inhibit excitatory transmission and enhance inhibitory transmission [131]. This is consistent with the enhanced permeability of GABA and the reduced permeability of aspartic acid, as observed in these experiments (Table 24). Electrophysiological evidence has also indicated [131] that GABA has a major role in barbiturate actions. Receptor binding studies have, however, failed to detect any interactions between GABA and barbiturates [132]. On this basis it has been concluded [131] that barbiturates do not affect the post-synaptic binding of GABA, even though GABA mimetic actions have been observed electrophysiologically. These studies appear to offer an explanation for these observations. The data on enhancement in the transport of GABA (Table24) suggest that access to a GABA receptor is likely to be facilitated by the presence of the liquid membrane generated by the barbiturates in association with membrane lipids at the receptor site.

The anticonvulsant activity of phenobarbitone, which has been used in the treatment of epilepsy, is ascribed to its ability to produce an increased concentration of GABA in the brain. Phenobarbitone is reported to be most effective when the brain GABA content has been depleted [133]. The enhancement in the permeability of GABA in the presence of the lecithin-cholesterol-sodium phenobarbital liquid membrane (Table 24) is consistent with these clinical observations.

The two barbitals presently studied are reported to have the following gradation in onset of action [134]: sodium pentobarbital > sodium Phenobarbital. This gradation in onset of activity of the barbitals is consistent with the observation on the concentration of the barbitals required to saturate the lecithin-cholesterol liquid membrane (Table 23): sodium phenobarbital > sodium pentobarbital. Thus sodium pentobarbital, which crosses the blood brain barrier the fastest [134], is required at the lower concentration to saturate the lecithin-cholesterol liquid membrane. Since modification in the permeability of the biological membrane would be maximum when the lipid bilayer is saturated with barbiturate, leading to maximum biological effect, the gradation in the onset of biological action appears to be a consequence of both factors, i.e. how fast it crosses the blood-brain barrier and how small is the concentration of drug required to saturate the lipid bilayer. The gradation in the onset of barbiturate action is also consistent with the conclusion that the CMC is a good indicator of the potency of surface-active drugs-the lower the CMC the more potent is the drug [64]. The CMC of sodium phenobarbital is higher than that of sodium pentobarbital.

Barbiturates are known to disturb the balance of the phases of sleep-the initial effect is that of reducing the proportion of REM (rapid eyeball movement) sleep in comparison to NREM (non-rapid eyeball movement) sleep [135]. This observation can also be explained in terms of the enhanced permeability of serotonin and reduced permeability of noradrenaline in the presence of the liquid membrane generated by the lecithin-cholesterol-barbiturate mixture (Table 24). It is documented that *raphe nuclei*, which are rich in serotonin, are responsible both for NREM sleep and for the transition to and onset of REM sleep. When a system of neurons in the pons known as the *locus ceruleus* (rich in noradrenaline) is destroyed, animals previously deprived of REM sleep fail to take the usual rebound excess of REM sleep when undisturbed [136]. The data (Table 24) indicate that the liquid membranes likely to be formed in the synaptic cleft by the barbiturates in association with the membrane lipids may enhance the access of serotonin to its site of action in the *locus ceruleus*, which may also contribute to the causation of imbalance in the phases of sleep by barbiturates.

Barbiturates are known to depress the respiratory drive and to disturb the rhythmic character of respiration [137]. It is also documented that iontophoretically applied GABA and glycine in the bulbar respiratory units have been found to inhibit medullary respiratory neurons [138,139], and glutamic and aspartic acids to excite the ongoing phasic neural activity of both inspiratory and expiratory neurons [139]. Thus the rhythmic character of respiration has been postulated to be a consequence of the actions of inhibitory amino acids like GABA and excitatory ones like aspartic acid [140]. The data (Table 24) indicate that enhancement in permeability of GABA and glycine and reduction in the permeability of excitatory neurotransmitters like aspartic acid due to the liquid membranes formed by barbiturates in association with the membranes lipids in the synaptic cleft of the respective neurons, may also be a factor responsible for disturbance in the rhythmic character of respiration.

Thus these studies [24] on the modification in the permeability of relevant permeants in the presence of the liquid membranes indicate effects, which are worthy of further investigation with natural membranes. The pH, which is likely to influence the ionization of the barbiturates and the permeants, is different from its physiological value in the present experiments for solute permeability ω measurements. This fact, however, may not alter the conclusions because qualitatively the pH of he experimental solutions are close to the pH of the solutions in the corresponding control experiments, at least in the order of magnitude.

6.2.8 Antihistamines –H_1 antagonists [8]

Surface-activity of antihistamines is documented in the literature [141,142]. Antihistamines have been shown to generate liquid membrane at the interface. Because antihistamines are known to be competitive antagonists of histamine [143, 144], data have been obtained on the transport of histamine through liquid membranes, which were generated by the antihistamines, in series with a supporting membrane and discussed in the light of the mechanism of their action. A Sartorius cellulose acetate microfiltration membrane/aqueous interface has been deliberately chosen as site for the formation of liquid membranes so that the role of passive transport through the liquid membranes is highlighted.

Three structurally dissimilar antihistamines (H_1-antagonists) namely chlorpheniramine maleate, diphenhydramine hydrochloride and tripelennamine hydrochloride, have been chosen for the study. The choice of structurally dissimilar drugs with in one pharmacological category makes the role of the liquid membranes in the mechanism of their action conspicuous.

Literature values of CMC of antihistamines are also recorded in Table 25. In the case of diphenhydramine hydrochloride it has been concluded [145] that the drug show aggregation beyond 0.05 M concentration. Hence, the CMC should be above this concentration. The CMC chlorpheniramine maleate is not documented in literature. The CMC values determined by Bhise et al [8], though lower than the literature values, were found to be consistent with the hydraulic permeability data. The resistance to volume flux in presence of the drugs increased with increasing concentration of the drugs up the CMC values beyond which it became more or less constant. This implies that these are the concentrations at which a complete liquid membrane is generated in series with the supporting membrane, in accordance with the liquid membrane hypothesis [32].

The hydraulic permeability data were utilized to demonstrate the existence of the liquid membrane in series with the supporting membrane. For the measurement of the solute permeability (ω) of histamine, as described in the earlier sections, two sets of experiments were performed. In the first set of experiments the permeant faced hydrophilic surface of the drug liquid membrane whereas in the second set of experiments it faced the hydrophobic surface. For details of the experiments and procedures the original paper may be consulted [8]. All measurements were made at constant temperature using a thermostat set at 37° ± 0.1°C. For solute permeability (ω) measurements the concentrations of the drugs taken were always higher than their CMCs to ensure that the supporting membrane was completely covered with the liquid membranes generated by the drugs.

Table 25. Critical micelle concentration data of antihistamines in aqueous solutions.

	CMC		
	$mol \cdot l^{-1}$	$mol \cdot kg^{-1}$	$mol \cdot l^{-1}$
Chlorpheniramine maleate	(1×10^{-4})*	-	-
Diphenhydramine hydrochloride	(1×10^{-3})*	0.122**	0.05***
Tripelennamine hydrochloride	(1×10^{-3})*	~ 0.20**	-

* Ref. 8, values are at 37°C.
** Ref. 144, values are at 30°C.
*** Ref. 145, values are at 25°C.

Antihistamines are known [143] to occupy histamine receptors causing exclusion of histamine from its site. The action is known K_o be competitive and reversible [146]. The antagonism is considered entirely on account of the specific interaction of antihistamine with the receptor. The data obtained by Bhise et al [8] however, indicate that the liquid membranes generated by the drugs also contribute to the antihistaminic action.

The data on the solute permeability (ω) of histamine in the presence of the antihistamine drugs are recorded in Table 26. The values are expressed as arithmetic mean ± standard deviation-based on the 15 repeats for each value of ω. The differences between the various ω-values in Table 26 were found to be statistically significant. The data in Table 26 clearly indicate that the liquid membranes generated by antihistamines themselves impede the transport of histamine to a notable extent. Chlorpheniramine maleate is known to be most potent [146] amongst all 3 antihistamines studied. This is consistent with the observation that the CMC of chlorpheniramine maleate is the lowest (Table 25) implying that it forms a complete liquid membrane at a much lesser concentration than the other two drugs. This, *prima, facie* indicates that the liquid membrane generated by the antihistamines at the site of action may play a role in the mechanism of their action.

Chlorpheniramine maleate which is known to be most potent of all the three drugs [146], impedes the transport of histamine more or less to the same extent in both the orientations-when the permeant faces the hydrophilic or the hydrophobic surface of the drug liquid membrane. The rest of the drugs, however, impede the transport of histamine more when the drug liquid membranes present their hydrophobic surface to the permeant than when the permeant faces the hydrophilic surface of the drug liquid membranes. Since, in the histamine receptor, existence of both hydrophilic and hydrophobic sites has been indicated [147], it appears that chlorpheniramine maleate gets attached to both hydrophilic and hydrophobic sites in the formation of liquid membrane, while the other two drugs get attached only to the hydrophilic sites. According to ward [148] 'potency may imply selectivity'. In other words, the more potent the drug is, the more selective it may be to the receptor. Thus the tendency of chlorpheniramine maleate to attach with both hydrophilic and hydrophobic sites implies its selectivity to histamine receptors, which is in keeping with Waud's statement.

Table 26. Permeability of histamine [b] (ω) in presence of antihistamines [a] (Ref. 8).

Drug	$\omega_1 \times 10^{10}$ (mol. s^{-1}. N^{-1})	$\omega_2 \times 10^{10}$ (mol. s^{-1}. N^{-1})	$\omega_3 \times 10^{10}$ (mol s^{-1} N^{-1})
Chlorpheniramine maleate	5.1855	3.0620	3.3460
	±0.6379	±0.4604	±0.3398
Diphenhydramine hydrochloride	5.1855	3.1063	1.5250
	±0.6379	±0.3251	±0.1424
Tripelennamine hydrochloride	5.1855	3.1579	2.6607
	±0.6379	±0.6379	±0.3145

Note: values of ω reported as arithmetic mean of 15 repeats ± S.D.

ω_1 = control value – when no drug was used; ω_2 = drug and histamine in the compartment C and water in the compartment D: permeant histamine facing hydrophilic surface of the drug liquid membrane.

ω_3 = drug in the compartment D and histamine in the compartment C: permeant histamine facing hydrophobic surface of the drug liquid membrane.

[a] The concentrations of chlorpheniramine maleate, diphenhydramine hydrochloride, tripelennamine hydrochloride were: 2 x 10^{-4} M, 2 x 10^{-3} M and 2 x 10^{-3}M, respectively.

[b] Initial concentration of histamine 10 μg /ml.

It is interesting to note that structure-activity studies of H_1-antagonists have exhibited a relationship with partition characteristics [149, 150] and association phenomena [151,152] both of which are related to surface-activity.

Although the reduction in the permeability of histamine on account of liquid membranes generated by the antihistamines is passive in nature, it is likely to be accompanied by consequent reduction in the active transport as well. This is because the presence of the liquid membrane generated by antihistamines is likely to reduce the access of histamine to its receptors.

The multiple effects [143] associated with antihistamines, viz. anticholinergic effects, local anesthetic effects or sedation, may also be explained by modification in the transport of relevant permeants. The liquid membrane generated by antihistamines may offer a varying degree of resistance to the transport of relevant permeants. Detailed investigations, however, are called for to assess the validity of the proposition. Thus liquid membranes generated by the antihistamines at the site of action also seem to contribute to the mechanism of their action.

6.2.9. H_2-anlagonist and histamine release blocker [9,10].

The studies on antihistamines have been conducted to include cimetidine, ranitidine, famotidine and disodium chromoglycate. The first three drugs are histamine H_2-receptor antagonist [153] while disodium cromoglycate in known to act by inhibiting release of histamine from mast cells [146]. All drugs cited have been found to be surface active in nature: their critical micelle concentrations as determined from the variation of surface tension with concentration are recorded in Table 27.

Table 27. Critical micelle concentrations (Ref. 9,10).

Drug	CMC
Cimetidine	5.1024 x 10^{-6}M
Ranitidine	1.0188 x 10^{-6}M
Famotidine	4.0000 x 10^{-6}M
Cromoglycate disodium	1.5925 x 10^{-6}M

All drugs have been shown to form liquid membrane in series with the supporting membrane by themselves and also in association with membrane lipids i.e. lecithin and cholesterol. Data on the solute permeability (ω) of histamine in the presence of liquid membrane have been obtained in the two orientation of the liquid membrane: orientation 1 where the permeant faces eth hydrophilic surface of the liquid membrane and orientation 2 where the permeant faces the hydrophobic surface of the liquid membrane. The data on ω for histamine are recorded in Table 28. The details of the experiments and procedures are given in section 6.1 and in original papers [9,10].

Both Cimetidine and ranitidine are known to be H_2-antagonists. The data on histamine permeability (ω) in presence of these drugs (Table 28) reveals that the liquid membranes, which are likely to be formed at the site of action of the respective drugs, may contribute to their biological action. A perusal of Table 28 reveals that permeability of histamine is reduced to a greater extent in the first set of experiments in which the permeant, histamine, faces the hydrophilic surface of the drug liquid membrane. This trend appears to indicate that the H_2-receptors are oriented in such a manner that their hydrophobic moieties are available to get attached with the hydrophobic moieties of the H_2-antagonists-cimetidine and ranitidine, leaving hydrophilic moieties of the drugs to face histamine molecules. This is in contrast to our earlier observation [8] in the case of H_1-antagonists, where the antagonists impeded the transport of histamine more when histamine faces hydrophobic surface of the liquid membrane generated. These observations, therefore, appear to indicate that orientation of H_1- and H_2 receptors for histamine may be opposite to each other. Similar opposing orientations of H_1- and H_2- receptors are already indicated in literature [154]

Ranitidine is known to be a more potent H_2-antagonist than cimetidine [152, 153]. This fact can be rationalized on the basis of CMC values (Table 27) of the two drugs. Since the CMC of ranitidine is less than that of cimetidine, the former would form the complete liquid membrane offering maximum resistance to the transport of histamine, at a lower concentration than cimetidine would require, thus making ranitidine more potent than cimetidine.

Table 28. Solute permeability (ω)[a] of histamine [b] in presence of drugs [c] (Ref. 9)

Drug	$\omega_1 \times 10^{10}$(mol. s^{-1}. N^{-1})	$\omega_2 \times 10^{10}$ (mol. s^{-1}. N^{-1})	$\omega_3 \times 10^{10}$(mol s^{-1} N^{-1})
Cimetidine	5.1855 ± 0.6379	3.0379 ± 0.3531	3.6516 ± 0.2716
Ranitidine	5.1855 ± 0.6379	1.6630 ± 0.3205	2.6741 ± 0.4347
Cromoglycate disodium	5.1855 ± 0.6379	1.3139 ± 0.1952	2.4207 ± 0.2631

ω_1 = Control value-when no drug was used.

ω_2 = Drug and histamine in compartment C and water in compartment D: orientation 1 of the liquid membrane.

ω_3 = Drug in compartment D and histamine in compartment C: orientation 2 of the liquid membrane.

[a] The values of ω expressed as arithmetic mean of 15 repeats ± standard deviation.

[b] Initial concentration of histamine 10 μg/ml.

[c] The concentrations of Cimetidine, ranitidine and cromoglycate disodium were 2.0410 x 10^{-5} M. 4.0756 x 10^{-6}M and 6.3700 x 10^{-6} M. respectively.

In the case of disodium cromoglycate also, which is a histamine release blocker, the transport of histamine is impeded most when the drug liquid membrane presents its hydrophilic surface of the permeant (Table 28). It appears, therefore, that a similar orientation of the liquid membrane with hydrophilic moieties of disodium cromoglycate molecules facing histamine molecules may be necessary even on mast cells. However, more information on the nature an orientation of the actual site of action on mast cells is called for.

In a recent study [10] by Pandi et al., solute permeability (ω) of histamine, acetylcholine and ions in the presence of liquid membranes generated by ranitidine and famotidine alone and also in association with lecithin-cholesterol mixture has been measured to throw light on their biological actions. It is reported that ranitidine and famotidine are competitive antagonist at the parietal cell H_2-receptor. Gastric acid secretion is complex and continuous process controlled by multiple control (neuronal) and peripheral endocrine and paracrine factors. Each factor attributes to the secretions of H^+ ions by parietal cells which are located in the body and fundus of the stomach. Acetylcholine (neuronal), histamine (paracrine) and gastrin (endocrine) acts on their specific receptors M_1, H_2 and CCK_2, that have been anatomically and/or pharmacologically localized to basolateral membrane of the parietal membrane of the parietal cell. The histamine is synthesized and secreted by enterochrommaffin-like cells, which are adjuvant to basolateral membrane of the parietal cells. These drugs reduce basal secretion of acid and also secretion stimulated by food, neural and hormonal influences [155].

For solute permeability measurements, the concentrations of the drugs taken were always higher than their CMCs (=2CMC to be precise). This was done to ensure that the supporting membrane was completely covered with liquid membrane generated by the drugs. The composition of lecithin-cholesterol mixture used in the solute permeability measurements was 1.175×10^{-6} M with respect to cholesterol and 1.919×10^{-5} M with respect lecithin because in an earlier study its had been shown [90,130] that at this composition the supporting membrane is completely covered by lecithin liquid membrane and is also saturated with cholesterol. In these studies also a cellulose acetate microfiltration membrane has been chosen as supporting membrane to highlight the role of passive transport through the liquid membrane in the biological action. For details of experiments and procedure the original paper may be consulted [10]. All experiments were done at constant temperature using a thermostat set at 37±0.1°C. The values of solute permeability (ω) of different permants are recorded in Table 29. Since in the earlier studies [9] on H_2-antagonists it was observed that permeability of histamine is reduced to a greater extent in the first set of experiments in which the permeant faces the hydrophilic surface of the drug liquid membrane in these experiments also the value of (ω) were obtained in the specific orientation, hydrophilic surface of the liquid membrane facing the approaching permeant Table 29.

From the perusal of solute permeability data (Table 29) it can be said that both RNT and FMT reduced the permeation of sodium, potassium, calcium and chloride ions, histamine and acetylcholine (Table 29) while enhanced permeation of bicarbonate ions. The above observations are consistent with the physiological role of RNT/FMT. The regulation of acid secretion by parietal cells is especially important in peptic ulcer and constitutes a particular target for drug action. Physiologically Cl^- ion is actively transported into canaliculi in the parietal cells. K^+ accompanies the Cl^- and is exchanged for H^+ from within the cell by a K^+/H^+.ATPase [156]. Liquid membrane likely to be formed by H_2-antagonists on the parietal cells reduces the passive transport of K^+ as well as Cl^- ions, which leads impairment of K^+/H^+ exchange. Due to the fact that availability of H^+ and Cl^- ions is reduced in the lumen, the formation of gastric hydrochloric acid is decreased, H_2CO_3, formed from CO_2 and H_2O, dissociates in a reaction catalysed by carbonic anhydrase to form H^+ and HCO_3^-. HCO_3^- exchanges across the basal membrane for Cl^-. The liberated HCO_3^- is combined with mucus to form a cytoprotective layer (pH 7) on the gastric lumen [156]. Transport of HCO_3^-ions is enhanced in presence of liquid membranes formed by H_2-antagonists. This fact may also contribute for its cytoprotective action of these drugs. Acetylcholine is released from neurons and stimulates specific muscarinic receptors on the surface of parietal cells and on surface of histamine containing cells leading to the activation of H^+/K^+ATP-ase via Ca^{2+}-dependent pathway from basolateral membrane [156]. Transport of acetylcholine in presence of liquid membranes generated by H_2 antagonists is reduced, which may impede H^+/K^+ATP-ase via Ca^{2+}-dependent pathway. It is known that parietal cell has H_2-receptors and is sensitive to histamine, responding to amounts that are below the threshold concentration that acts on H_2-receptors in blood vessels. In man the histamine is derived from mast cells or histamine containing cells similar to mast cells, which lie close to the parietal cell [156]. Stimulation of H_2-receptors increases cAMP and these second messengers synergies to produce acid secretions. Transport of histamine in presence of liquid membranes generated by H_2-antagonists is decreased which will reduce the availability of histamine for H_2-receptors.

Table 29. Solute permeability (ω) of various permeants in presence of liquid membrane generated by ranitidine (RNT) and famotidine (FMT) alone and in presence of lecithin-cholesterol mixture (Ref. 10).

Permeant	Initial concentration	RNT (2 CMC) (x 10^{-6}) (mol $s^{-1}N^{-1}$)				FMT (2 CMC) (x 10^{-6}) (mol $s^{-1}N^{-1}$)			
		ω_0	ω_1	ω_2	ω_3	ω_0	ω_1	ω_2	ω_3
Histamine	10.0 mg/ml	0.123 ± 0.026	0.043± 0.045	0.053± 0.0341	0.014± 0.024	0.120 ± 0.036	0.046± 0.054	0.048± 0.072	0.012± 0.074
Acetylcholine	1.0 µg/ml	1.121 ± 0.055	0.423 ± 0.042	0.412 ± 0.082	0.214 ± 0.082	1.080 ± 0.064	0.513 ± 0.053	0.552 ± 0.048	0.342 ± 0.078
Cl^- ions	500 µg/ml	88.00 ± 0.074	10.01 ± 0.071	41.00 ± 0.024	4.123 ± 0.043	87.01 ± 0.04	10.92 ± 0.056	12.42 ± 0.078	7.34 ± 0.092
HCO_3^- ions	500 µg/ml	102.00± 0.052	7.0 ± 0.089	130.00 ± 0.078	7.162 ± 0.034	101.14± 0.058	10.00 ± 0.063	150.12 ± 0.081	12.00 ± 0.091
Na^+ ions	5.382 mg/ml	340.23± 0.037	110.53± 0.042	200.01± 0.073	100.42± 0.036	344.04± 0.042	100.43± 0.033	150.42± 0.018	74.43± 0.024
K^+ ions	10.43 mg/ml	800.24± 0.056	640.41± 0.075	710.17± 0.013	617.83± 0.087	824.21± 0.027	592.12± 0.048	670.14± 0.081	400.12± 0.073
Ca^{2+}ions	10.0 mg/ml	480.0 ± 0.082	305.36± 0.082	372.14± 0.075	271.26± 0.048	488.36± 0.064	290.24± 0.058	350.25± 0.039	250.42± 0.092

Values of ω are reported as arithmetic mean often repeats ± S.D., ω_0 : when no drug was used; ω_1: in the presence of lecithin-cholesterol mixture;

ω_2: in presence of RNT/FMT;

ω_3 : in presence of RNT/FMT and lecithin-cholesterol mixture.

CMC of RNT : 1×10^{-6} M , CMC of FMT : 4×10^{-6} M

It has been found that, RNT and FMT are antagonists of the histamine H_2-receptors. When analysed by the classical Schild method pA2-values*of RNT an FMT are found to be 6.8 and 7.7, respectively with dimaprit as agonist and 6.5 and 7.7 respectevely with histamine as agonist [157]. It could be seen from Table 29, that relative reduction in transport of the histamine is more by FMT than that of RNT (difference in ω_0 and ω_2 values), which shows close alliance with the pA2-values. The reduction in the transport of the histamine in presence of lecithin-cholesterol mixture and lecithin-cholesterol-drug mixture shows further evidence of the findings of he transport studies (difference in ω_1 and ω_3 values).

This study [10] in no way refutes the already established mechanism of action of H_2-antagonists. However it provides a more rational and dynamic approach to their mechanism of action by highlighting the role played by the liquid membranes generated by these drugs. Further *in vivo* studies are required to be designed and done for further confirmation of the hypothesis.

6.2.10 Steroids [15,17]

Steroids are known to be surface active [158] in nature; CMCs shown in Table 1 and their interactions with constituents of biological membranes is documented in literature [159]. Three representative steroidal drugs, namely testosterone propionate an androgen, ethynylestradiol, an estrogen and hydrocortisone acetate, a glucocorticoid have been chosen for the study of the role of liquid membranes generated by them in their biological action. The drugs and their mixtures with lecithin have been shown to generate liquid membranes in series with a supporting membrane. In this case also, as already explained, a cellulose acetate microfiltration membrane (Sartorius cat no.11107) has been used as supporting membrane to highlight the role of passive transport through the liquid membrane in their biological actions. Data on solute permeability (ω) of relevant permeants were obtained in the presence of liquid membranes generated by the drugs alone and also in association with lecithin. The data on solute permeability (ω) were obtained in both the orientation of the liquid membrane i.e. hydrophilic surface of the liquid membrane facing the approaching permeants and hydrophobic surface of the liquid membrane facing the approaching permeants. Details of the experiments and procedures are given in the original paper [15]. All experiments were carried out at constant temperature using a thermostat set at $37 \pm 0.1^{\circ}C$.

The normalized values (r) of solute permabilities obtained by dividing the experimental values by the values for the corresponding control experiments ($r = \omega_{exp}/\omega_{con}$) recorded in Table 30 reveal the following trends. The permeability of glucose is enhanced in both orientations of the liquid membranes generated by the steroidal drugs or the lecithin-steroidal drugs mixture. The permeability of amino acids however is enhanced only in the specific orientation of the liquid membranes with their hydrophilic surface facing the permeants. The cause of the enhancement of permeabilities is difficult to determine at this stage, Nevertheless, trends observed in the permeability of amino acids (Table 30) appear relevant to some of the biological actions of the steroidal drugs.

The permeability of amino acids in the presence of lecithin-steroidal drug mixtures is enhanced more in the case of testosterone than ethinyl estradiol. In the absence of lecithin the trend in permeability of amino acids was reversed (Table 30). The anabolic effect of steroids involves increased mobilization of amino acids, which implies their enhanced permeability and also increased accessibility near the relevant biological membrane. Since, in biological cells, androgens are known to be more anabolic than estrogens [160], it is expected that androgens are more likely to increase amino acid permeability than are estrogens. The same trend is observed in these experiments (Table 30). Hence the role of liquid membranes, generated by a lecithin-steroidal drug mixture, in influencing the permeability of amino acids in indicated. Possibly, the incorporation of steroidal drugs in the phospholipids, with their hydrophilic ends specifically oriented to face the approaching amino acids, may be necessary for their anabolic action.

Table 30. Values of normalized permeability (r) of glucose, leucine, histidine and tryptophan in the presence of steroidal drugs and lecithin-steroidal drug mixtures[a] (Ref. 15).

Permeants	Steroidal drugs		
	Testosterone propionate	Ethinyl estradiol	Hydrocortisone acetate
Glucose			
r^b	1.639	1.070	1.290
r^c	1.650	1.224	1.484
r^d	1.185	1.081	1.022
r^e	1.073	1.144	1.206
Leucine			
r^b	1.171	1.373	1.210
r^c	0.535	0.585	0.619
r^d	1.178	1.162	1.095
r^e	0.990	0.980	0.852
Histidine			
r^b	1.226	1.855	1.222
r^c	0.545	0.778	0.490
r^d	1.274	1.200	1.197
r^e	0.930	0.773	0.715
Tryptophan			
r^b	1.337	1.448	1.160
r^c	0.675	0.611	0.629
r^d	1.400	1.258	1.143
r^e	1.000	0.691	0.681

[a] Lecithin concentration in he mixtures, 15.5 ppm.

[b] Permeants and the steroidal drugs in compartment C and water in compartment D.

[c] Perments in compartment C and the steroidal drugs in compartment D.

[d] Lecithin-steroidal drug mixture and the permeants in compartment C and water in compartment D.

[e] Permeants in compartment C and lecithin-steroidal drug mixture in compartment D.

Glucocorticoids such as hydrocortisone are known to mobilize amino acids from a number of acids from a number of tissues [161]. In these investigation it was also observed (Table 30) that hydrocortisone, either by itself or in association with lecithin, enhanced the permeability of the amino acids. Since for this action the specific orientation of the steroids with their hydrophilic ends facing the permeants was observed to be necessary (Table 30), it is tempting to suggest that a similar orientation of glucocorticoids such as hydrocortisone may also be necessary in biological cells.

The lecithin-steroid liquid membrane at the surface of cellulose acetate may be structurally different from the lipid bilayer characteristic of biological membranes. Nevertheless, since the trends observed in the present experiments are similar to those that occur on biological cells, these studies are indicative of the possible role of liquid membranes in the action of steroidal drugs.

Unlike the effect on amino acids, the uptake of glucose is reduced in the presence of glucocorticoids [161]. These experiments, however, showed an increase in permeability of glucose in the presence of steroids in both orientations-the hydrophilic ends facing the permeants and the hydrophobic ends facing the permeants. Hence these observations on the increase in permeability of glucose (Table 30) do not appear to be relevant to the biological effects of glucocorticoids on glucose transport.

Steroids are known to exert their anabolic action by combing with a soluble receptor present in the cytoplasm [162] of the target cells. Since these actions are accompanied by increased mobilization of amino acids and the synthesis of proteins, these studies indicate that the liquid membranes generated by the steroidal drugs in association with phospholipids such as lecithin may have a role to play in the mechanism of action of the drugs.

Since the entire hypothalamus is under the influence of neurotransmitters in the release of hypophysial hormones [163-165] including gonadal steroid hormones responsible for a variety of physiological function, in another study [17] transport of neurotransmitters, viz. adrenaline, noradrenaline, dopamine and serotonin, through the liquid membranes generated by the steroid hormones in association with sphingomyelin, which is the relevant phospholipid in brain, has been studied. The data obtained on the modification in the permabilities of neurotransmitters in the presence of the liquid membranes have been discussed in the light of the various physiological actions of the steroid hormones. Incidentally suggestions to the effect that modifications in the permeability of cell membranes brought about by steroid hormones may play significant roles in their physiological actions, have already been made in literature [166, 167].

Hydraulic permeability data were utilized to demonstrate the formation of liquid membranes by sphingomyelin-steroid mixtures. The value of solute permeability (ω) of relevant permeants were determined in the presence of liquid membranes generated by sphingomyelin –steroid mixture. The composition of sphingomyelin-steroid mixtures used for (ω) measurements were those at which the sphingomyelin liquid membrane was shown to be completely saturated with the steroids. The details of experiments and procedures are given in the original paper [17].

Table 31. Solute permeability (ω)[a] of various permeants in presence of sphingomyelin gonadal steroid mixtures (Ref. 17).

	$\omega_0 \times 10^{10}$ (mol $s^{-1}N^{-1}$)	$\omega_1 \times 10^{10}$ (mol $s^{-1}N^{-1}$)	$\omega_2 \times 10^{10}$ (mol $s^{-1}N^{-1}$)	$\omega_3 \times 10^{10}$ (mol $s^{-1}N^{-1}$)
Adrenaline [b]	2.050 ± 0.043	0.855 ± 0.053	1.151 ± 0.022	1.481 ± 0.01
Noradrenaline[b]	4.467 ± 0.192	2.775 ± 0.062	4.275 ± 0.056	3.160 ± 0.087
Dopamine [b]	3.266 ± 0.097	2.825± 0.014	4.394 ± 0.069	2.892 ± 0.088
Serotonin[b]	3.189 ± 0.089	2.579 ± 0.181	3.771 ± 0.128	3.612 ± 0.159

[a] Values of are reported as arithmetic mean of 10 repeats ± S.D.
[b] Initial concentrations: adrenalin = 3.000×10^{-5} mol/lit. noradrenaline = 5.889×10^{-5} mil/lit., dopamine =5.273×10^{-5} mol/lit. Serotonin = 2.466×10^{-5} mol/lit.
ω_0 Control value when spingomyelin alone was used (spingomyelin concentration = 18ppm).
ω_1 Values in the Presence of sphingomyelin-ethinyl estradiol mixture of composition 18 ppm with respect to sphingomyelin and 2.0×10^{-6} M with respect to ethinyl estradiol.
ω_2 Values in the presence of sphingomyelin-progesterone mixtures of composition 18 ppm with respect to sphingomyelin and 0.5×10^{-6} M with respect to progesterone. ω_3 Values in the presence of sphingomyelin-testosterone propionate mixtures of composition 18 ppm with respect to sphingomyelin and 2.0×10^{-6} M with respect to testosterone propionate.

The values of solute permeability (ω) of the various biogenic amines in the presence of liquid membranes generated by sphingomyelin-steroid hormone mixtures are recorded in Table 31. The modifications in the values of solute permeabilities (ω) of the various neurotransmitters (Table 31) appear relevant to various physiological functions of the steroid hormones.

It is well known that in response to a hypophysiotropin, adenohypophysial hormones and the gonadotropins are released which stimulates their target tissues. Target tissue stimulation results in increased secretion of target tissue hormones such as thyroid hormones, adrenal glucocorticoids and gonadal steroid hormones. These hormones then in addition to acting on their respective target tissues to mediate their action also act on higher brain centers, hypothalamus and pituitary and exercise a negative feedback control.

Biogenic amines particularly dopamine, noradrenaline and serotonin, have been implicated in the feedback mechanism [163,168-170]. For example, dopamine has been shown to cause the release of LH/FSH-RH and P-RIH. The release of LH/FSH-RH produced by intraventricularly injected dopamine is blocked by the previous intraventricular injection of estradiol [163]. The impediment in the transport of dopamine in the presence of the liquid membranes generated by the sphingomyelin-ethinyl estradiol mixture as observed in this study (Table 31) appears to be a contributing factor to the negative feedback mechanism. It is also documented [163] that patients treated with drugs like reserpine and chlorpromazine, which have been shown [2,3] to impede the transport of biogenic amines including dopamine display evidence of altered pituitary functions, e.g. failure to ovulate. This observation is

Abbriviation-used : LH/FSH-RH Lenutinizing hormone/Follicle stimulating hormone relasing hormone, P-RIH prolactin release inhibiting hormone P-RH, prolaction releasing hormone

consistent with the conclusion that impediment in the transport of dopamine due to the ethinyl estradiol liquid membrane formed in association with sphingomyelin (Table31) may be a contributing factor in the negative feedback mechanism. It may be pointed out that there is evidence [168] of dopamine being directly released into paretial vessels and acting on pituitary. It is also documented that implantation of testosterone in the median eminence of rats inhibits pituitary gonadotropin secretion [171] by decreasing the level of gonadotropin-releasing hormones [164]. The reduced permeability of biogenic amines like dopamine in the presence of sphingomyelin-testosterone mixture (Table 31) could be a plausible explanation for this observation.

At hypothalamic level, the inhibitory hormone P-RIH controls the secretion of prolactin in mammals and possibly by a prolactin-releasing hormone, P-RH, -the role if any, of P-RH is only of secondary importance. The release of P-RIH from neuroendocrine transducer cells in the median eminence is controlled by hypothalamic dopamine. The drugs like reserpine, chlorpromazine and haloperidol, which are known to reduce the permeability of dopamine [1-3] and lower its concentration in hypothalamic region, are known to decrease the P-RIH release which, in turn, promotes prolactin release [172]. Since ethinyl estradiol was found to reduce the permeability of dopamine (Table 31), it should have effects similar to that of reserpine, chlorpromazine and haloperidol, which indeed is the case [172]. It is reported that disorders like galactorrhea or gynaecomastia may also arise from estrogen secreting tumors and also as a side effect of oral contraceptives [163].

MSH secretions by the pars intermedia of the pituitary gland are reported to be under the control of catecholamines, viz. adrenaline, noradrenaline and dopamine [173]. Drugs such as, reserpine, haloperidol and chlorpromazine, which block the actions of catecholamines [1-3], are reported to stimulate MSH secretion [173]. Karkun and Sen [174] reported increased pituitary levels of MSH in ovariectomaized rats after estrogen treatment for thirty days. These finding of Karkun and Sen, also corroborated by later workers [173], are consistent with the reduced permeability of catecholamines in the presence of ethinyl estradiol as observed in this study (Table 31).

Dopamine, noradrenaline and serotonin are reported to increase growth hormone release in animals and in man [169], the role of serotonin, however, is controversial. Estrogens have been used to treat acromegalics [175]. It is also documented that excess androgen secretion can lead to shortened stature [176]. Reduced permeability of biogenic amines, particularly dopamine and noradrenaline, in the presence of the liquid membranes generated by the steroids in association with the membrane lipid (Table 31) is consistent with these observations.

Neurohypophysial secretions containing ADH, oxytocin and neurophysins are controlled by both acetylcholine and noradrenaline, the former has stimulatory effect while the latter has an inhibitory effect [163, 177]. Ovarian hormones are reported to facilitate the release of neurohypohysial hormones [177]. The data on the reduced permeability of noradrenaline (Table 31) appear to suggest that the reduced access of noradrenaline to the postsynaptic receptor due to the steroid-phospholipid liquid membrane may reduce its inhibitory effect and thus contribute to the increased releases of neurohypophysial hormones.

Abbriviation-used : MSH melanocyte stimulating hormone, ADH antiddirutic hormone

Estrogen possesses neuroleptic like quality and potentiates neuroleptic-induced parkinsonism [178,179]. Reduced concentrations of dopamine and serotonin in brain have been linked with neuroleptic actions and symptoms like parkinsonism arising from it [1,50,180]. The impediment in the transport of dopamine and serotonin in the presence of estrogen-sphingomyelin liquid membrane (Table 31) appears to be one of the contributing factors to these effects.

Antidepressant drugs, e.g. Imipramine are known to act by reducing the uptake of biogenic amines [172,181]. The reported antidepressant effects of estrogen [167,182] are, therefore, consistent with the reduced permeability of noradrenaline and serotonin in the presence of the sphingomyelin-estrogen liquid membranes (Table 31). Similarly enhanced permeability of serotonin in the presence of progesterone-sphingomyelin liquid membrane (Table 31) appears consistent with the reported depressant effects of progesterone [183].

Reduction in concentration of serotonin at the postsynaptic receptor has been implicated in migraine. Premenstrual migrainous headache is aggravated by oral contraceptives and mitigated by switching to progestrogen only preparations [183-186]. These observations appear consistent with the reduced and enhanced permeabilities of serotonin observed respectively in the presence of the liquid membranes generated by estrogen and progesterone in association with sphingomyelin. (Table 31).

Noradrenaline and serotonin have profound effect on body temperature [185]. When injected into the anterior hypothalamus or the cerebral ventricles in experimental animals, serotonin produces a rise in body temperature whereas noradrenaline produces a fall [185]. Change in body temperature, which occurs after ovulation is ascribed to increased progesterone levels in blood [183,185,187,188]. The enhancement and the reduction in the permeabilities of serotonin and noradrenaline respectively in the presence of progesterone – sphingomyelin liquid membrane (Table 31) appear to be a contributing factor to the thermogenic effects of progesterone.

Thus it appears that modification in the permeability of neurotransmitter molecules in the presence of gonadal steroid hormones-brain phospholipid liquid membranes may also play a significant role in the physiological functions of the steroid hormones.

6.2.11. Fat soluble vitamins-vitamin E, A and D. [19,20,27]

6.2.11.1. Vitamin E: Studies on α-tocopherol [19]

α-Tocopherol is the most important tocopherol because it comprises about 90% of the tocopherols in animal tissues and exhibits maximum biological activity. It is distributed throughout the tissues of animals and man and its deficiency causes a variety of syndromes in the animal organism. Just by looking at the structure of α-tocopherol one suspects it to be surface-active in nature. In fact it is: CMC value is given in Table 1. In view of Kesting's hypothesis [32] as is likely that the phenomenon of liquid membrane formation at the interface may play a role in the actions of α-tocopherol.

Prompted by this conception investigations were carried out to explore the role of liquid membrane phenomenon in the actions of α-tocopherol. Critical micelle concentration of α-tocopherol in water has been determined. The data on hydraulic permeability have been obtained to demonstrate: (1) the formation of a liquid membrane; and (ii) the incorporation of α-tocopherol in the lecithin-cholesterol liquid membrane existing in series with the supporting membrane. Transport of relevant permeants in presence of the liquid membrane generated by the lecithin-cholesterol-α-tocopherol mixture has been studied and the data obtained have been discussed in the light of the various syndromes caused by vitamin E deficiency.

For details of experiments for determination of the data on hydraulic permeability and solute permeability ω the original paper may be consulted [19]; it is also given in section 6.1 in a generalized way.

For solute permeability (ω) measurements, lecithin-cholesterol-α tocopherol mixtures of composition 1.919×10^{-5} M with respect to lecithin, 1.175×10^{-6}M with respect to cholesterol and 3.75×10^{-8}M with respect to α-tocopherol was used because it was demonstrated [19] that at this composition the lecithin liquid membrane which completely covers the supporting membrane is saturated with both cholesterol and α-tocopherol.

Data on the solute permeability (ω) of several permeants, namely estrogen, progesterone, cystine, methionine and cations (Na^+ , K^+ and Ca^{2+} ions), in the presence of the liquid membranes generated by the mixture of lecithin, cholesterol and α-tocopherol in series with the supporting membrane are recorded in Table 32. The data appear to be relevant to causation of various syndromes in animal organisms due to deficiency of vitamin E, i.e. α-tocopherol.

Except for the work cited by Wagner and Folkers [189] there is enough evidence to indicate that vitamin E is essential for normal reproduction in several mammalian species [190,191] and its deficiency is known to cause habitual abortions. The fundamental mechanism by which vitamin E deficiency interferes with reproduction is obscure [191]. The data (Table 32) on the impedimenting in the transport of oestrogen and progesterone in the presence or α-tocopherol may offer an explanation for occurrence of habitual abortions caused by vitamin E deficiency.

Table 32. Solute permeability (ω) of various permeants in presence of lecithin-cholesterol-α-tocopherol mixture[a] (Ref. 19).

	$\omega^b \times 10^9$ (mol $s^{-1}N^{-1}$)	$\omega^c \times 10^9$ (mol $s^{-1}N^{-1}$)
Methionine [d]	8.79 ± 0.27	6.25 ± 0.24
Cystine [e]	2.62 ± 0.09	4.66 ± 0.56
Creatinine [f]	0.27 ± 0.03	0.29 ± 0.02
Ethinyl oestradiol [g]	5.50 ± 0.40	4.14 ± 0.24
Progesterone [h]	4.90± 0.38	4.11 ± 0.20
Sodium (chloride) [i]	0.12 ± 0.01	0.12 ± 0.01
Potassium (chloride) [j]	0.13 ± 0.01	0.14 ± 0.01
Calcium (chloride) [k]	0.16 ± 0.03	0.18 ± 0.01

[a] Lecithin concentration, 1.919×10^{-5} M ;cholesterol concentration, 1.175×10^{-6}M; α-tocopherol concentration, 3.75×10^{-8}M. [b] Control value when no α-tocopherol was used.
[c] Lecithin-cholesterol-α-tocopherol mixture in compartment C of the transport cell (Fig. 2 Chapter 5) together with the permeant. [d] Initial concentration 100 mg /l. [e] Initial concentration 100 mg/l.
[f] Initial concentration 1 g/l. [g] Initial concentration 50 mg/l. [h] Initial concentration 100 mg/l.
[i] Initial concentration 5.382 g/l. [j] Initial concentration 10.430 g/l. [k] Initial concentration 0.222 g/l.

It is not only the high concentrations of oestrogen and progesterone but also a proper ratio of their concentrations, which is essential for the maintenance of pregnancy [192]. As the data in Table 32 indicate, the deficiency of vitamin E in the membranes of the uterus would enhance the outflow of oestrogen and progesterone to an unequal extent. This outflow would disturb the oestrogen-progesterone ratio resulting tin the failure of pregnancy.

In many species, deficiency of vitamin E leads to the development of muscular dystrophy. Metabolic disturbances during muscular dystrophy include increased water content of the tissues, changes in electrolyte pattern and increased excretion of creatine in urine-creatinurea [193]. The values of solute permeability, ω for the cations and also for creatinine in the presence of α-tocopherol do not show any significant difference in comparison to the values obtained from the control experiment where no α-tocopherol was used. The data on hydraulic permeability (Table 33), however, appear relevant to causation of increased water content of the tissues and creatinurea. The data in Table 33 imply that the cell membranes deficient in vitamin E are likely to be more permeable to water which may be one of the factors responsible for the increased water content of the tissue. It has been suggested [193,194] that creatinurea in nutritional muscular dystrophy might be due to hydration of creatinine to creatine due to increased water content of the tissues-creatinine is formed inside the cells as a result of creatine metabolism. The alteration in the normal water balance of tissues is a consistent finding in the biochemical and histological examinations of tissues affected by vitamin E depletion [193]. Nitowsky et al. [195] have shown that tocopherol can decrease the elevated creatine excretions of children with cystic fibrosis.

Table 33. Values of L_p at various concentrations of α-tocopherol in lecithin-cholesterol -α-tocopherol mixtures [a] (Ref. 19).

Concentration$\times10^8$M	0.00	1.25	2.50	3.75	5.00	10.00
L_p x 10^8 ($m^3 . s^{-1} . N^{-1}$)	1.575 ± 0.084	1.402 ± 0.045	1.296 ± 0.012	1.178 ± 0.011	1.185 ± 0.031	1.164 ± 0.033

[a] Lecithin and cholesterol concentrations kept constant at 1.919×10^{-5} M and 1.175×10^{-6} M, respectively.

Dam and associates [196] have shown that supplementing the diet with either vitamin E or cystine could prevent muscular dystrophy in chicks. Later Machlin and Shalkop [197] showed that cystine and methionine were equally effective in prevention of dystrophy. However, Scott and Calvert [198] have reported that cystine is more effective than methionine. The data (Table 32) show that the permeability of cystine is enhanced and that the amount of methionine was reduced in the presence of α-tocopherol. This observation is consistent with the inference drawn by Scott and Calvert that cystine is more effective than methionine in the prevention of dystrophy.

Certain diets low in protein and especially in the sulfur-containing amino acids, particularly cystine, have been found to produce an acute massive hepatic necrosis in experimental animals. Vitamin E deficiency is reported to enhance the effects of such diets, whereas added vitamin E exerts a preventive action upon the necrosis [199]. The enhanced permeability of cystine in the presence of α-tocopherol (Table 32) could be a plausible explanation for these observations on the causation and prevention of hepatic necrosis.

Thus the studies reported above indicate that phenomenon of liquid membrane formation may also play a notable role in the causation and prevention of various syndromes due to vitamin E deficiency. It may be emphasized once again that since the supporting membrane chosen in this study was a non-specific, non-living membrane, the present study highlights the role of passive transport in biological action.

6.2.11.2 Vitamin A-retinol acetate [20]

Vitamin A is surface-active in nature [200]. Vitamin A is therefore expected to generate a liquid membrane at the interface. The cytoplasmic membrane consists of phospholipids and proteins. The phospholipids molecules are arranged in a bimolecular layer with polar groups directed toward both sides. The nonpolar part of vitamin A is likely to be placed across the hydrophobic core of the membrane, consisting of a phospholipids bilayer and polar part may get oriented toward either the exterior or the interior of the cell because they have primarily aqueous content. At the critical micellar concentration (CMC) of vitamin A the cytoplasmic membrane may be completely covered by vitamin A and likely to form a liquid membrane. Because of the liquid membrane formed by vitamin A on the cytoplasmic membrane of the cell, transport of essential amino acids and cations required for various physiological functions may be altered. In the model studies by Nagappa et.al [20], experiments on vitamin A are reported and the data are viewed in the light of the liquid membrane hypothesis of drug action [64]. Data on the hydraulic permeability have been obtained to demonstrate the formation of liquid membranes by vitamin A in series with the supporting membrane. The data obtained on the modification in the permeability of relevant amino acids such as serine, threonine, arginine, and histidine and various ions such as calcium, sodium, and potassium in the presence of the liquid membrane have been discussed in the light of the various physiological functions of vitamin A. Details of the experiments and procedures are described in the original paper [20].

The CMC of aqueous vitamin A as determined from the variation of surface tension with concentration was found to be 6.0×10^{-9} M. The hydraulic permeability data, which were utilized to demonstrate the formation of liquid membrane by aqueous vitamin A, were obtained using a cellulose acetate microfiltration membrane (Sartorius Cat No.11107) as supporting membrane. For measurement of solute permeability (ω) the compartment C of the transport cell (Fig. 2 Chapter 5) was filled with aqueous solution of vitamin A along with the desired concentration of the permeant and the compartment D was filled with deionzed water. The concentration of vitamin A used in solute permeability measurements was always higher than its CMC to make sure that the interface was completely covered by the liquid membranes. All measurements were made at $37 \pm 0.1^{\circ}$C.

The data on solute permeability (ω) of different relevant permeants in presence of the vitamin A liquid membrane are recorded in Table 34.

Table 34. Solute permeability (ω) of various permeants in the presence of liquid membranes generated by vitamin A (ω_l), along with control values (ω_0) when no vitamin A was used (Ref. 20).

Permeants	Initial concentration (mg ml^{-1})	ω_0 x 10^5 (moles $s^{-1}N^{-1}$)	ω_l x 10^5 (moles $s^{-1}N^{-1}$)
Serine	0.2	498.29 ± 4.15	238.18 ± 1.85
Threonine	0.2	219.93 ± 1.58	105.56 ± 1.23
Arginine	0.2	38.22 ± 0.50	90.05 ± 0.56
Histidine	0.2	7.14 ± 0.03	8.23 ± 0.01
Calcium (chloride)	10.0	10.71 ±0.01	16.01 ±0.01
Sodium (chloride)	5.4	2.38 ± 0.02	2.73 ± 0.01
Potassium (chloride)	10.4	3.20 ± 0.03	3.80 ± 0.04

[a] Values of ω_0 and ω_l are reported as the arithmetic mean of 10 repeats ± SD in each case. Vitamin A concentration used: 12 x 10^{-9}M (CMC).

The solute permeability data (Table 34) show that the transport of the amino acids serine and threonine is reduced and that of arginine and histidine enhanced. It is reported that vitamin A promotes the synthesis of fibronectin and inhibits the synthesis of keratin [200]. Serine and threonine are neutral amino acids, whereas arginine and histidine are basic in nature [201]. Serine and threonine are essential for the synthesis of fibronectin whereas arginine and histidine are essential for keratin synthesis [200]. The vitamin A is known to cause a transformation of membrane lipids from the bimolecular leaflet configuration to the micellar configuration, which changes the permeability properties of the cell membrane [202]. The retinol acetate molecule's hydrophobic part is likely to bind the hydrophobic part of the cellulose acetate membrane (supporting membrane) and the hydrophilic part is drawn outward away from it. Arginine and histidine, which are basic in nature, may be repelled by the hydrophilic part of the retenol acetate liquid membrane leading to an impediment in the permeation of these amino acids. On the other hand, serine and threonine, the neutral amino acids, are not affected by the hydrophilic part of retinol, leading to enhancement in the permeation of these amino acids by passive transport.

Solute permeability data (Table 34) show that there is an increase in the transport of sodium, potassium and calcium ions when compared with control. The enhanced transport of sodium, potassium and calcium in the presence of liquid membranes of vitamin A may be due to the possibility of formation of hydrophilic pathways through which ions can move more freely in comparison to the control, where no vitamin A is used. For further confirmation of this, transport studies through lecithin-cholesterol liquid membranes in the presence of vitamin A are called for.

6.2.11.3 Vitamin D_3- Cholecalciferol [27]

In an earlier study [130] cholesterol has been shown to generate liquid membrane in series with a supporting membrane. Since vitamin D_3, which is an essential dietary requirement, plays vital role in the homeostasis of mineral metabolism, and has structural similarity with cholesterol (structures shown in Fig 5), should also be surface active in nature and hence capable of generating liquid membrane at the interface (Kesting's hypothesis).

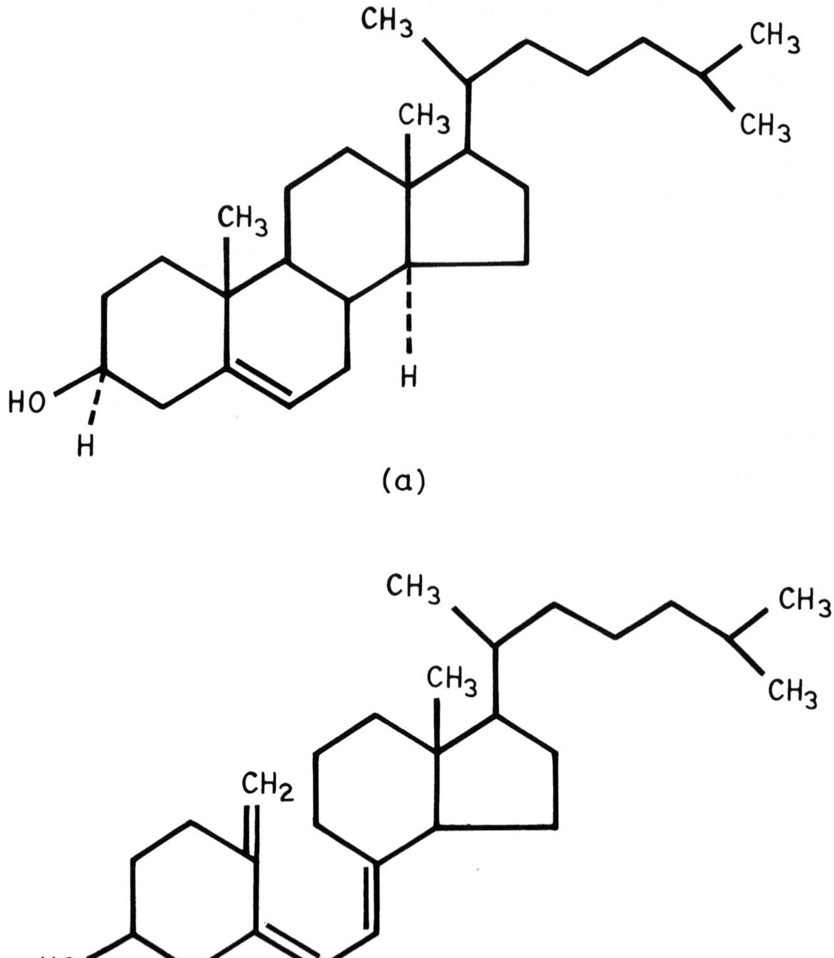

Fig. 5 Structures of (a) cholesterol and (b) vitamin D_3 (Cholecalciferol)

In the light of these facts it was considered [27] desirable to study the transport of cations, phosphate and glucose across the liquid membrane generated by vitamin D_3 to gain information about its biological actions.

Vitamin D_3 was found to the surface active (CMC=8 x 10^{-9}M) and using the hydraulic permeability data it was shown to generate a liquid membrane, which completely covered the supporting membrane at its CMC [27]. Solute permeability (ω) of relevant permeants in the presence of the liquid membrane was determined. The data on solute permeability of different permeants in the presence of the liquid membrane generated by vitamin D_3 are recorded in Table 35. In the solute permeability measurement the concentration of vitamin D_3 was always higher than its CMC to ensure that the interface is completed covered by the liquid membrane. All measurements were made at 37 ± 0.1°C.

Table 35. Solute permeability (ω_l)[a] of various permeants in presence of liquid membrane generated by vitamin D_3 along with the control values (ω_0) when no vitamin D_3 was used. (Ref. 27).

Permeants	Initial Concentration (mg/ml)	ω_0 x 10^5 (mol s^{-1} N^{-1})	ω_l x 10^5 (mol s^{-1} N^{-1})
Calcium ions ($CaCl_2$)	10.000	12.60 ± 0.03	22.31 ± 0.03
Phosphate ions (KH_2PO_4)	0.050	1.49 ± 0.01	5.30 ± 0.02
Sodium ions (NaCl)	5.382	3.50 ± 0.02	4.40 ± 0.02
Potassium ions (KCL)	10.430	5.10 ± 0.04	6.70 ± 0.04
Glucose	20.000	2.29 ± 0.05	2.94 ± 0.06

[a] Values of ω_0 and ω_l are reported as the arithmetic mean of 10 repeats ± SD in each case. Vitamin D_3 concentration, 16 x 10^{-9} M (2 x CMC).

Data on the solute permeability of permeants such as calcium, phosphate, sodium and potassium ions and glucose in the presence of liquid membrane likely to be generated by vitamin D_3 in series with the supporting membrane are recorded in Table 35. The trends in the modification in the values of solute permeability observed in these model studies are consistent with various biological actions of vitamin D_3. The trend observed in these studies is that the permeability of calcium, sodium, potassium, glucose and phosphate are all enhanced in the presence of vitamin D_3 liquid membrane. The main function of vitamin D_3 is mineralizing of the skeleton [203]. The hardness of the bone is achieved by the deposition of calcium and phosphorus ions as calcium carbonate, calcium fluoride and magnesium phosphate. In natural condition, bone is calcified structure [204] and it contains calcium, phosphorus and magnesium in large amounts (45%) and potassium, sodium and chloride in small amounts.

6.2.12 *Autacoids –Prostaglandin E_I and $F_{z\alpha}$ [21,22]*

The prostaglandins are among the most prevalent autacoids and have been detected in almost every tissue and body fluid; they produce, in minute amounts, a remarkably broad spectrum of effects that embrace practically every biological function. No other autacoids show more numerous and diverse effects than do prostaglandins. Just by looking at the structure of prostaglandins, their surface-active nature becomes apparent *prima facie*. The CMC values are given in Table 1. One can, therefore, suspect that prostaglandins, when added to an aqueous phase, according to Kesting's liquid membrane hypothesis would generate surfactant layer liquid membranes at the interface.

In these studies [21,22], the data on the hydraulic permeability in the presence of various concentrations of the prostaglandins have been obtained to demonstrate the formation of the surfactant layer liquid membrane in series with a supporting membrane. The data on the hydraulic permeability in the presence of varying concentrations of the prostaglandins in a mixture of lecithin and cholesterol of fixed composition, have been utilized to demonstrate the incorporation of the prostaglandins into the liquid membrane generated by the lecithin-cholesterol mixture. Transport of several relevant permeants through the liquid membranes, generated by the lecithin-cholesterol-prostaglandin mixtures, in series with a supporting membrane, has been studied, and the data obtained have been discussed in the light of the reported biological effects of the prostaglandins.

The hydraulic permeability data at various concentrations of prostaglandins, both PGE, and $PGF_{2\alpha}$, were found to be represented by the proportional relationship, $J_v = L_p \Delta P$. The trend in the values of L_p in case of both the prostaglandins, E_I and $F_{2\alpha}$ was found to be in accordance with Kesting's hypothesis indicating the formation of liquid membranes in series with the supporting membrane. Agreement of the experimental values of L_p with those calculated using the mosaic model, lent further support to the formation of liquid membrane.

Information on the incorporation of prostaglandins into the liquid membrane generated by the lecithin-cholesterol mixture was obtained from the hydraulic permeability data at varying concentrations of the prostaglandins in the lecithin-cholesterol mixture of fixed composition, 15.542 ppm with respect to lecithin and 1.175×10^{-6} M with respect to cholesterol. The data revealed that as the concentration of the prostaglandins is increased, holding the concentration of lecithin and cholesterol constant, the value of L_p, which measures the reciprocal of the resistance to volume flow, decreases. This decreasing trend in the values of L_p continues up to PGE_1 concentration equal to 0.6×10^{-8} M and a $PGF_{2\alpha}$ concentration equal to 6.97×10^{-8} M, and thereafter, the values of L_p becomes more or less constant. This trend in the values of L_p is indicative of the strengthening of the hydrophobic core of the liquid membrane generated by the lecithin-cholesterol mixture at the interface due to incorporation of the prostaglandins in it. It is also apparent from the trend that at a concentration equal to 0.6×10^{-8} M, the lecithin-cholesterol liquid membrane is saturated with PGE_1 and at a concentration equal to 6.97×10^{-8} M, the lecithin-cholesterol liquid membrane is saturated with $PGF2_{\alpha}$. In order to ascertain whether the added prostaglandin reaches straight to the interface or not, surface tensions of solutions of various concentrations of the prostaglandins –both PGE_1 and $PGF_{2\alpha}$ prepared in the aqueous solutions of lecithin-cholesterol mixtures of composition 15.542 ppm with respect to lecithin and 1.175×10^{-6}M with respect to cholesterol were measured. The surface tension of the aqueous solution of the lecithin cholesterol mixture showed a further decrease upon addition of the prostaglandins. The decreasing trend of the surface tensions continued up to 0.6×10^{-8}M concentration in the case of PGE_1, and up to 6.97×10^{-8} M concentration in the case of $PGF_{2\alpha}$. This trend indicates that the added prostaglandin, both PGE_1 and $PGF_{2\alpha}$, reach deep into the interface of the liquid membranes generated by the lecithin-cholesterol mixtures in series with the supporting membrane.

For solute permeability (ω) measurements for the relevant permeants the method outlined in section 6.1 was used. Compartment C of the transport cell (Fig. 2 Chapter 5) was filled with the solution of known concentration of the permeant prepared in the aqueous solution of lecithin, cholesterol, and one of the prostaglandins (PGE_1 and $PGF_{2\alpha}$) under study and compartment D was filled with distilled water. The composition of the aqueous solution of the lecithin-cholesterol-prostaglandin mixture used in the solute permeability experiments was such that the liquid membrane generated by lecithin, in series with the supporting membrane, was completely saturated with both cholesterol and the prostaglandin under study. Since lecithin, cholesterol, and prostaglandins are all surface active in nature, i.e., they have both hydrophilic and hydrophobic parts in their structure, it is obvious that in the liquid membranes generated in the solute permeability experiments, the hydrophobic tail of these molecules will be oriented preferentially toward the hydrophobic supporting membrane and the hydrophilic moieties will be drawn away from it. All measurements were made at 37 ± 0.1°C.

The data, recorded in Table36 on the solute permeability (ω) of various permeants in the presence of the liquid membranes generated by PGE_1 and $PGF_{2\alpha}$ in association with the lecithin cholesterol mixtures, appear relevant to the various reported pharmacological actions of the prostaglandins. The data (Table 36) show that the solute permeability (ω) for glucose is increased in presence of PGE_1 and $PGF_{2\alpha}$, the increase in the presence of PGE_1 being much larger than the increase in the presence of $PGF_{2\alpha}$. This observation on the increase in permeability of glucose is consistent with the literature reports, particularly in the case of PGE_1. It is documented [205, 206] that in isolated adipose tissues PGE_1 stimulates glucose uptake.

Cardiac output is generally increased by the prostaglandins of E and F Series [207]. It is also known [208] that adrenaline is a powerful cardiac stimulant and enhances cardiac output by acting on β_1 receptors. The data obtained in this study (Table 36) indicate that transport of adrenaline is increased in the presence of the liquid membranes generated by the prostaglandins. This observation suggests that the increased permeability of adrenaline due to the prostaglandins present in the membranes of myocardial cells facilitating interaction with β_1 receptors may also be a contributing factor to the reported increase in cardiac output by the prostaglandins.

Prostaglandins of E series are known to inhibit the gastric acid secretion stimulated by feeding histamine [209, 210] and this has raised the possibility of the therapeutic utility of certain methylated analogs of prostaglandins for peptic ulcers [211]. The gastric acid secretion by histamine is exerted through H_2-receptor antagonists [212]. It has been indicated [9] that an impediment in the transport of histamine due to the liquid membranes, which are likely to be generated by the H_2-receptor antagonists, drugs like Cimetidine and ranitidine, at the site of action may also contribute to their H_2-antagonistic action. The data on the transport of histamine in the presence of the PGE_1 liquid membrane (Table 36) appear relevant to the reported [209, 210] inhibition of gastric acid secretion by the PGE_1. It appears that the resistance offered by the PGE_1 to the transport of histamine impeding its access to the H_2-receptors, may also be a cause of the inhibition of histamine-induced gastric acid secretion from the parietal cells. Although histamine transport is also impeded in the presence of $PGE_{2\alpha}$ (Table 36.), the relevance of this observation in the context of gastric acid secretion is not clear.

Table 36. Solute permeability (ω)[a] of various permeants in presence of liquid membranes generated by prostaglandin E_1 (ω_1) and prostaglandin $F_{2\alpha}$ (ω_2) in lecithin-cholesterol mixtures[b] along with the control values (ω_0) when no prostaglandin was used (Ref. 21, 22).

Permeants	Initial Concentration (mg liter $^{-1}$)	$\omega_0 \times 10^9$ (mole $s^{-1}N^{-1}$)	$\omega_1{}^c \times 10^9$ (mole $s^{-1}N^{-1}$)	$\omega_1{}^d \times 10^9$ (mole $s^{-1}N^{-1}$)
Glucose	10	0.288 ± 0.030	0.412 ± 0.070	0.360 ± 0.011
Histamine	10	0.392 ± 0.019	0.244 ± 0.007	0.229 ± 0.001
Adrenaline	100	1.592 ± 0.038	1.697 ± 0.026	2.216 ± 0.009
Ethinyl estradiol	50	2.280 ± 0.046	3.135 ± 0.035	3.424 ± 0.068
Progesterone	100	0.198 ± 0.032	0.697 ± 0.093	0.279 ± 0.006
Glycine	100	1.517 ± 0.061	2.284 ± 0.059	1.279 ± 0.034
γ-Amino butyric acid	200	0.916 ± 0.012	1.170 ± 0.024	1.153 ± 0.039
Sodium chloride	5.382	0.133 ± 0.008	0.196 ± 0.012	0.197 ± 0.002
Potassium chloride	10.430	0.122 ± 0.008	0.055 ± 0.001	0.101 ± 0.003
Serotonin	10	1.442 ± 0.038	1.135 ± 0.057	2.222 ± 0.021
Dopamine	10	0.462 ± 0.025	0.193 ± 0.002	0.553 ± 0.041
Noradrenaline	10	0.443 ± 0.071	0.328 ± 0.014	0.711 ± 0.005

[a] Values of ω are reported as arithmetic means of 15 repeats ± S.D..
[b] Lecithin concentration 15.542 ppm; cholesterol concentration, 1.175×10^{-6}M.
[c] Prostaglandin E_1 concentration, 0.65×10^{-8}M.
[d] Prostaglandin $F_{2\alpha}$ concentration, 8.5×10^{-8}M.

Table 36. reveals that in the case of PGE_1 the transport of both glycine and GABA is enhanced, whereas in the case of $PGF_{2\alpha}$ the transport of GABA is enhanced and that of glycine is impeded. The enhancement in the transport of glycine and of GABA leading to their increased concentration in the brain could also be the reason for reported [213-215] effects such as sedation, stupor, catatonia, etc., induced by the administration of prostaglandins, particularly PGE_1, in animals.

It is reported [213] that in the intact central nervous system of a chloroluse-anesthetized chick, intravenous administration of $PGF_{2\alpha}$ potentiates the crossed extensor reflex while PGE_1 inhibits it. The opposing trends observed in the transport of glycine (Table 36.), which is known [216] to be utilized by the inhibitory interneurons of the spinal cord, may be relevant to the reported potentiation and inhibition of the crossed extensor reflex in chicks.

Prostaglandins have been used as abortive agents [217]. The data on the permeability of estrogen and progesterone (Table 36) appear relevant to their abortive action. No only high concentration of estrogen and progesterone but also a proper ratio of their concentrations is essential for the maintenance of pregnancy [192]. As these data indicate, the presence of high concentrations of prostaglandins in the membranes of uterus would enhance the outflow of estrogen and progesterone to an unequal extent. This out flow would not only decrease the concentrations of estrogen and progesterone but also disturb the estrogen progesterone ratio resulting in failure of pregnancy.

The data on the permeability of estrogen and progesterone also appear relevant to the causation of primary dysmenorrhea. There is substantial evidence to indicate that prostaglandin is a major causal factor in primary dysmenorrhea [218]. Drugs having prostaglandin synthetase inhibitory activity have been reported to be effective in the treatment of dysmenorrhea. The effectiveness of oral contraceptive in the treatment of dysmenorrhea is also well established [218]. These observations appear to indicate that the enhanced permeability (outflow) of estrogen and progesterone in the presence of the increased concentration of prostaglandins, particularly $PGF_{2\alpha}$, in the endometrium may also be a factor responsible for dysmenorrhea.

Prostaglandins of E and F series are present in the renal medulla. Renal prostaglandins have been implicated in antihypertensive action [219]. It is suggested that prostaglandins may exert an antihypertensive action, acting either as peripheral vasodilators or by promoting diuresis with sodium loss, i.e., natriuresis [219]. The enhanced permeability to sodium ions in the presence of prostaglandins as observed in these experiments appears consistent with the latter mechanism. Sodium reabsorbation in proximal tubule is active in nature and is mediated by carbonic anhydrase [220]. Besides forces moving Na^+ ion and water out of the proximal tubule, there is component of leakage back across the tubular epithelium into the lumen of the proximal nephron [221]. The back leak is passive in nature and its amount is influenced by peritubular osmotic pressure [221]. The increased passive transport of Na^+ ions in the presence of prostaglandins (Table 36) may, thus offer an explanation for the diuretic and natriuretic effects of the prostaglandins due to the back-leak mechanism leading to their antihypertensive action.

The toxin *Vibrio cholerae* affects electrolyte handling by the epithelial cells of the intestinal mucosa in such a way that there is hypersecretion into the gut resulting in the profuse watery stools that characterize cholera. It has been suggested that the toxin acts by stimulating prostaglandin synthesis [222]. The enhanced permeability of Na^+ ions in the presence of the prostaglandins, as observed in this study (Table 36), suggests that a back-leak mechanism similar to the one proposed in the case of natriuretic and diuretic effects of the prostaglandins [221], may also explain the hypersecretion into the lumen of the intestines due to the increased concentration of the prostaglandins in the epithelial cells of the intestinal mucosa.

$PGF_{2\alpha}$ does not affect the transport of K^+ ions significantly (Table 36). In the presence of PGE_1, however, a decrease in the transport of K^+ ions is observed (Table 36). The observation of the decreased permeability of K^+ ions may be relevant to the causation of Barter's syndrome. Barter's syndrome, an unusual and complex disorder, which is characterized by, among other symptoms, hypokalemia, i.e., excessive loss of potassium, has been associated with excessive production of renal prostaglandins [223]. This is obvious from the fact that Barter's syndrome has been successfully treated with drugs like indomethacin and aspirin [224-227], which have prostaglandin synthetase inhibitory activity. Although potassium reabsorption in proximal tubules is active in nature, these data suggest that impediments in the transport of K^+ ion due to the increased concentration of the prostaglandins in the tubular cells, may also contribute to the urinary potassium wasting, leading to hypokalemia.

On a macroscopic level inflammation is usually accompanied by the familiar clinical signs of erythema, edema, hyperalgesia and pain [228]. Prostaglandins are always released when cells are damaged and have been detected in increased concentrations in inflammatory exudates [228]. During inflammation chemical mediators like histamine, serotonin etc. are also liberated locally, which stimulate sensory nerve endings and cause pain [228,229]. Prostaglandins by themselves are not known to act directly to stimulate sensory receptors subserving pain [230]. It is documented that histamine or serotonin antagonists have little therapeutic effect in inflammation whereas aspirin-like drugs which have little or no effect upon the release of activity of histamine or serotonin but are well known for their prostaglandin synthase inhibitory activity are therapeutically important in the treatment of inflammation [228].

The data on the reduced volume flow in the presence of prostaglandins [22] indicate that the liquid membranes formed by prostaglandins released in the interstitial fluid offering resistance to the volume flow may be a contributing factor to the causation of edema. Similarly, impediment in the transport of histamine and serotonin in the presence of the liquid membranes generated by the prostaglandins, as observed in these studies [21, 22] (Table 36), may lead to the accumulation of histamine and serotonin in the interstitial region causing hyperalgesia and pain. Thus it appears that the phenomena of liquid membrane formation by the prostaglandins may be a contributing factor to the causation of edema, hyperalgesia and pain in inflammation and its cure by the prostaglandin synthase inhibiting drugs like aspirin. It may be mentioned that intradermal, intravenous and intra-arterial injections of prostaglandins produce effects strongly reminiscent of inflammation [228].

Transport through liquid membrane bilayers generated by prostagland in E_1 has been studied in the presence of hydrocortisone. The data have indicated the formation of aqueous pores when hydrocortisone is added on both the sides of the prostaglandin E_1 liquid membrane bilayer [16]. The phenomenon of aqueous pore formation has been utilized to explain the therapeutic action of hydrocortisone in the treatment of inflammation. A detailed discussion is presented in Chapter 5 Section 5.4.2.

The suggestion that prostaglandins, particularly prostaglandin E_1, may be implicated in migraine has been made in literature [231,232]. This suggestion has been prompted by the observation that intravenous injection of prostaglandin E_1 in non-migrainous subjects consistently resulted in vascular headache that bore migrainous features. Reduction in the concentration of serotonin at post-synaptic receptor resulting in defective neurotransmission has also been implicated in migraine [231]. Since serotonin is a prostaglandin releasing factor, it has been suggested [231] that hypotheses implicating either of these agents are not mutually exclusive. The data on the reduced permeability of serotonin in the presence of prostaglandin E_1 (Table 36) suggest that the reduced access of serotonin to the postsynaptic receptor due to the prostaglandin liquid membrane formed at the receptor site could also be a contributing factor to the causation of migraine by prostaglandins.

Shock is considered essentially to be an inadequate tissue perfusion that impairs normal organ functions [233]. Elevated levels of circulating prostaglandins have been observed in several shock states [234-236], though the exact significance of prostaglandins in various shock models remains unclear [233]. Aspirin like drugs, which inhibit prostaglandin synthesis, are reported to have beneficial effects in several shock states [233]. The data on the reduction in volume flow in the presence of prostaglandins [22] appear to indicate that the liquid membranes formed by prostaglandins in the blood capillaries offering resistance to the volume flow into the interstitial region resulting in impaired tissue perfusion could be a plausible explanation for these observation.

A similar explanation can be offered in the case of secondary glaucoma due to inflammation. Anterior uveitis particularly iridocyclites results in an increased intraocular pressure because of the swelling and the increased rate of fluid formation including the inflammatory exudates. This condition is reported to respond to non-steroidal anti-inflammatory drugs [237]. The ability of prostaglandins to raise intraocular pressure in rabbit eyes is well known [238]. The outflow of aqueous humor in humans and primates occurs primarily through the conventional drainage pathway through the angle of anterior chamber via canal of Schlemm [239]. The data on reduced volume flow in the presence of prostaglandin [22] appear to suggest that blockade of the drainage pathway by the prostaglandin liquid membranes could be a contributing factor to the increased intraocular pressure in to secondary glaucoma due to inflammation.

It is reported [240,241] that prostaglandins often modify sympathetic neuroeffector junctions in exceedingly low concentration. For example, prostaglandins of E series inhibit noradrenaline output from adrenergic nerve endings and depress the response of the

noradrenaline output from adrenergic nerve endings and depress the response of the innervated structures whereas contrary effects leading to increased output of noradrenaline or heightened responsiveness of the effector organ have been noted with prostaglandins of F series. These observations appear consistent with the findings [22] that the permeability of noradrenaline is reduced in the presence of prostaglandin E_1 and enhanced in the presence of prostaglandin $F_{2\alpha}$ (Table 36).

Prostaglandins of E series are known [242] to relax bronchial smooth muscle and produce bronchodilation in the lungs *in situ*. Bronchoconstrictor responses to histamine, serotonin and other bronchospasmogens are counteracted by prostaglandins of E series [242]. The data (Table 36) therefore suggest that the liquid membrane formed by prostaglandin E_1 offering resistance to the transport of histamine and serotonin to their sites of action could be one of the contributing factors to the observed bronchodilation effects of prostaglandin E_1 and also to the observe counteraction of the bronchoconstrictor respone to histamine and serotonin.

Prostaglandins of E series are reported [240, 241] to inhibit water reabsorption induced by antidiuretic hormone in toad bladder. The reduced values of L_p, as observed in these studies [22] may also contribute to the reported inhibition of water reabsorption in the presence of prostaglandins of E series leading to diuresis.

In epileptic patients marked increase in prostaglandin $F_{2\alpha}$ levels in cerebrospinal fluid has been detected [243]. It is also documented that prostaglandins of E series antagonize convulsions induced by pentylenetertrazol, penicillin and picrotoxins [244]. These observations appear consistent with the trends observed in the solute permeability data for excitatory neurotransmitters, viz. dopamine, serotonin and noradrenaline, and inhibitory neurotransmitters, i.e. glycine and γ-aminobutyric acid. (GABA), in the presence of prostaglandin E_1 and prostaglandin $F_{2\alpha}$ (Table 36). Transport of the excitatory neurotransmitters is impeded in the presence of prostaglandin E_1 and enhanced in the presence of prostaglandin $F_{2\alpha}$. Transport of the inhibitory neurotransmitter GABA though enhanced in the case of both, prostaglandin E_1 and prostaglandin $F_{2\alpha}$, the transport of glycine is enhanced in the case of prostaglandin E_1 and impeded in the case of prostaglandin $F_{2\alpha}$ (Table 36).

The nerve cell bodies of the paravantricular and supraoptic nuclei have both cholinergic and noradrenergic nerve endings impinging on them. Thus the activity in the neurosecretory cells is perhaps also controlled by noradrenaline. It is reported that noradrenaline injected into carotid circulation inhibits the release of antidiuretic hormone [245]. The data on the reduced permeability of noradrenaline in the presence of prostaglandin E_1, (Table 36) appear to indicate that access of noradrenaline to the postsynaptic receptor may be reduced due to the resistance offered by the liquid membranes generated by prostaglandin E_1 in association with the membrane lipids and thereby stimulate antidiuretic hormone release. It is interesting to point out that prostaglandins of E series when injected into common carotid artery or the cerebral ventricles also stimulate the release of antidiuretic hormone [245].

At hypothalamic level, the secretion of prolactin in mammals in controlled by the prolactin release-inhibiting hormone (P-RIH) and possibly by a prolactin-releasing hormone (P-RH). The role, if any, of P-RH is of secondary importance [245]. The release of P-RIH from neuroendocrine transducer cells in the median eminence is controlled by hypothalamic dopamine and it has been shown that the drugs; like reserpine, chlorpromazine and haloperidol which reduce the permeability of dopamine [1-3] and thereby lower its concentration in the hypothalamic region, decrease the release of P-RIH and hence promote prolactin release [245]. Systemic administration of prostaglandins, particularly prostaglandin E_1 has been shown to stimulate prolactin release [246]. This observation is consistent with the decreased permeability of dopamine in the presence of prostaglandin E_1 as observed in this study (Table 36). It appears that the reduced access of dopamine to its site of action in hypothalamus, which also is reported to be the site of action of prostaglandins [246], may be a contributing factor in the prolactin release stimulating action of prostaglandin E_1.

Thus it appears that modification in the transport of relevant permeants to their respective sites of action due to the liquid membrane formed by prostaglandins in association with membrane lipids may also contribute to their biological actions.

6.2.13. Antidepressant drugs [4]

Tricyclic antidepressant drugs like imipramine are known to be surface active in nature [247, 248]. Explanation of the antidepressant action is based on the fact that these drugs reduce the uptake of catecholamines in the nervous tissue [249]. Data on hydraulic permeability has been obtained and utilized to demonstrated [4] the formation of liquid membrane by imipramine hydrochloride in series with supporting membrane cellulose acetate microfilatration membranes Cat No.11107. Measurements of solute permeability (ω) of biogenic amines and cations in the presence of the liquid membrane generated by imipramine hydrochloride have been made. For the measurements of solute permeability (ω) two sets of experiments have been performed: in the first set of experiments the permeants face the hydrophilic surface of the liquid membrane whereas in the second set of experiments permeants face the hydrophobic surface of the liquid membrane, in the control experiments, however, no imipramine hydrochloride was used. In the solute permeability experiments the concentration of imipramine hydrochloride has always higher than it CMC, to ensure that the supporting membrane was completely covered with the liquid membrane. All measurements were made at $37 \pm 0.1^{\circ}C$. The data on solute permeability (ω) are recorded in Table 37.

The values of ω recorded in Table 37 indicate that in the first set of experiments, where the Imipramine liquid membrane presents a hydrophilic surface to the permeants, permeability of biogenic amines is increased. In the second set of experiments, however, where the Imipramine liquid membrane presents a hydrophobic surface to the approaching permeant, a marked decrease in the permeability of biogenic amines is observed. Since *in vivo* imipramine is known to act by reducing the uptake of biogenic amines [249] this study indicates that the specific orientation of Imipramine with hydrophobic ends facing the permeants would also be necessary in the vicinity of nerve terminals. This reduction in the passive transport of biogenic amines and cations (Table 37) is also likely to be accompanied by a consequent reduction in their active transport because access of the permeants to the active site located on the nerve membrane is likely to be effectively reduced due to the resistance offered by the liquid membrane interposed in between. Thus although neuronal uptake of biogenic amines is by active process [250] the formation of liquid membrane by imipramine seems to have a contribution in the mechanism of its action.

Table 37. Solute permeability ω of biogenic amines and cations in presence of imipramine hydrochloride (Ref. 4).

	ω_1 (moles $N^{-1}sec^{-1}$)	ω_2 (moles $N^{-1}sec^{-1}$)	ω_3 (moles $N^{-1}sec^{-1}$)
Dopamine	2.657×10^{-10}	1.048×10^{-10}	4.337×10^{-10}
Noradrenaline	9.893×10^{-11}	4.820×10^{-11}	1.210×10^{-9}
Adrenaline	4.625×10^{-10}	1.768×10^{-10}	6.887×10^{-10}
5-Hydroxytryptamine	2.272×10^{-10}	0.973×10^{-10}	9.541×10^{-10}
Sodium (chloride)	0.862×10^{-10}	0.469×10^{-10}	0.712×10^{-10}
Potassium (chloride)	4.757×10^{-10}	1.383×10^{-10}	1.861×10^{-10}
Calcium (chloride)	0.566×10^{-10}	0.102×10^{-10}	0.208×10^{-10}

Note: Imipramine hydrochloride concentration = 5.92×10^{-4}M
ω_1 –Control value when no Imipramine was used.
ω_2- Imipramine in compartment D of the transport cell-the second set of experiments.
ω_3-Imipramine in compartment C of the transport cell-the first set of experiments.

In certain tissues imipramine is known to increase outflow of noradrenaline [250,251]. This may be because of the specific orientation of imipramine with its hydrophilic ends facing the catecholamines. Presumably even on adrenoceptors similar orientation of imipramine molecule with respect to the relevant biological membrane is necessary. However more understanding in terms of orientation of imipramine molecule with respect to relevant biological membrane is necessary.

The reduced permeability to cations in both orientations of imipramine can be explained on the basis of hydrophilicity of the ions. This observation may have relevance to the effect of imipramine on nerve conduction. The tricyclic antidepressants (TCA) are found to cause postural hypotension [252]. Calcium ion when enters the inside the vascular smooth cells causes their excitation and contraction and thereby increases the total peripheral resistance and this causes increased blood pressure. It was found [253] that the liquid membrane formed by TCA drugs decreased the permeability of calcium, sodium and potassium ions. This observation if viewed in the light of permeation of ions might explain the blood pressure lowering effect of TCAs. The liquid membranes formed around the vascular smooth muscle cells may decrease the permeation of extra cellular calcium and sodium into the cell.

Decrease in intracellular sodium concentration in the vascular smooth muscle may decrease stiffness of vessel wall, increase their compliance and dampen responsiveness to constrictor stimuli (by nor-adrenaline, angiotensin II). Quite recently Nagappa et. Al [253] has investigated the influence of membrane lipids in the actions of TCA drugs. However, no clear-cut conclusions can be drawn due to inadequate design of experiments.

6.2.14. Antiepileptic drugs. [31]

Antiepileptic drugs are known to stabilize biological membranes [254] after interacting with them. They are known to contain both hydrophilic and hydrophobic moieties in their structure [255]. The antiepileptic drugs therefore, are expected to be surface active in nature (CMCs given in Table 1) and hence capable of generating liquid membrane at the interface in accordance with Kesting's hypothesis. Depressant drugs, in general, are reported to populate at the air-solution interface [256]. In these studies, existence of liquid membranes generated by the antiepileptic drugs, at a cellulosic microfiltration membrane /aqueous interface has been demonstrated. Data on the modification in the transport of γ-aminobutyric acid (GABA) in the presence of liquid membranes have been obtained and discussed in the light of the mechanism of faction of drugs. Three structurally dissimilar antiepileptic drugs, namely diphenylhydantoin, carbamazepine and valproate sodium, have been chosen for the present study. A Sartorius cellulose acetate microfiltration membrane was, as in other cases cited in the foregoing sections, deliberately chosen as supporting membrane for the liquid membranes so that the role of passive transport through the liquid membrane is highlighted.

Hydraulic permeability data in the presence of antiepileptic drugs, in case of all three drugs, were found to be in accordance with the relationship $J_v = L_p \Delta_p$. These data were utilized as in the in the other cases discussed above to demonstrate the formation of liquid membrane in series with the supporting membrane. The values of hydraulic conductivity coefficient L_p show a progressive decrease with increase in concentration of the drugs up to their CMCs, beyond which they become more or less constant. The normalized values of hydraulic conductivity coefficient –the values of (L_p/L^0_p) where L^0_p is the value of L_p when no drug was used, are plotted against drug concentrations in Fig 6 for all the three drugs.

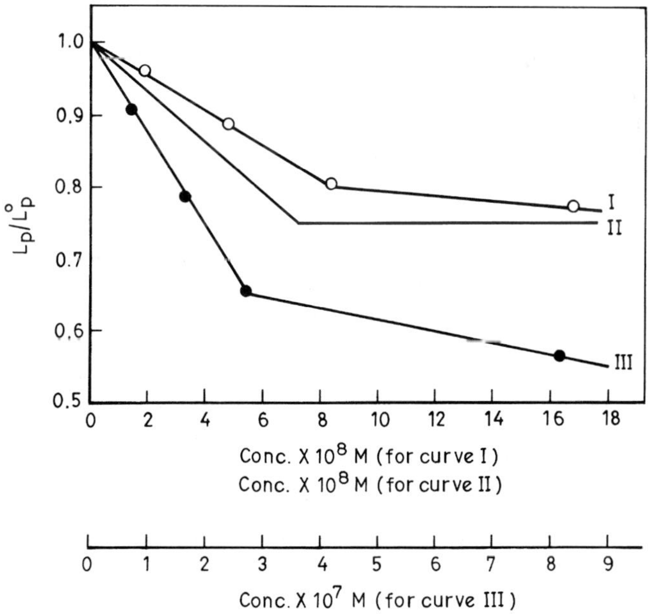

Fig. 6 Variation of (L_p/L^0_p) with concentration of the drugs. Curves I, II and III represent data in presence of carbamazepine, valproate sodium and diphenylhydantoin, respectively (Ref. 31).

The progressive decrease in the values of hydraulic conductivity with increasing concentrations of the drugs up their CMCs (Fig.6) is indicative of the progressive coverage of the supporting membrane with the liquid membrane generated by the drug, in accordance with the liquid membrane hypothesis [32]. At the CMC, coverage of the supporting membrane with the liquid membrane is complete. The slight decrease in the values of (L_p/L^0_p) beyond the CMCs particularly in the case of diphenylhydantoin and carbamazepine may be due to densing of the liquid membrane, which is completely developed at the CMC of the drugs, as postulated by Kesting in the liquid membrane hypothesis [32].

Analysis of the flow data in the light of mosaic model [43-45] was utilized to further confirm the existence of liquid membrane in series with the supporting membrane.

Since antiepileptic action is determined by the concentration of γ-aminolintyric acid (GABA) in brain, data on the modification of permeability of GABA in the presence of liquid membrane generated by the antiepileptic drugs have been obtained and are recorded in Table 38. The data in Table 38 are for the two orientation of the liquid membrane: one in which the permeant GABA faces the hydrophilic surface of the liquid membrane and the other in which it faces the hydrophobic surface of the liquid membrane. Details of the experiment are given in the original paper[31] and also in section 6.1.

Table 38. Permeability of GABA[a] (ω)[e] in presence of antiepileptic drugs[b] (Ref. 31)

	ω_1 x 10^{10} (mol $s^{-1}N^{-1}$)	ω_2 x 10^{10} (mol $s^{-1}N^{-1}$)	ω_3 ẋ 10^{10} (mol $s^{-1}N^{-1}$)
Diphenylhydantoin	1.1682 ± 0.1532	1.8487 ± 0.6815	1.1501 ± 0.1078
Carbamzepine	1.1682 ± 0.1532	4.8175 ± 0.4424	1.0680 ± 0.1545
Valproate Sodium	1.1682 ± 0.1532	2.0039 ± 0.3782	1.6624 ± 0.1534

ω_1 : control value – when no drug was used. ω_2 : The value of ω when permeant GABA facing hydrophilic surface of the liquid membrane.

ω_3 : The value of ω when the permeant GABA facing hydrophobic surface of the drug liquid.

[a] Initial concentration of GABA is 200μg/ml

[b] The concentration diphenylhydantoin, Carbamzepine and valproate sodium are 8×10^{-7}M, 1.6×10^{-7} and 1.6×10^{-4}M respectively.

[c] Values of ω are reported as arithmetic mean of 10 repeats ± standard deviation.

In the solute permeability measurements the concentration of the drugs was always higher than their CMCs. A perusal of Table 38 reveals that in the first set of experiments, where the permeant, GABA, faces the hydrophilic surface of the drug liquid membrane, the permeability of GABA ins enhance considerably in case of all the three drugs. In the second set of experiments, however where the permeant GABA faces hydrophobic surface of the drug liquid membrane there is a distinct reduction in the permeability of GABA, except in case of sodium valproate, where an increase in the permeability is observed (Table 38). Even in the case of sodium valproate, the increase in the permeability of GABA is much more in the first set of experiments than in the second set. These observations on the increase in the permeability of GABA appear relevant to the antiepileptic action.

The antiepileptic drugs which, when administered, exert stabilizing effect [254] on excitable cell membranes, are known to increase the concentration of GABA in brain. These experiments appear to indicate that increased permeability of GABA in presence of the drug liquid membranes, which are likely to be formed at the site of action, may be responsible for the increased concentration of GABA in brain. Since enhancement in the permeability of GABA was observed to be maximum in the first set of experiments, it appears that the specific orientation of the drug molecules in the liquid membrane, with their hydrophilic ends facing the permeant may be necessary even at the actual site of action. To substantiate this conjecture, detailed investigations of the nature of the site of action are called for.

Another indication of the possible role of the liquid membrane phenomenon in antiepileptic action is obtained from the gradation in values of CMCs of the drugs (Table1) vis-à-vis the gradation in the concentrations of these drugs in plasma. The CMC values of the three drugs are in the following order (Table 1):

valproate > diphenylhydantoin > carbamazeine

which also the gradation in their concentrations in plasma [257]. Concentrations of the drugs in plasma can be taken to be a measure of their concentrations at the site of action. The reported concentrations of these drugs in plasma [257] are far higher than their respective CMCs. Hence complete liquid membranes can be generated by the drugs at the site of action. Since modification in the permeability of GABA due to the presence of the drug liquid membranes is responsible for the antiepileptic action, the concentrations of the drugs required to produce maximum biological response may be related to their CMCs. CMC is the concentration at which the interface is completely covered by the liquid membrane and therefore, modification in the permeability of biomembranes to GABA will be maximum at this concentration. Hence agreement between the gradation in the concentration of the drugs in plasma and the gradation in their CMCs is also indicative of the contribution of the liquid membrane generated by these drugs, to their antiepileptic action.

This study, thus, indicated that the formation of liquid membrane at the site of action, by the drugs, modifying the transport of GABA, may be an important step common the mechanism of action of all the three drugs, namely diphenylhydantoin, Carbamzepine and valproate sodium.

6.2.15 Hypnotic and sedative (7, 26)

Three drugs namely, diazepam, nitrazepam and chlordizepoxide belonging to the category of benzodaizepines have been investigated for the role of liquid membrane phenomena in the biological action of these drugs. All three drugs are reported to be surface active [7, 26, 258]. The CMC are given in Table 1 and hence these drugs should be capable of forming liquid membranes in accordance with kestings hypothesis [32].

Existence of a liquid membrane generated by diazepam at the interface has been demonstrated. The transport of gamma-aminobutyric acid (GABA) and glycine through the diazepam liquid membrane has been studied. In all drugs, which act by modifying the permeability of biomembranes, it is relevant to study the interaction of the drugs with membrane lipids. In fact, studies on interaction of several drugs with phospholipids monolayers are available in the literature [247]. The studies have therefore been extended to the liquid membranes generated by lecithin-cholesterol-diazepam mixture. The transport data indicate that he liquid membrane phenomena may make a notable contribution to the actions of diazepam.

In these experiments also a cellulosic microfiltration membrane (Sartorius Cat No.11307) has been used to highlight to contribution of passive transport through the liquid membrane.

For hydraulic permeability measurements two sets of experiments were performed. In one set of experiments aqueous solutions of diazepam of various concentrations ranging from 0 to 2×10^{-4} M filled in compartment C of the transport cell (Fig.2 Chapter 5) whereas compartment D was filled with water. The concentration range from 0 to 2×10^{-4}M was chosen to get data on both the lower and the higher side of the CMC of diazepam. In another set of experiments solutions of various concentrations of diazepam prepared in an aqueous solution of lecithin-cholesterol mixture which was 15.542 pip with respect to lecithin and 1.175×10^{-6} M with respect to cholesterol, filled in compartment C, whereas compartment D of the transport cell was filled with water. This particular composition of the lecithin-cholesterol mixture was chosen because it has been shown in an earlier study [90] that at this composition the liquid membrane generated by lecithin is saturated with cholesterol and completely covers the supporting membrane.

Solute permeabilities (ω) for glycine and GABA were measured in presence of both diazepam and the lecithin-cholesterol-diazepam mixture. For measurements of ω in presence of diazepam two sets of experiments, solution of the permeant-glycine or GABA, prepared in aqueous solution of known concentration of diazepam, filled in the compartment C, whereas compartment D was filled with water. In the second set of experiments, compartment D was filled with the aqueous solution of diazepam, and compartment C was filled with the aqueous solution of the permeant. However, in control experiments no diazepam was used. The concentration of diazepam used in these experiments was 1.6×10^{-4} M which is well above its CMC. Similar sets of experiments were carried out ω measurements in presence of lecithin-cholesterol-diazepam mixtures. The diazepam concentration in these experiments was 0.75×10^{-4}M –the concentration at which the liquid membrane generated by lecithin-cholesterol mixture becomes saturated with diazepam. The details of the procedure adopted for hydraulic permeability measurements and (ω) measurements are described in the original paper [7] (See also section 6.1)

The hydraulic permeability data at all concentration of diazepam were found to obey the relationship $J_v = L_p \Delta_p$. The variation of L_p with concentration in the presence and in the absence of lecithin-cholesterol mixture is shown in Fig.7. The trend in curve I of Fig 7 is indicative of progressive coverage of the supporting membrane with diazepam liquid membrane, in accordance with Kesting's hypothesis [32]. At the CMC the coverage of the supporting membrane with diazepam liquid membrane is complete.

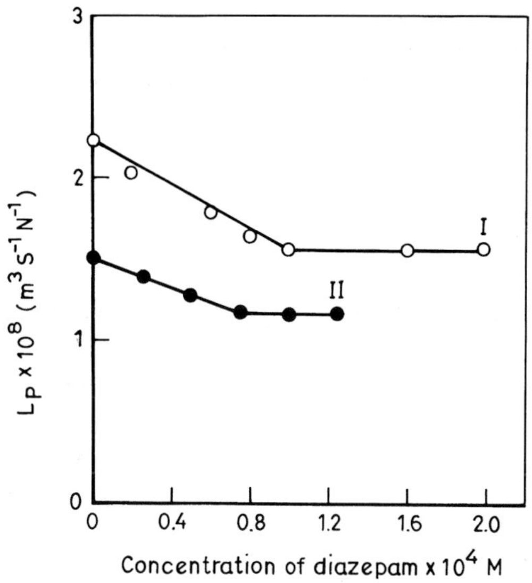

Fig. 7 Variation L_p with diazepam concentration. Cureve I represents data in the presence of diazepam alone, while curve II represents data in the presence of lecithin-cholesterol –diazepam mixtures (Ref. 7).

The hydraulic permeability data for the other set of experiments, where compartment C of the transport cell (Fig.2 Chapter 5) contained aqueous solutions of various concentrations of diazepam prepared from lecithin-cholesterol mixtures and compartment D control mixtures and compartment D contained distilled water, were also found to obey the linear relationship; $J_v = L_p \Delta P$. the values of L_p show a decrease with increasing concentration of diazepam up to 0.75×10^{-4}M beyond which they become more of less constant (Fig. 7 curve II).This indicates that diazepam is incorporated within the liquid membrane generated by the lecithin-cholesterol mixture and that when its concentration equal 0.75×10^{-4}M, the lecithin-cholesterol liquid membrane is saturated with diazepam. To ascertain whether or not diazepam is found at the interface, surface tensions of solutions of lecithin-cholesterol mixtures of fixed composition-15.542 ppm with respect to lecithin and 1.175×10^{-6} M with respect to cholesterol were measured. The surface tension of the aqueous solutions of the lecithin cholesterol mixture showed a further decrease with the increase in concentration of diazepam up to 0.75×10^{-14}M. This indicates that the diazepam penetrates the liquid membrane generated by the lecithin-cholesterol mixture and is found at the interface.

The values of ω for glycine and GABA in presence of diazepam indicate (Table 39) that the diazepam liquid membrane in both the sets of experiments offers resistance to the transport of the aminoacids. Because diazepam is surface active [258], it consists of both hydrophobic and hydrophilic moieties. In the first set of experiments where diazepam and the permeants are present in the same compartment, the hydrophobic ends of the diazepam molecules will be preferentially oriented toward the hydrophobic supporting membrane. Thus in the first set of experiments the permeant will face the hydrophilic surface of the liquid membrane generated by diazepam. In the second set of experiments, however where diazepam is present in compartment D and the permeants in compartment C (Fig.2 Chapter 5), the liquid membrane will present a hydrophobic surface to the permeant. The data in Table 39 reveal that in both the orientations the diazepam liquid membrane impedes the transport of amino acids.

Table 39. Solute permeability ω of amino acids in presence of diazepam[a] and lecithin-cholesterol mixture[b] (Ref. 7).

	$\omega_1^c \times 10^{10}$	$\omega_2^d \times 10^{10}$	$\omega_3^e \times 10^{10}$	$\omega_4^f \times 10^{10}$	$\omega_5^g \times 10^{10}$	$\omega_6^h \times 10^{10}$	$\omega_7^i \times 10^{10}$
Glycine	18.145	15.681	5.269	13.297	14.827	2.007	2.330
GABA	16.438	14.479	8.731	10.345	12.942	3.781	1.783

Note. All values given in moles N^{-1} sec^{-1}.
[a] Diazepam concentration, 1.6 x 10^{-4}M.
[b]Lecithin concentration, 15.542 ppm; cholesterol concentration, 1.175 x 10^{-6}M; diazepam concentration, 0.75 x 10^{-4}M.
[c] Control value when no diazepam was used.
[d] Diazepam in compartment C.
[e] Diazepam in compartment D.
[f]Control value when lecithin-cholesterol mixture was taken in compartment C.
[g] Lecithin-cholesterol-diazepam mixture in compartment C together with the permeant.
[h] Control value when lecithin-cholesterol mixture was in compartment D.
[i] Lecithin-cholesterol-diazepam mixture in compartment D and permeant in compartment C.

The values of ω in presence of lecithin-cholesterol-diazepam mixtures (Table 39) appear to have some bearing on the mode of drug action. When lecithin-cholesterol-diazepam are present in compartment C (Fig. 2 Chapter 5) together with the permeants, the liquid membrane generated by the lecithin-cholesterol-diazepam mixture, i.e. the composite liquid membrane, presents a hydrophilic surface to the permeants. Similarly, when the permeants are present in compartment C and the lecithin-cholesterol-diazepam mixture in compartment D, the composite liquid membrane presents a hydrophobic surface to the permeants. In the former case the permeability of both glycine and GABA is enhanced considerably in comparison to the permeability values for blank experiments where no diazepam was used.

The observation of increase permeability of GABA appears to have biological relevance because in case of benzodaizepines, biochemical [259, 260] and neurophysiological [261-263] evidence has suggested that facilitating synaptic action of GABA in brain may exert the antianxiety action of diazepam. Displacement of an endogenous modulator protein forming part of macromolecular complex constituting GABA receptor ionophore has been suggested [264] as one possible mechanism of such an action. However, events at cellular and molecular level resulting in GABA potentiation are completely unknown [265]. The increased permeability of GABA through the lecithin-cholesterol-diazepam composite liquid membrane in the specific orientation of hydrophilic ends facing the permeants, as indicated in these experiments, can also be an explanation for facilitation of GABA action leading to antianxiety action of diazepam. Benzodaizepines are also known [266] to bind to sites that have high affinity for strychnine, which is an antagonist of glycine. It has been suggested, therefore, that some actions of benzodaizepines may result from their interaction with glycine receptor [267]. Hence, the observed increase in a permeability of glycine when it faces the hydrophilic surface of the composite liquid membrane generated by lecithin-cholesterol-diazepam mixture appears relevant.

The decrease in permeability of GABA (Table 39) when it faces the hydrophobic surface of the composite liquid membrane generated by the lecithin-cholesterol-diazepam mixture does not appear to be relevant to the antianxiety action of diazepam. Nevertheless, it can offer some insight into the cause of certain other effects of diazepam. Diazepam has been shown to inhibit the uptake of GABA into synaptosomes prepared from mouse brain [268] and also from rat cortical slices [269]. Calcium-dependent release of GABA bas also been shown to be inhibited by diazepam [268]. The observed decreased in permeability of GABA when it preferentially faces the hydrophobic end of the lecithin-cholesterol-diazepam composite liquid membrane may help to explain these effects.

Thus the ability of diazepam to become incorporated within the liquid membrane, which is generated by lecithin-cholesterol mixtures, and to modify the permeabilities of the inhibitory neurotransmitter amino acid molecules-glycine and GABA, appears to be related to the modality of drug action.

Since benzodaizepines in addition to anxiolytic action are also known to exert myorelaxant and anticonvulsant actions [270-272] involving multiplicity of neurotransmitter systems [270] including catecholamines, serotonin, γ-aminobutyric acid (GABA) and glycine, a more detailed study has been conducted by Raju et al. [26]. The study has been conducted [26] on two benzodiazepines, namely nitrazepan and Chlordiazepoxide. Data on hydraulic permeability have been obtained to demonstrate the formation of liquid membrane and also the incorporation of these drugs into the liquid membranes generated by the lecithin-cholesterol mixtures. Transport of the relevant permeants, viz. glycine, GABA, noradrenaline, dopamine and serotonin, through the liquid membrane generated by the lecithin-cholesterol-benzodiazepine mixtures has been studied and the data obtained have been utilized to throw light on the role of liquid membrane phenomenon in the biological actions of these drugs.

In experiments for determining ω, a solution of desired concentration of the permeant prepared in the aqueous solution of the lecithin-cholesterol-benzodiazepine mixture of known composition was filled in compartment C and water in compartment D of the transport cell. The details of the method are described in the original paper [26]. The composition of the lecithin-cholesterol-benzodiazepine mixture used in the experiments for ω measurements was derived from the hydraulic permeability data in the presence of varying concentrations of benzodiazepines in the aqueous solution of the fixed composition of the lecithin-cholesterol mixture. The composition of the lecithin-cholesterol-benzodiazepine mixtures used in the solute permeability (ω) measurements were those at which the liquid membrane generated by lecithin completely covers the interface and is saturated with both cholesterol and the benzodiazepine under study.

All measurements were made at constant temperature using a thermostat set at 37° ± 0.1°C. Solute permeability data recorded in Table 40 appear relevant to the reported biological actions of the benzodiazepines.

Table 40. Solute permeability (ω)[a] of various permeants in presence of lecithin-cholesterol benzodiazepine mixtures[b] (Ref. 26).

Permeants	Initial concentration x 10^3 (mole lit^{-1})	(ω_0) x 10^9 (mol s^{-1}N^{-1})	(ω_1) x 10^9 (mol s^{-1}N^{-1})	(ω_2) x 10^{10} (mol s^{-1}N^{-1})
Glycine	1.333	1.584 ± 0.022	2.476 ± 0.071	2.185 ± 0.105
γ-Aminiobutyric Acid (GABA)	1.940	0.974 ± 0.051	3.151 ± 0.116	2.817 ± 0.082
Noradrenaline	0.059	0.351 ± 0.039	0.197 ± 0.077	0.516 ± 0.019
Dopamine	0.0527	0.473 ± 0.062	0.342 ± 0.039	0.278 ± 0.051
Serotonin	0.0247	0.764 ± 0.016	1.109 ± 0.027	0.837 ± 0.107

[a] Values of ω are reported as arithmetic mean of 15 repeats ± S.D.
[b] Lecithin concentration (15.542 ppm) and cholesterol concentration (1.175 x 10^{-6}M).
ω_0- Control values when no drug was used.
ω_1 Nitrazepam concentration (7.5 x10^{-6}M).
ω_2 – Chlordiazepoxide concentration (1.668 x 10^{-5}M).

Biochemical and neurophysiological evidences recorded in literature [259-265] have suggested that facilitating synaptic action of GABA in the brain may exert antianxiety action of benzodaizepines. Enhanced permeability of GABA through the liquid membrane, as observed this study (Table 40), could also facilitate GABA potentiation leading to the antianxiety action of benzodaizepines.

Glycine present in relatively high concentration in the gray matter of the spinal cord is known to cause muscle relaxation by depressing the excessive motor activity [273,274]. The enhanced permeability of glycine through the lecithin-cholesterol-benzodaizepines composite liquid membrane (Table 40) may facilitate its access to the glycine receptor in the central nervous system and thus may also contribute to the reported muscle relaxant action of benzodaizepines.

Use of benzodaizepines in the treatment of epilepsy is documented [273,275]. Electrophysiological and biochemical evidences have linked the actions of benzodaizepines to their ability to potentiate the effects of exogeneous GABA or to enhance GABA mediated presynaptic and post-synaptic inhibitory pathways [275,276]. The enhanced permeability of GABA as observed in the present study (Table 40), may contribute to the reported anti-epileptic effects of benzodaizepines.

The benzodaizepines are believed to suppress the ability of the limbic system to activate the reticular formation and thus induce sleep in cases of insomnia due anxiety [277]. This effect appears to be due to the GABA potentiation to which the enhanced permeability of GABA (Table 40) may be a contributing factor.

Nitrazepam, like barbiturates, is known to disturb the balance of the phases of sleep [277]. The initial effect is that of reducing the proportion of REM (rapid eyeball movement) sleep in comparison to NREM (non rapid eyeball movement) sleep [277]. Raphe nuclei, which are rich in serotonin, are responsible both for NREM sleep and for the transition to and onset of REM sleep [278]. When the locus ceruleus, which is rich in noradrenaline, is destroyed, animals previously deprived of REM sleep fail to take the usual rebound excess of REM sleep when undisturbed [278]. The data in Table 40 indicate that the liquid membrane may be formed by the nitrazepam in association with the membrane lipids in the synaptic cleft and may enhance the access of serotonin to its site of action in the raphe nuclei and reduce the access of noradrenaline to its site of action in the locus ceruleus causing imbalance in the phases of sleep.

It is documented that patients treated with benzodaizepines also show failure to ovulate [279] like those treated with drugs like reserpine and chlorpromazine [280] which impede the transport of dopamine. The data in Table 40 indicate that impediment in the transport of dopamine due to the liquid membranes of the benzodaizepines in association with membrane lipids, which acts at the level of median eminence [280] to stimulate the release of LH/FSH-RH (lueteinzing hormone/follicle stimulating hormone), could also be a factor responsible for this side effect of the benzodaizepines.

One side effect of benzodaizepines is reported to be weight gain due to renewed appetite [279-281]. Although the pharmacology of eating behavior is complex and is governed by several factors [281], broadly speaking GABA and noradrenaline acting at the level of hypothalamus are known to act as feeding enhancers and feeding inhibitors respectively [282]. The data recorded in Table 40 on the enhanced permeability of GABA and the reduced permeability of noradrenaline appears consistent with these observations particularly in the case of nitrazepam. The observation that permeability of noradrenaline is enhanced in the case of chloridazepoxide appears consistent with the report that it is less toxic than nitrazepam [283].

According to the liquid membrane hypothesis of drug action [64] the CMC of the drug is a good indicator of its potency-lower the CMC more potent is the drug. Since the CMC value of Nitrazepam is lower than that of Chlordiazepoxide it should be more potent than Chlordiazepoxide, which indeed is the case [284].

Thus it appears that the phenomenon of liquid membrane formation may contribute to the biological actions of Nitrazepam and Chlordiazepoxide also.

6.2.16. β-Blockers [14]

About 17 different β-blockers are being used clinically through out the world. Of the three β-blockers namely propranolol hydrochloride, atenolol and metoprolol have been investigated recently for the role of liquid membranes in their action. All three drugs mentioned above were found to be surface active (CMCs shown Table 1). These drugs have been shown to generate liquid membranes by themselves and also in association with lecithin and cholesterol in series with a cellulosic supporting membrane (Sartorius Cat no.11107).

Transport of biogeneic amines, namely, adrenaline, non-adrenaline, dopamine and important cations like sodium (Na^+), potassium (K^+) and calcium (Ca^{2+}) ions, through the liquid membranes generated by the lecithin-cholesterol and β-blockers mixture in series with a supporting membrane has been studied. The data indicate that modification in the transport of relevant permeants due to the liquid membranes generated by β-Blockers in association with lecithin and cholesterol may also contribute to the biological actions of these drugs [14].

For solute permeability (ω) measurements two sets of experiments were performed. In one set of experiments the compartment C of the transport cell (Fig. 2 Chapter 5) were filled with the mixture of lecithin cholesterol and one β-blocker drug along with the desired concentration of the permeant and the compartment D was filled with the water alone. In the other set of experiments lecithin and cholesterol were not used only the β-Blocker drug and the permeant were used. In both the sets of the experiments pH of the two compartments were fixed at 7.4 using a phosphate buffer. In the first set of experiments the composition of the lecithin-cholesterol β-blocker mixture was the one at which the liquid membrane generated by lecithin was fully saturated by cholesterol and the β-blockers drug. In the other set of experiments the concentration of β-blockers drugs was always higher than their CMCs to ensure that the supporting membrane is completely covered by the drug liquid membrane. All measurements were made at $37 \pm 0.1^{o}C$.

Propranolol amongst the three β-blocker drugs studied [14] is reported to be the most potent. This is consistent with the fact that the CMC of propranolol in the lowest (Table 1): According to liquid membrane hypothesis of drug action [64], lower the CMC more potent is the drug.

Data on the solute permeability (ω) of biogenic amines and relevant cations in the presence of liquid membranes generated by β-blocker drugs and the β-blocker drug in association with lipids-lecithin and cholesterol are recorded in Table 41.

The data in Table 41 indicate the solute permeability (ω) of catecholamines viz adrenaline, noradrenaline and dopamine and of relevant cations all show a decrease, in case of all the three β-blocker drugs, in the presence of lipids-lecithin and cholesterol. This indicates that probably interaction of β-blocker drugs with membrane lipids may be necessary, *in vivo,* for their biological action. Adrenaline, nor-adrenaline and dopamine are important catecholamines, which are known to have a variety of receptors. The solute permeability of catecholamines, viz. adrenaline, nor-adrenaline and dopamine indicate that the transport of nor-adrenaline and adrenaline is impeded. However, not much difference is observed in case of dopamine. The liquid membranes likely to be generated at the cell membrane may be impeding the transport of nor-adrenaline and adrenaline and thus slowing down the rate of formation of catecholamine receptor complex. This may also be a contributing factor to anti-hypertensive action of β-blockers.

Gupta has reported that β-blockers can act on myocardial cell membrane producing cadiodepressant effects via changes in basic electrophysiological properties of the membrane such as automaticity, excitability, conductivity and refractoriness [285]. The transport of ions is altered in the presence of liquid membrane generated by β-blockers and particularly β-blockers in association with lecithin-cholesterol. The observed electro-physiological changes may have bearing with the solute permeability data observed in these studies [14].

Saitta et al. [286] have reported that β-blockers causes a decrease in the sodium flux through passive permeability. This is evident from the solute permeability data (Table 41), which show that the transport of the sodium ions is impeded in presence of liquid membranes generated by lecithin-cholesterol and β-blocker mixture.

Skeberdis et al. [287] have reported that metoprolol antagonizes L-type Ca^{2+} current induced by isoprenaline, dobutamine and salbutamol in frog ventricular myocytes. This is consistent with the decrease in calcium transport (Table 41) in the presence of liquid membranes generated by β-blockers-lecithin-cholesterol mixture.

Thus the role of liquid membrane in the biological action of β-blocker is indicated by these studies.

6.2.17 Antibecterials [28]

Studies on two representative drugs of this class, ciprofloxacin and norfloxacin, were undertaken for the role of surface activity *vis-à-vis* liquid membranes in their biological actions. These drugs have both hydrophilic and hydrophobic groups in their structure and hence they are likely to be surface-active [288]. In fact they are: CMC values are given in Table 1. Both these drugs were shown to generate liquid membranes in series with the supporting membrane and modify transport of relevant permeants. All measurements were made at 37 ± 0.1°C. The details of the experiments are described in the original paper [28].

Data on the solute permeability (ω) of relevant permeants such as dextrose, K^+ Ca^+, Mg^{2+},NH_4^+ and PO_4^3 ions in presence of liquid membranes generated by these drugs in series with the supporting membrane are shown in Table 42. The permeability of all the permeants is diminished in the presence of ciprofloxacin and norfloxacin liquid membranes. Ciprofloxacin and norfloxacin, which contain carboxylic acid group at third position, in aqueous medium. Anionic surfactants are electrolytes and a surface ion is an anion, when surfactants dissociates in water [289]. Hence these molecules may act as anionic surfactant molecules. Due to their surface activity, molecules may self-aggregate or bind with the supporting membrane. The non-polar part of these drugs is likely to be associated with non-polar part of cellulose acetate membrane, a supporting membrane. In such an event the polar part (-anionic) is expected to be projected out wards away from the supporting membrane. In this study, transport of both anions and cations are impeded (Table 42), which may be due to repulsion between cations and surface anionic charge. So, in both cases, the ions cannot permeate the supporting membrane freely. Likewise, dextrose molecules permeation is impeded in presence of liquid membrane formed by these drugs.

In an earlier study, alteration of the hydrophobicity of bacterial membrane by antibiotics, such as ciprofloxacin and norfloxacin was reported [290]. The most significant reduction of bacterial cell surface hydrophobicity was found after treatment at 1/16 of MICs (to 20.3% for both drugs, compared with control values) [291]. The reduction in hydrophobicity may be due to accumulation of these drugs on bacterial membranes [292]. In this study, these antibiotics are found to interact with the hydrophobic surfaces of cellulose acetate supporting membrane and form a liquid membrane.

Table 41. Solute permeability (ω) of various permeants in presence of liquid membrane generated by β-blockers alone and β-blockers in presence of lecithin-cholesterol mixture (Ref. 14).

Permeants	Initial concentration	$\omega_0 \times 10^6$ (moles $s^{-1}N^{-1}$)	Atenolol			Metoprolol			Propranolol		
			$\omega_l \times 10^6$ (moles $s^{-1}N^{-1}$)[a]	$\omega_a \times 10^6$ (moles $s^{-1}N^{-1}$)[a]	$\omega_b \times 10^6$ (moles $s^{-1}N^{-1}$)[a]	$\omega_l \times 10^6$ (moles $s^{-1}N^{-1}$)[a]	$\omega_a \times 10^6$ (moles $s^{-1}N^{-1}$)[a]	$\omega_b \times 10^6$ (moles $s^{-1}N^{-1}$)[a]	$\omega_l \times 10^6$ (moles $s^{-1}N^{-1}$)[a]	$\omega_a \times 10^6$ (moles $s^{-1}N^{-1}$)[a]	$\omega_b \times 10^6$ (moles $s^{-1}N^{-1}$)[a]
Potassium (as chloride)	10.430 mg/ml	729 ± 0.16	737 ± 0.93	746 ± 0.14	744 ± 0.77	814 ± 0.11	714 ± 0.13	702 ± 0.12	612 ± 0.43	592 ± 0.46	561 ± 0.73
Sodium (as chloride)	5.382 mg/ml	373 ± 0.72	341 ± 0.17	340 ± 0.17	321 ± 0.15	380 ± 0.08	340 ± 0.72	317 ± 0.92	460 ± 0.24	442 ± 0.62	414 ± 0.64
Calcium (as chloride)	10 mg/ml	449 ± 0.32	441 ± 0.23	407 ± 0.97	398 ± 0.74	486 ± 0.08	485 ± 0.42	471 ± 0.06	367 ± 0.08	313 ± 0.04	296 ± 0.04
Adrenaline	10 μg/ml	1.312 ± 0.06	1.016 ± 0.08	0.984 ± 0.02	0.870 ± 0.07	1.402 ± 0.02	1.394 ± 0.07	1.359 ± 0.04	1.329 ± 0.08	1.303 ± 0.06	1.264 ± 0.08
Non-adrenaline	10 μg/ml	0.974 ± 0.07	0.964 ± 0.06	0.890 ± 0.01	0.873 ± 0.05	1.010 ± 0.06	0.874 ± 0.09	0.817 ± 0.08	1.408 ± 0.06	1.367 ± 0.08	1.323 ± 0.02
Dopamine	10 μg/ml	0.787 ± 0.08	0.781 ± 0.07	0.753 ± 0.03	0.759 ± 0.03	0.709 ± 0.06	0.706 ± 0.04	0.707 ± 0.05	0.608 ± 0.03	0.606 ± 0.01	0.594 ± 0.02

ω_0 = Values of ω when no drug was used and no lecithin-cholesterol mixture was used

ω_l = Values of ω when only lecithin-cholesterol mixture was used

ω_a = Values of ω when only β-blockers drug was used

ω_b = Values of ω when only lecithin-cholesterol and β-blockers drug was used

Concentration of Lecithin = 1.919×10^{-5} M

Concentration of Cholesterol = 1.175×10^{-6} M

Propranolol Concentration = 8.0×10^{-5} M

Atenlol concentration = 8.0×10^{-3} M

Metoprolol concentration = 12.0×10^{-3} M

Table 42. Solute permeability (ω) of various permeants in the presence of liquid membranes generated by norfloxacin and ciprofloxacin (Ref. 28).

Permeants	Initial conc. (g/l)	Norfloxacin (6×10^{-4}M)		Ciprofloxacin (6×10^{-4}M)	
		$(\omega_0) \times 10^6$	$(\omega_1) \times 10^6$	$(\omega_0) \times 10^6$	$(\omega_1) \times 10^6$
		Membrane 1		Membrane 2	
Dextrose	10.0	10.29 ± 0.02	5.18 ± 0.06	14.84 ± 0.11	6.71 ± 0.04
K^+	0.02	27.93 ± 0.31	5.56 ± 0.30	28.20 ± 0.28	5.09 ± 0.13
Ca^{2+}	0.02	31.82 ± 0.93	14.32 ± 0.96	28.64 ± 0.23	10.91± 0.16
Mg^{2-}	0.2	49.44 ± 0.96	26.99 ± 0.85	54.17 ± 0.65	28.03± 0.56
NH_4^-	2.5	38.93 ± 0.04	12.43 ± 0.14	32.57 ± 0.49	6.70 ± 0.05
PO_4^{3-}	0.05	7.35 ± 0.04	2.74 ± 0.02	6.60 ± 0.05	2.33 ± 0.02

Values of ω (moles $s^{-1}N^{-1}$) are reported as arithmetic mean of 10 repeats ± SD

ω_0 When no drug was used; and ω_1 in the presence of norfloxacin/ciprofloxacin. In case of nor floxacin and ciprofloxacin a new membrane was used each time.

6.2.18 ACE inhibitors [29].

About sixteen different angiotensin converting enzyme (ACE) inhibitors are employed worldwide. All ACE inhibitors effectively block the conversion of angiotensin I to angiotensin II and have similar therapeutic effects, adverse effect profiles and contraindications. Apart from captopril and lisnopril, all other ACE inhibitors are prodrugs [293], hence hey were selected for investigation. Both these drugs contain hydrophilic and hydrophobic part in their structure [293] and are likely to be surface active in nature. In fact they are: the CMC values are given in Table 1. Hence these drugs are likely to be generate liquid membrane at interface. The transport of biogeneic amines, cations such as Na^+, K^+, Ca^{2+} and neutral molecule like glucose in the presence of the liquid membrane generated by captopril and lisnopril has been studied, and data have been discussed in the light of the biological action of the drugs.

Both these drugs have been shown to generate liquid membrane in series with the supporting membrane. Data on solute permeability of relevant permanents in the presence of drug liquid membranes, have been obtained and are recorded in Table 43. All measurements have been made at 37 ± 0.1°C. For details of the experiments original paper should be consulted [29].

The solute permeability for sodium ions is enhanced in the presence of liquid membrane generated by captopril and lisnopril (Table 43). These observations are consistent with the reports that ACE inhibitors produce natriuresis [293] (excretion of abnormal amounts of sodium in urine). So the liquid membrane generated by captopril and lisnopril in kidney may also be contributing by increase the transport of sodium that leads to natriuresis.

The observed trend is also in correlation with the substantial lowering of blood pressure in sodium depleted (hyponatremic) patients than the sodium replete patients with single dose of captopril [294].

Table 43. Solute permeability (ω) of various permeants in presence of liquid membrane generated by captopril and lisnopril (ω_l), along with the control values of ω_0, when no inhibitor was used (Ref. 29).

Permeants	Initial concentration	$\omega_0 \times 10^6$ (mole s^{-1} N^{-1})	$\omega_l \times 10^6$ (mole s^{-1} N^{-1})	
			Captopril	Lisnopril
			(12×10^{-4}M (2CMC)	(14×10^{-4}M (2CMC)
Potassium (chloride)	10.430 mg/ml	708.18 ±17	627.43 ±23	576.65 ±16
Sodium (chloride)	5.382 mg/ml	188.50 ± 16.4	249.22 ± 13.2	239.22 ± 13.20
Calcium (chloride)	10 mg/ml	444.61 ± 28	335.50 ± 23	324.60± 21
Adrenaline	10 μg/ml	0.763 ± 0.063	0.818 ± 0.021	0.782 ± 0.055
Noradrenaline	10 μg/ml	1.100 ± 0.050	1.187 ± 0.031	1.912 ± 0.034
Dopamine	10 μg/ml	1.200 ± 0.028	Undetectable	Undetectable
Glucose	20 mg/ml	17.62 ± 1.4	21.36 ± 1.11	22.04 ± 1.036

Values of ω are reported as arithmetic mean of 10 repeats ± S.D

ACE inhibitors may rarely cause hyperkalemia in patients with renal insufficiency or in patient taking potassium-sparing diuretics, potassium supplements, β-adrenoceptor blockers or NSAID. Also, significant retention of potassium is encountered with ACE inhibitors in patients with normal renal function who are not taking other drugs that cause potassium retention [293]. So, retention of potassium ion in the blood that leads to hyperkalemia is inconsistent with solute permeability observations (Table 43) that transport of potassium ions is reduced when compared to control. So, the liquid membrane generated by both the ACE inhibitors captopril and lisnopril may reduce the transport of potassium into the urine by kidney, which may lead to hyperkalemia.

The depolarization of vascular smooth muscle is primarily dependent on the influx of calcium ions. An increase in cytosolic calcium results in the constriction of smooth muscle through myosin light chain [295]. ACE inhibitors lower systemic vascular resistance and mean diastolic and systolic blood pressure in various hypertensive states. They dilate both veins and arterioles [296]. These reports are consistent with the observation (Table 43) that both captopril and lisnopril reduce the transport of calcium ion compared to control. So, by reducing the transport of calcium ions to the vascular smooth muscle by the liquid membrane generated by captopril and lisnopril may also contribute to the vasodilatation effect of the drugs.

A reversible side effect of ACE inhibitors is spillage of glucose into the urine in the absence of hypoglycemia, the mechanism of which is unknown [297]. These studies have shown enhanced transport of glucose across the liquid membrane generated by captopril and lisnopril (Table 43) compared to control. The observed glycosuria may have bearing with the liquid membrane phenomenon of the drugs. It has already been shown that lisnopril treatment is not associated in diabetic patients and it does not affect glycemic control [298]. Also, it has been shown in a few case reports that hypoglycemia results from the combination of an ACE inhibitor and an oral hypoglycemic agent [299]. In these cases, an enhanced transport because of the liquid membrane may be influencing the glucose influx into the cell.

Dopamine is the immediate metabolic precursor of the noradrenaline and adrenaline. The cardiovascular effects of dopamine are mediated by several distinct types of receptors that vary in their affinity for catecholamines. At high concentrations, dopamine; activates vascular alpha-one adrenergic receptors, leading to vasoconstriction and increase in systolic blood pressure [300]. Although the transport of noradrenaline and adrenaline are not affected much, dopamine transport is highly affected in presence of liquid membrane generated the captopril and lisnopril (Table 43). These trends of catecholamines transport are consistent with antihypertensive effect of ACE inhibitors. The impediment in catecholamines transport may be contributing to the transport of circulating catecholamines especially dopamine an observed antihypertensive effect may be due to the combination of ACE inhibition and decreased dopaminergic activity.

Having consolidated the experimental studies on the role of liquid membranes in drug action to different pharmacological categories, on a wide variety of drugs belonging we proceed to make a critical assessment of the liquid membrane by hypothesis of drug action in the next chapter.

REFERENCES

[1] S. B. Bhise, P.R. Marwadi, S.S. Mathur and R.C. Srivastava, J. Pharm. Sci., 71(1982)526.

[2] S. B. Bhise, P.R. Marwadi, S.S. Mathur and R.C. Srivastava, J.Pharm. Sci., 71 (1983)599.

[3] S. B. Bhise, P.R. Marwadi, S.S. Mathur and R.C. Srivastava, Biophys. Chem., 17(1983) 187.

[4] R.C. Srivastava, R.P.S. Jakhar and S.B. Bhise. J. Colloid Interface Sci., 87 (1982) 56.

[5] A.N. Nagappa. P.L. Kole, P.V. Pandi, 1. Shanmuka K.Girish and P.K. Rahul Mishra, Colloids and Surfaces, B(Bio-interfaces)32 (2003), 169.

[6] R.C. Srivastava, R. K. Sharma, R. Srinivasan and S.B. Bhise, J. Colloid Interface Sci., 94(1983) 456.

[7] R.C. Srivastava, R. K. Sharma and S.B. Bhise, J. Colloid Interface Sci.,93(1983) 72.

[8] S.B. Bhise, C.V.S. Subrahmanyam and R.C. Srivastava, Int. J. Pharm., 17(1983) 263.

[9] S.B. Bhise, C.V.S. Subrahmanyam and R.C. Srivastava, Int. J. Pharm., 21(1984) 27.

[10] P.V. Pandi, P.L. Kole, Rahul Mishra, I. Shanmukha, K. Girish and A.N. Nagappa, Colloids and Surfaces B(Bio-interfaces) 35(2004)243.

[11] S. B. Bhise, D. Bhattarcharjee and R.C. Srivastava, J.Pharm. Sci., 73 (1984) 1579.

[12] R.C. Srivastava, M.N.A. Rao, Rajesh Sood and S.B. Bhise, Colloids and Surfaces, 19 (1986) 83.

[13] S. B. Bhise, C.V.S. Subrahmanyan, R.K. Sharma and R.C. Srivastava, Int. J. Pharm, 28 (1986) 145.

[14] A. N. Nagappa, P.L. Kole, P.V. Pandi, K. Girish, P.K. Rahul Mishra and I. Shanmukha, J. Membr. Sci., 211 (2003) 349.

[15] R.C. Srivastava, R.K.Sharma and S.B. Bhise, Colloids and Surfaces, 14(1985) 1.

[16] R. C. Srivastava, G.S.K. Velu, A.N. Nagappa and D.B. Raju, J.Colloid Interface Sci., 131 (1989) 584.

[17] A.N. Nagappa, K.U. Bhaskar, M. Shankar, S. Mitra and R.C. Srivastava, Indian J.Biochem. Biophys., 25(1988) 350.

[18] R.C. Srivastava, M. Shankar, A. Tandon and A.N. Nagappa, Int. J. Pharm, 38 (1987) 239.

[19] A. Tandon, A.N. Nagappa and R.C. Srivastava, Int J. Pharm, 32(1986) 39.

[20] A.N. Nagappa, R.T. Patil, P.V. Pandi, P.L. Kole, K. Ziyauddin and I. Shanmukha, J. Colloid Interface Sci. 271 (2004) 416.

[21] R. C. Srivastava, A. Tandon, A.N. Nagappa and R.K. Sharma, J.Colloid Interface Sci., 117 (1987) 375.

[22] A.N. Nagappa, D.B. Raju and R.C Srivastava, Indian J. Biochem, Biophys, 26(1989) 172.

[23] R.C. Srivastava, D.B. Raju, V. Singh and S. Upadhyay, J. Membr, Sci., 58 (1991) 211.

[24] A.N. Nagappa, D.B. Raju and R.C. Srivastava, Colloids and Surfaces, 35(1989) 17.

[25] A. N. Nagappa, P.V. Pandi, P.K. Misra, K. Girish and I. Shanmukha, Indian J. Biochem. Biophys. 39(2002) 406.

[26] D.B. Raju, A.N. Nagappa and R.C. Srivastava, Indian J. Biochem. Biophys., 26(1989) 104.

[27] A.N. Nagappa, K. Zeeyauddin, P.V. Pandi, R.T. Patil and K. Girish, J. Surface Sci. Technol. 18(2002) 43.

[28] A. N. Nagappa, P.L. Kole, P.V. Pandi, R.T. Patil K. Zeeyauddin and I. Shanmukha, Indian J. Biochem. Biophys, 41(2004) 48.

[29] A.N. Nagappa, R. T. Patil, P.V. Pandi and K. Zeeyauddin, Indian J. Biochem. Biophys., 38(2001)412.

[30] V. Singh, B. Malhotra, D.B. Raju, A.N. Nagappa and R.C. Srivastava, Indian J. Biochem. Biophys., 28(1991) 34.

[31] S.B. Bhise, C.V.S. Subrahmanyam, A.K. Molhotra and R.C. Srivastava, Int. J. Pharm., 24 (1985) 297.

[32] R.E. Kesting, W.J. Subcaskey and J.D. Paton, J. Colloid Interface Sci, 28 (1968) 156.

[33] P.A.J. Jansen, Int. J. Neurophychiat., S(1967)10.

[34] M.A. Spirtes and P.S. Guth, Biochem. Pharmacol 12(1963) 37.

[35] P.S. Guth, Fed. Proc. 21(1962) 1100.

[36] R.S. Stacey, Brit. J. Pharmacol., 16(1961)284.

[37] H. Well-Malherbe and H.S. Posner, J. Pharmacol.Exp. Therap. 140(1963) 93.

[38] E.T. Eckhardt and W.M. Govier, Proc. Soc. Exp. Biol.Med., 97 (158)124.

[39] K.F. Gey and J. Pletscher, J. Pharmacol. Exp. Therap. 133(1961) 18.

[40] J. Axelord, L.G. Whitby and G.Hertling, Science, 133(1961)18.

[41] L.L.M. Van Deenen and R. A. Demel, Biochim. Biophys. Acta. 94(1965) 314.

[42] G. Zografi and D.E. Auslander J. Pharm. Sci., 54(1965) 313.

[43] K.S. Spiegler and O. Kedem., Desalination 1(1966)311.

[44] T.K. Sherwood, P.L.T. Brian and R.E. Fischer, Ind. Eng. Chem. (Fundam), 6(1967)2.

[45] F.L. Harris, G.B. Humphreys and K.S. Spiegler, In (P. Meares, ed.), Membrane Separatem Processes, , Elsevier, Amsterdam, 1976, Chapt 4 p.126.

[46] I.D. Kuntz, In (M.E. Wolff ed.), Burger's Medicmal Chemistry, Part I ,John Wiley and Sons, Newyork, 1980, p.291.

[47] P.A.J. Janssen, Int. Rev. Neurobiol. 8(1965) 221.

[48] M. Grabowska, Pol. J. Pharmacol. 28(1976) 253.

[49] R. Byck, In L.S. Goodman and A. Gilman (eds); Pharmacological Basis of Therapentics, Macmillan, New york, 1975, p.172.

[50] O. Hornykiewicz, Pharmacol. Rev., 18(1966) 925.

[51] T.N. Chase and D.L. Murphy, Annu. Rev. Pharmacol, 13(1973) 188.

[52] S.H. Snyder, Am. J. Psychiatry, 133 (1976) 197.

[53] N. Weiner, In A.G. Gilman, L.S. Goodman and A. Gilman (eds.), Pharmacological Basis of Therapeutics, Macmillan, Newyork, 1980, p.202.

[54] J. Raggendal and A. Dahlshram. J.Pharm. Pharmecol, 24 (1972) 565.

[55] D. Palm, H. Grobecker and I.J. Bak, In H. J. Shumann and G. Kroneberg (eds.) "Bayer Symposium II, New Aspects of Storage and Release Mechanisms of Catecholamines" Springer Verlag, Berlin, 1970, pp 188-198.

[56] J.G. Ashwin and L.B. Jacques, Thromb. et. Diath. Hemorrhag, 5(1961) 543.

[57] B.C. Bose and R. Vijayavargirja, Arch. Int. Pharmacodyn., 127(1960) 27.

[58] S.M. Kirpekar and J.J. Lewis, Anesthesia, 15(1960) 175.

[59] A. Soulairac and M.L. Soulairac, Soc. Biol., 154(1960) 510.

[60] D.R. Laurence, Clinical Pharmacology, 4th Edn.., ELBS and Churchil Livingstone, Edinburgh, 1973, pp 18.15.

[61] B.N. Premchandra and C.W. Turner, Proc. Soc. Exp. Biol. Med., 104(1960) 306.

[62] W.C. Bowman and M.J. Rand, Text Book of Pharmacology, Blackwell Scientific Publication, Oxford, 1980, pp 11.13, 12.19.

[63] J. Knoll, Arch. Exp. Path. Pharmak., 238(1960) 114.

[64] R.C. Srivastava, S.B. Bhise and S.S. Mathur, Adv. Colloid Interface Sci., 20(1984) 131.

[65] M. Ligo, A. Hoshi, A. Nakamura and K. Kunetani, J. Pharm. Dyn., 1(1978)49.

[66] S.O. Zaki, Y. Ike, H. Mizuno, K. Ishikawa and H. Mori, Bull. Chem. Soc. Jpn., 50(1977) 2406.

[67] W.C. Bowman and M.J. Rand. Text Book of Pharmacology Blackwell, Oxford 1980, pp 38.16.

[68] P. Calabrsi and R.E. Parks Jr., In A.G. Gilman, L.S. Goodman and A. Gilman (eds.), The Pharmacological Basis of Therapeutics, 6th edn., Macmillan New york, 1980, pp 1256-1313.

[69] R. S. Hilman, in A. Gilman, L.S. Goodman and A. Gilman (eds), The Pharmacological Basis of Therapeutics, 6th edn., Macmillan, New york, 1980, pp 1331-1346.

[70] W.C. Bowman and M.J. Rand., Text Book of Pharmacology, 2nd edn. Blackwell Scientific Publications, Oxford, 1980 pp 38.9.

[71] M.M. Burger, in B.F. Horeoker and E.R. Stadlman (eds.). Current Topics in Cellular Regulations, Vol 3, Academic Press, New york, 1971, pp 135-193.

[72] P. Emmelot, in E.J. Ariens (ed.), Molecular Pharmacology, Vol II Academic Press, Newyork 1964, pp. 133-134.

[73] W.C. Bowman and M.J. Rand, Text Book of Pharmacology, Blackwell Scientific. Publications, Oxford, 1980, pp 3.12.

[74] P. Emmelot, in E.J. Ariens (ed.) Molecular Pharmacology, Vol. II, Academic Press New York, 1964, p 72-73.

[75] A.R. Horler, In D.M. Davies (ed.), Text Book of Adverse Drug Reactions, 2nd Edn., Oxford University Press, Oxford, 1981, pp 514-515.

[76] P. Calabresi and R.E. Parks Jr., In A.G. Gilman, L.S. Goodman and A.Gilman (eds.) The Pharmacological Basis of Therapeutics, 6th edn., Macmillan, New york, 1980, pp 1280.

[77] W.C. Bowman and M.J. Rand., Text Book of Pharmacology, Blackwell Scientific publications Oxford, 1980, pp 21.40.

[78] C.A. Finch, C.A. Colman, A.G. Motrulsky, D.M. Donahue and R.H. Rciff., Blood, 11(1956) 807.

[79] R. Stibbins, J. Scott and V. Herbert, Semin-Haematol, 10(1973) 235.

[80] R. Stibbins and J.R. Bertino, Clin. Haematol., 5(1976) 619.

[81] G.H. Mudge, in A.G. Gilman, L.S. Goodman and A. Gilman (ed.) The Pharmacological Basis of Therapeutics, Macmillan New York 1980, pp 903, 994.

[82] L.Z. Benet, J. Pharmcokinet. Biopharm., 7(1979) 1.

[83] U. Gundert-Remy, D. Vankenne, E. Weber, H.E. Geissler, B. Grebian and E. Mutschler, Eur. J. Clin. Pharmacol., 16(1979) 39.

[84] G.H. Mudge, in A.G. Gilman, L.S. Goodman and A. Gilman (eds.) The Pharmacological Basis of Therapeutics 6th Ed.; Macmillan New York, 1980, p. 908.

[85] S. Erlinger, D. Dhumeaux, P. Bertholof and D. Mount., Am. J. Physiol, 219(1970)416.

[86] M.J. Dunn, In A.F. Lant and G.M. Wilson (eds.), "Modern Diuretic Therapy in the Treatment of Cardiovascular and Renal Diseases" Experta Medica, Ansterdam, The Netherlands, 1973, p.196.

[87] S. Ahmed, Br. Med. J., 3(1974) 657.

[88] A Fine, I.S. Henderson, D.R. Morgan and W.J. Tilstone, Br. Med. J., 2(1977) 1061.

[89] W.C. Bowman and M. J. Rand, Text Book of Pharmacology, Blackwell Scientific Publications Oxford, 1980 pp. 18.30.

[90] R.C. Srivastava and R.P.S. Jakhar, J. Phys. Chem. 86(1982) 1441.

[91] F.H. Brain and B.J. Thomas (Jr.), In A.G. Gilman, L.S. Goodman, T.W. Rall and F. Murad (eds.), The Pharmacological Basis of Therapeutics, Macmillan New York, 1985, pp 715-747.

[92] H. T. Roland and B.C. Leonard In George de Stevens (ed.), Cardiac Stimulant Substances-Medicinal Chemistry, A Series of monographs, Vol 6, Academic Press New York, 1967 pp 103-136.

[93] T. Akera, F.S. Lerson and T.M. Brody, J. Pharmcol. Exp. Ther. 173(1970) 145.

[94] T.J. Hougen and T.W. Smith, Ure. Res., 42 (1978) 856.

[95] G.N. Winifred, in E.E. Bittar (Ed), Membrane and Ion Transport, Academic Press, New York, 1970 pp 75-94.

[96] R.D. Tanz, J. Pharmacol. Exptl. Therap., 144 (1964) 205

[97] W. C. Bowman and M.J. Rand, Text Book of Pharmacology, Blackwell, London, 1980, pp 22.77.

[98] W.C. Bowman and M.J. Rand, Text Book of Pharmacology, Blockwell, London, 1980, pp 22.54.

[99] W.C. Bowman and M.J. Rand, Text Book of Pharmacology, Blackwell, London, 1980, pp 22.83.

[100] R.D. Gillis and A.Q. John, Pharmacol Rev., 31(1979) 19.

[101] P. Seeman, Pharmacol. Rev., 32 (1980) 229.

[102] W.C. Bowman and M.J. Rand., Text Book of Pharmacology, Bloackwell London, 1980, pp 11.28.

[103] W.C. Bowman and M.J. Rand, Text Book of Pharmacology, Blackwell, London, 1980 pp 22.84, 11.28.

[104] W.C. Bowman and M.J. Rand, Text Book of Pharmacology, Blackwell, London, 1980 pp 22.84, 25.7

[105] J.M. Ritchie and N.M. Greene, in A.G. Gilman, L.S. Goodman and A. Gilman (eds.), The Pharmacological Basis of Therapeutics, Macmillan, New York, 1980, pp 301.

[106] H. Small, J.Am. Chem. Soc., 90(1968) 2217.

[107] J.C. Skou, Acta Pharm. Toxicol, 10(1954) 280,317,325.

[108] B. Hille, J. Gen. Physiol 69(1977) 497.

[109] S. Wiedling and C. Tegner, Prog. Med. Chem., 3(1963) 332.

[110] J.C. Skou, J. Pharm. Pharmacol., 13(1961) 204.

[111] R. Straub, Experientia, 12(1956) 182.

[112] J.M. Ritchie and P. Greengard, Annu. Rev. Pharmacol., 6(1966)405.

[113] A.M. Shanes, J. Gen. Physiol., 33(1950) 729.

[114] N.L. Gershfield, J. Phys. Chem., 66(1962) 1923.

[115] J. Buchi and X. Perlia, In E.J. Ariens (ed.) Drug Design, Vol. III Academic Press, New York, 1972, pp 375.

[116] A.G. Lee, Nature, (London), 262(1976) 545.

[117] D. Papahadgopoulos, K. Jacobson, G. Poste and G. Schepherd, Biochim. Biophys. Acta, 394(1975) 504.

[118] C. Tanford, The hydrophobic Effect: Formation of Micelles and Biological Membrane, 2nd Edn., Wiley Inter Science, New York, 1980, pp 131.

[119] M. Montal, Annu, Rev. Biophys, Bioeng., 5(1976)119.

[120] R.E. Thomas, In M.E. Wolff (ed.) Burger's Medicinal Chemistry Part III, 4th Edn., Cardiac Drugs, John Wiley, New York, 1981 pp 85-87.

[121] W.C. Bowman and M.J. Rand, Text Book of Pharmacology, Blackwell, Oxford, 1980 pp 22.55, 22.58.

[122] R.J. Kuffner, M.T. Bush and L.T. Bircher, J.Am, Chem. Soc. 79(1957) 1587.

[123] A. Giacalone, Gazz Chem. Ital, 70(1940) 656

[124] A. Bornsherin, R. Voigt and H. Florstedt, Pharmazie, 28(1973) 756.

[125] F.A. Villalonga and E.N. Philips, J. Pharm. Sci., 69(1980) 102.

[126] P. Seeman, Pharmacol, Rev., 24(1972) 583.

[127] A.G. Lee, Bichim. Biophys. Acta, 455(1976) 102.

[128] P.M.G. Broughton, G. Huggins and J.R.P. O'Brien, Lancet, 1(1956) 180.

[129] S. Sun and A.H.C. Chun, J. Pharm. Sci, 66(1977) 477.

[130] R.C. Srivastava and R.P.S. Jakhar, J. Phys. Chem., 85(1981) 1457.

[131] J.K. Ho and R.A. Harris, Annu. Rev. Pharmcol. Toxicol., 21(1981)88.

[132] R.W. Olsen, M.K. Ticku, D. Greenlee and P. VanNess, In H. Kofod, P. Krogsgaard Larson and J. Scheel-Kruger (eds.), GABA Neurotransmitter, GABA receptors and Ionosphore binding site: Interaction with various Drugs., Munksgard, Copenhagen, 1979, pp 165-178.

[133] W.C. Bowman and M.J. Rand, Text Book of Pharmacology, Blackwell, Oxford 1980, pp 14.15.

[134] S.C. Harvey, in L.S. Goodman, A.Gilman and A.G. Gilman (eds.), The Pharmacological Basis of Therapeutics, 5th edn., Macmillan New York, 1975, pp 101-123.

[135] W.C. Bowman and M.J. Rand, Text Book of Pharmacology, Blackwell, Oxford, 1980, pp 8.7.

[136] W.C. Bowman and M.J. Rand, Text Book of Pharmacology, Blackwell, Oxford, 1980, pp 6.26.

[137] S.C. Harvey, In L.S. Goodman, A Gilman and A.G. Gilman (eds.), The Pharmacological, Basis of Therapeutics, Macmillan New York, 1980, p 355.

[138] S. M. Denavit and V. Champagnat, Brain Res., 97 (1975) 356.

[139] J.R. Toleikis L. Wang and L.L. Boyarsky, J. Neuro. Sci. Res., 4(1979) 225.

[140] S.M. Denavrt, J. Champagnat, and J.C. Vellute, In C. Van Euler and H. Lagercrantz (eds.), Central Nervous Control Mechanisms in Breathing, Pergamon, New York, 1979, pp 256-260.

[141] D. Attwood, J. Pharm. Pharmacol., 24(1972)751.

[142] D. Attwood and O.K. Udeala, J. Phys. Chem., 79(1975) 889.

[143] W.W. Douglos, In A.G. Gilman, L.S. Goodman, A.Gilman (eds.), The Pharmacological Basis of Therapentics, 6th edn., Macmillan, New York, 1980, pp 623-625.

[144] D. Attwood and O.K. Udeala, J. Pharm. Pharmacol., 27(1975)754.

[145] R.G. Johnson, F.M. Goyan and L.D. Tuck., J. Pharm. Sci., 54(1965)1176.

[146] D.T. Witiak and R.C. Cavestry, In D.W. Foye (ed.) Principles of Medicinal Chemistry, Lea and Febiger, Philadelphia, 1981, pp 479-485.

[147] A. Korolkovas, Essentials of Molecular Pharmacology, Wiley-Inter science, New York, 1970 247.

[148] D.R. Waud, Pharmacol. Rev. 20(1968) 49.

[149] D. Brick, F G Van and E.J. Lien, Eur. J. Pharmacol., 44(1977) 251.

[150] B. Testa and L. Rossetti-Murset., Helv. Chim, Acta, 61(1978) 2530.

[151] D. Attwood and O. K. Udeala, J. Pharm. Sci., 65(1976) 1053.

[152] J. Gettins, R. Greenwood, J. Rassuig and E. Wyn-Jones, J.Chem. Soc., Chem. Comm., 24(1976) 1030.

[153] W. Domschke, G.Lux and S. Domschke, Lancet, 1(1979) 320.

[154] C.R. Ganellin and G.J.Durant, In M.E. Wolff (ed.), Burger's Medicinal Chemistry, Part III, John Wiley and Sons, New York 1981, p.524.

[155] C.T. Richardson, Gastroenterology, 74 (1978) 366.

[156] H.P. Rang, M.M. Dale and D.M. Ritter, Pharmacology, Churchill Livingston, New York, 1995.

[157] M.J. Krielaart, D.M. Veenstra and K.J. Van Bvuren, Agents Actions 31(1990) 23.

[158] H.V. Bergmeyer and W.Dirscher., Biochem. Z., 323(1952) 104.

[159] E.N. Willnier, Biol. Rev., 36(1968) 368.

[160] F. Murad and R.C. Haynes Jr., In AG Gilman, L.S. Goodman and A. Gilman (eds.), The Pharmacological Basis of Therapeutics, 6 edn., Macmillan New York, 1980, p.1425.

[161] R.C. Hynes Jr. and R. Murad, In A.G. Gilman, L.S. Goodman, A Gilman (eds.), The Pharmacological Basis of Therapeutics, 6th edn, Macmillan, New York, 1980, p 1474.

[162] F. Murad and R.C. Haynes. Jr., In A. G. Gilman, L.S. Goodman A Gilman (eds.), The Pharmacological Basis of Therapeutics, 6th edn, Macmillan, New York, 1980, p. 1453.

[163] W.C. Bowman and M.J. Rand, Text Book of Pharmacology, Blackwell, London, 1980, pp 19.1-19.70.

[164] A.V. Schally, Science, 202(1978) 18.

[165] C.A. Wilson, Adv. Drug. Res., 8(1974)119.

[166] L.L. Engel, In International Series of Monographs on Pure and Applied Biology, Pergamon Press, Oxford, 1961, pp 1-13.

[167] S.B. McEwen and B. Parson, Annu. Rev. Pharmacol Toxicol, 22(1982) 555.

[168] J. Meites and W.E. Sonntag, Annu. Rev. Pharmacol. Toxicol., 21(1981)295.

[169] J. Meites, J. Simpkins, J. Bruni and J. Advis., IRCS J.Med. Sci., 5(1977) 1.

[170] R.I. Weiner and W.F. Ganong, Physiol. Rev., 58(1978) 905.

[171] J.M. Davidson, In W.E. Ganon and L. Martini (ed.), Frontiers in Neuroendocrinology, Oxford University Press, New York, 1969, pp 343-388.

[172] W.C. Bowman and M.J. Rand, Text Book of Pharmacology, Second edition, Blackwell Scientific Publications, London 1980, pp 15.1-15.31.

[173] A. J. Thody, In A.B. Simonds, (ed.) Advances in Drug Research Vol 11, Academic Press New York, 1977, pp 23-74.

[174] J.N. Karkun and D.P. Sen, Indian J. Med. Res., 53(1965) 226.

[175] L.S. Philips and R. Vassilopulou-Sellin, New Engl. J. Med. 302(1980) 371,438.

[176] E. Mac Hadley, Endocrinology, Prentice Hall Inc., Englewood Cliffs, New Jersey, 1984, pp 286.

[177] S.M. Sief and A.G. Robinson, Annu. Rev. Physiol., 40(1978) 345.

[178] P. Seeman, Pharmacol. Rev., 32(1980) 229.

[179] P. Bedard, J. Dankova, R. Boucher, and R. Langelier, Can. J. Physiol. Pharmacol., 56(1978)538.

[180] N.T. Chase and D.L. Murphy, Annu. Rev Pharmacol., 13(1973) 181.

[181] P.A. Shore, Annu. Rev. Pharmacol., 12(1972)209.

[182] E.L. Klaiber, D.M. Broverman, W. Vogel and Y. Kobayashi, J. Colloid Inteface Sci, 28(1979) 156.

[183] I. Rothchild, In R.S. Harris, I.G. Wool an J A. Loraine (eds.), Vitamins and Hormones, Academic Pren New York, 1965, pp 243-249.

[184] N.H. Raskin, Annu. Rev. Pharmacol. Toxicol., 21(1981)463.

[185] W.C. Bowman and M.J. Rand, Text Book of Pharmacology, Second Edn. Blackwell Scientific Phublications, London, 1980 pp 20.1-20.62.

[186] P.O. Lundberg, Acta Endocrinol., 40 (Suppl 68)(1962) 5.

[187] W.C. Bowman and M.J. Rand, Text Book of Pharmacology, Second Edition, Blackwell Scientific Publications, London 1980, pp. 31.3-31.5.

[188] F. Murad and R.C. Haynes, Jr., In A.G. Gilman, L.S. Good man and A Gilman (eds). The Pharmacological Basis of Therapeutics, Macmillan New York, 1980, pp 1420-1477.

[189] A.F. Wagner and K. Folkers, Perspect. Biol Med., 6(1963) 347.

[190] J. Marks, In R.S. Harris and I.G. Wool (Eds). Vitamins and Hormones, Vol 20, Academic Press, New York, 1962 pp 573-293.

[191] H.G. Mandel and W.H. Cohn, In A.G. Gilman, L. S. Goodman and A.Gilman, The Pharmacological Basis of Therapeutics 6th Edition, Macmillan New York, 1980, pp 1597.

[192] F.A. Knicl, In (M. Tansk, Section ed.), International Encyclopedic of Pharmacology and Therapeutics, section 48, Vol. I, Peramon Press, Oxford, 1971, pp 275-290.

[193] K.E. Mason, In R.S. Harris and K.V. Thimann (eds.), Vitamins and Hormones, Vol.2, Academic Press, New York, 1944, pp 107-154.

[194] H.H. Beard, Annu. Rev. Biochem., 10(1941)245.

[195] H.M. Nitowsky, J.T. Tildon, S. Levin and H.H. Gordon, Am. J. Clin, Nutr., 10(1962)368.

[196] H. Dam, I. Prange and E. Sondergaard, Acta Pathol. Mocrobiol. Scand. 31 (1952) 172.

[197] L.J. Machiln and W.T. Shalkop, J. Nutr., 60(1956)87.

[198] M. L. Scott and C.C. Calvert, J. Nutr., 77(1962)105.

[199] H.A. Harper, Review of Physiological Chemistry, 15th Edn., Lange Medical Publications, Maruzen, Japan, 1975 p 95.

[200] J.T. Dingle and J.A. Lucy, Biochem. J., 90(1963) 133.

[201] A.V.S.S. Ramarao Text Book of Biochemistry, UBS Publishers, New Delhi- 1999 p.158.

[202] J.G. Hardman and L.E. Limbird (eds.). In Goodman and Gilman's the Pharmacological Basis of Therapeutics, McGraw Hill, New York, 1996 p.1573.

[203] D.R. Laurence and M.J. Brown, Clinical Pharmacology, 8th Edn., Churchill Livingston, New York, 1977 p 671.

[204] A.V. S.S. Ramarao, Text Book of Biochemistry, 8th Edn. UBS Publishers, New Delhi, 1998, pp 206.

[205] U.S. Van Euler and R. Eliasson In Prostaglandins Medicinal Chemistry-A Series of Monographs, Vol. 8, Academic Press New York, 1967, p 125.

[206] S. Bergstrom, L.A. Carlson and J.R. Weeks, Pharmcol. Rev.,20 (1968) 1.

[207] S. Moncada, R.J. Flower and J.R. Vane, in A. G. Gilman, L.S. Goodman and A Gilman (eds.). The Pharmacological Basis of Therapeutics, 6th Edn., Macmillan, New York, 1980, pp. 673.

[208] N. Weiner In A.G. Gilman, L. S. Goodman, A. Gilman (eds.), The Pharmacological Basis of Therapeutics, 6th Edn, Macmillan, New York, 1980 p146.

[209] A. Roberts, In M. Silver, B.J. Smith and J.J. Koksis (eds.), Prostaglandins in Hematology, Spectrum Publications Inc., New York, 1977 p.187.

[210] B. J. R. Whittle, N.K. Broughton-Smith,S. Moncada and J. R. Vane, Prostaglandins, 15(1978) 955.

[211] S.M.M. Karim and W.P. Fung, Adv. Prostaglandin Thomboxane Res., 2(1976) 539.

[212] W.C. Bowman and M.J. Rand, Text Book of Pharmacology, Blackwell Oxford 1980, p. 25.14.

[213] E. W. Horton, Physiol. Rev., 49(1969) 122.

[214] E.W. Horton, In Prostaglandins-Monographs on Endocrinology, Vol.7, Springer Verlag, Berlin, 1972.

[215] F. Coceani, Arch. Intern. Med., 133(1974) 119.

[216] W.C. Bowman and M.J. Rand, Text Book of Pharmacology, Blackwell, Oxford, 1980 pp 18.1, 18.2.

[217] V.J. Goldberg and P.W. Ramwell, Physiol. Rev., 55(1975) 325 (see also Ref. 216 pp 12.37).

[218] W.Y. Chan, Annu. Rev. Pharmacol, Toxicol., 23(1983) 131.

[219] W.C. Bowman and M.J. Rand, Text Book of Pharmacology, Blackwell, Oxford, 1980, pp 12.30-12.39.

[220] F.C. Rector, N.W. Carter and D.W. Seldin, J. Clin. Invest., 44(1965) 278.

[221] D.C. Barter, Annu. Rev. Pharmacol. Toxicol., 23(1983) 45.

[222] W.C. Bowman and M.J. Rand, Text Book Of Pharmacology, Blackwell, Oxford, 1980, pp 2.24, 12.39,25.26.

[223] R.J. Flower, S. Moncada and J.R. Vane, In A G Gilman, L.S. Goodman and A Gilman (eds.), The Pharmacological Basis of Therpentics, 6th Edn Macmillan, New York, 1980, pp 687.

[224] R. Verbrckmoes, B. Van Damme, J. Clement, A. Amery and P. Michielsen, Kidney Int. 9(1976) 302.

[225] L. Zancan, F. Zacchello and F. Mantero, Lancet, 2(1976) 1354.

[226] P.V. Halnshka, H. Wohltmann, P.J. Privitera, G. Hurwitz and H.S. Margolius, Ann. Intern. Med., 87(1977)281.

[227] L. Norby, W. Flamenbanm, R. Lentz and P. Ramwell, Lancet, 2(1976) 604.

[228] J.F. Roderick, M. Salvador and R.V. John, In A.G. Gilman, L.S. Goodman and A Gilman (eds.), 6th edn., Macmillan, New York, 1980, pp 682.

[229] W.C. Bowman and M.J. Rand, Text Book of Pharmacology, Blackwell, Oxford, 1980 pp 13.16.

[230] W.C. Bowman and M.J. Rand, Text Book of Pharmacology, Blackwell, Oxford, 1980 pp.12.40.

[231] N.H. Ruskin, Annu. Rev. Pharmacol, Toxicol, 21(1981) 463.

[232] L.A. Carlson, A.G. Ekelund and L. Oro, Acta Med. Scand, 183(1987) 301.

[233] F. Giora and M.H. John, Anu. Rev. Pharmacol. Toxicol, 27(1987) 301.

[234] A. Bult and A.G. Herman, In P.M. Vanhutte, H. Denolin, A. Goosens and A.G. Herman (eds.), Cardiovascular Pharmacology of the Prostaglandins, Raven, New York, 1982, pp 327.

[235] J.R. Fletcher and P.W. Ramwell, Adv. Shock Res., 3(1980) 57.

[236] F.L. Anderson, W. Jubitz, T.J. Tsagarics and H. Kudia, Am. J. Physiol., 228(1975) 410.

[237] W.C. Bowman and M.J. Rand, Text Book of Pharmacology, Blackwell, Oxford, 1980, pp 29.11.

[238] I.H. Leopold and E. Duzman, Annu. Rev. Pharmacol. Toxicol., 26(1986)401.

[239] E. P. David, Pharmacol. Rev., 33(1981) 133.

[240] M. Salvador, J.F. Roderick and R.V. John, In A.G. Gilman, L.S. Goodman, A. Gilman, (eds.), The Pharmacological Basis of Theerapentics, 6the Edn. Macmillan, New York, 1980, pp 675.

[241] J. Nakano and M.C. Koss, South Med. J., 66(1973) 709.

[242] W.C. Bowman and M.J. Rand, Text Book of Pharmacology, Blackwell, Oxford, 1980. pp 12.39.

[243] L.S. Wolfe and F. Coceani, Annu Rev. Physiol., 41(1979) 669.

[244] M. Salvadon, J.F. Roderick and R. V. John, In A.G. Gilman, L.S. Goodman, T.W. Rall and F. Murad (eds.), The Pharmacological Basis of Therapentics, Macmillan, New York, 1985, pp 667.

[245] W.C. Bowman and M.J. Rand, Text Book of Pharmacology, Blackwell, Oxford, 1980, pp 19.13.

[246] H.R. Behrman, Annu. Rev. Physiol, 41(1979) 685.

[247] A. Felmeister, J. Pharm. Sci. 61(1972) 151.

[248] P. Seeman and H. S. Bialy, Biochem. Pharmacol., 12(1963) 1181

[249] P.A. Shore, Annu. Rev. Pharmacol., 12(1972)209.

[250] L.L. Iversen, In B.A. Collingham (ed.), Drugs and Transport processes, Macmillan, New York, 1974, pp 275-286

[251] J. Glowniski, In H.J. Shumann and G. Kronberg (eds.) New Aspects of Storage and Release Mechanism of Cateehomines, Bay Symposium II, Springer Verlag, Berlin, 1970, pp 237-248.

[252] D. Cloin, Therapentic Drugs, Chap 9, Churchill Livingstone, New York, 1992.

[253] A.N. Nagappa, P.L. Kole, P.V. Pandi, I Shankukha, K. Girish and Rahul P.K. Mishra, Colloids and Surfaces B: Biointerfaces, 32(2003) 169.

[254] W.C. Bowman and M.J. Rand, Text Book of Pharmacology, Blackwell, Oxford, 1980 pp 18.30.

[255] E. I. Isaacson and J.N. Delgado, In M.E. Wolff (ed.) Burger's Medicinal Chemistry, part III, John Wiley and Sons, New York, 1980, p 854.

[256] A. S. Buchanan, M.L. Glenda and A. Shulman, Eur. J. Pharmacol., 7(1969) 60.

[257] T.W. Rall and L.S. Schleifer, In A.G. Gilman, L.S. Goodman and A Gilman (eds.) The Pharmacological Basis of Therapentics, Macmillan New York, 1980, pp 455, 460, 463.

[258] J. Pawelek, R. Bichonski and B. Byrski., Pol. J. Pharmacol. Pharm., 28(1976) 449.

[259] E. Costa, A. Guidotti, C.C. Mao and A. Suria, Life Sci, 17(1975) 167.

[260] A. Guidotti, In M.A. Lipton, A. DiMascio and K.F. Killam (eds.) Psychopharmacology: A Genration of Progress, Vol. 18, Raven, New York, 1978 pp 1349-1357.

[261] P. Pole, H. Mohler and W. Haefely, Naunyn Schmiedeberg's Arch. Pharmacol., 248(1974) 319.

[262] D.W. Choi, D.H. Farb and G.D. Fisehbach, Nature, London, 269(1977) 342.

[263] R. Mac Donald and J.L. Barker, Nature, London, 271 (1978) 563..

[264] G. Toffano, A. Guidotti and E. Costa, Proc. Natl. Acad. Sci., USA, 75 (1978) 4024.

[265] W. Schallek, W.D. Horst and W. Schlosser, In S. Garattini, A. Goldin, F. Hawking and I.J. Kopin (eds)., Advances in Pharmacology and Chemotherapy, Vol. 16 Academic Press New York 1976, pp 57.

[266] S.C. Harvey, In A.G. Gilman, L.S. Goodman and A Gilman (eds.). The Pharmacological Basis of Therapeutics, 6th Edn., Macmillan, New York, 1980 pp 345.

[267] S.H. Snyder, S.J. Enna and A. B. Young, Am. J. Psychiatry, 134(1977) 662.

[268] R.W. Olsen, E.E. Lamar and J.D. Bayless, J. Neurochem., 28(1977) 299.

[269] M. Harris, J.M. Hopkin and M. J. Neal, Brit. J. Pharmacol., 47(1973) 229.

[270] E. Costa and A. Guidotti, Annu. Rev. Pharmacol. Toxicol., 19(1979) 531.

[271] G. Zbinden and L.O. Randall, Adv. Pharmacol. Chemother 5(1967) 213.

[272] L.O. Randall and W. Schallek, In, DHE from (ed.) Psychopharmacology, Public Health Ser. Publ. Washington D.C. 1968, pp 153-184.

[273] W.C. Bowman and M.J. Rand, Text Book of Pharmacology, Blackwell, London, 1980, pp 18.1-18.35.

[274] S.H. Snyder and S.J. Enna, In E. Costa and P. Greengard, (eds.), Mechanism of Action of Benzodiazepnies, Adv. Biochem. Phychopharmacol., Raven, New York, 1975, Vol. 14, pp 81-91.

[275] W.R. Theodore and S.S. Leonard, In A G Gilmanl L.S. Goodman and A Gilman (eds.), The Pharmacological Basis of Therapeutics, Macmillan, New York, 1980, pp. 448-474.

[276] R.W. Olsen, Annu. Rev. Pharmacol. Toxicol. 22(1982) 245 (See also. R.W. Olsen, In A.G. Gilman, L.S. Goodman, T.W. Rall and F. Murad (eds.), The Pharmacological Basis of Therapeutics, Macmillan, New York, 1985, Chap 20, p 465).

[277] W.C. Bowmn and M.J. Rand, Text Book of Pharmacology, Blackwell, London 1980, pp 8.1-8.16.

[278] W.C. Bowman and M.J. Rand, Text Book of Pharmacology, Blackwell, London, 1980, pp 6.1-6.33.

[279] R.J. Baldessorini, In A.G. Gilman, L.S. Goodman and A. Gilman (eds.), The Pharmacological Basis of Therapeutics, Macmillan, New York, 1980, pp 391-447.

[280] W.C. Bowman and M.J. Rand, Text Book of Pharmacology, Blackwell, London 19.1-19.70.

[281] E.M. John and S.L. Allen Annu Rev. Pharmacol. Toxicol., 25(1985)127.

[282] N.W. Norman, In A G Gilamn, L.S. Goodman and A Gilman (Eds), The Pharmacological Basis of Therapeutics, Macmillan, New York, 1980, pp 138-175.

[283] C.S. Harvey, In A.G. Gilman, L.S. Goodman, T.W. Rall and F. Murad (eds.). The Pharmacological Basis of Therapeutics, 7th Edn., Macmillan New York, 1985, pp 352-71.

[284] L.O. Randall, W. Schallek, In G. Maxwell (ed.) Psychopharmacological Agents, Vol. III, Academic Press, New York, 1974, pp 175-281.

[285] S.P. Gupta, Curr. Pharm. D, 4(1998) 455.

[286] A. Saitta, M.N. Saitta, M. Bonaiuto, M. Casteldo, A. Sardo, A. Messina and M. Cinquegrani, Clin. Ther., 148(1997) 237.

[287] V.A. Skeberdi, J. Jurevicius and R.Fischmeister, Br. J. Pharmacol., 121 (1997)1277.

[288] K. Tnsuji, Surface activity, Principle Phenomena and Applications Academic Press, 1988, p.211.

[289] K. Tnsuji, Surface activity, Principle Phenomena and Applications Academic Press, 1988, pp.20-30.

[290] J. Hernandez-Borrel and M.T. Monterro, Int. J. Pharm 252(2003) 149.

[291] A. Hostacka, Microbios, 91(1997) 137.

[292] M.A. Ramadan, A.F. Tawfik, T.A. Alkersh and A.M. Shibl., J. Infect. Dis 171(1995) 483.

[293] E.K. Jackson and J.C. Garrison, In J.G. Hardman and L.E. Limbird (eds.,), Goodman. Gilman's Health Professions Division, New York, 1996, pp 733-754.

[294] C. Dollery, Therapeutic Drugs, Churchill Liuingistone, Edinburgh, 1999 pp C38-C43.

[295] E.K. Jackson and J.C. Garrison, In J.G. Hardman and L.E. Limbird (eds.), Goodman and Gilman, The Pharmacological Basis of Therapeutics, McGraw Hill, Health Professions Division, New York, 1996 pp 767-772.

[296] D.W. Johns, C.R. Ayers and S.C. Williams, Hypertension 6(1984) 545.

[297] M.D. Cressman, D.G. Vidt and C. Acker, Lancet 2 (1982) 440.

[298] C. Dollery, Therapeutic Drugs, Churchill Livingstone, Edinburgh, 1999, pp L36-167.

[299] J. T. Dipro, R.L. Talbert, G.C. Yee and G.R. Matzke, Pharmacotherapy A pathphysiologic Approach, Appleton and Lange, Connecticut, 1999, pp.146.

[300] B.B. Hoffman and J.R. Lefkowitz In G. Hardman and L.E. Limbird (eds.), McGraw Hill, Health Professions Divisions, New York, 1996, pp 211.

Chapter 7

Assessment of the Hypothesis

The studies recorded in Chapter 6 indicate that the drugs, which act by altering the permeability of cell membranes after interacting with them, are surface active in nature and generate a liquid membrane at their site of action either by themselves alone or in association with the membrane lipids. The liquid membranes thus generated modify the transport of relevant permeates to their site of action which makes a definite contribution to drug action. In this chapter we proceed to make an assessment of the hypothesis: the liquid membrane hypothesis of drug action (section 4.2). For this assessment let us first see what are implications of the hypothesis and to what extent these are substantiated.

7.1. Implications of the hypothesis

Studies on the liquid membranes generated by surface-active drugs can provide a clue to their quantitative action. This explanation arises because the CMC of the drug indicates the concentration at which the interface will be completely covered by the drug liquid membrane. At this concentration (CMC), therefore, modification in the permeability of biological membrane would be maximum. This conclusion implies that at the CMC the magnitude of the biological effect would also be maximum. Hence, the lower the CMC of a drug, the lower the concentration required to alter the membrane transport and as a consequence, more potent would be the drug. Thus, CMC's of a series of drugs with the same pharmacological action can be a good indicator of their potency. This inference has been substantiated in almost all studies recorded in Chapter 6. For example let us take the case of haloperidol and chlorpromazine [1, 2]. The CMC of haloperidol is 1×10^{-6}M while that of chlorpromazine is 4.5×10^{-5}M. Haloperidol is known to be more potent than chlorpromazine on a milligram basis [3].

Another example substantiating this result is that of local anesthetic drugs: the lower the CMC the more potent the drug. In a series of local anesthetic, it was found [4], that CMC's and minimum blocking concentrations (MBC) for nerves (Table 19 of chapter 6) are identical. This finding indicates that formation of a liquid membrane between cations like sodium, potassium and the nerve membrane appears to be an important step in the mechanism of action of local anesthetics. It is proposed that interaction of local anesthetics with the lipid micro-environment of the sodium channel results in its fluidization, causing a blockade of sodium transport [5]. Thus, a physical mechanism can provide a satisfactory explanation for local anesthesia. Formation of liquid membranes by these drugs within sodium channels and polar head interaction of the drugs with the lipid micro-environment of the channels can, therefore, explain why nerve blocking concentration and CMC's are identical (Table 19 of chapter 6).

The liquid membranes generated by surface active drugs are expected to have two types of orientations with respect to the approaching permeants. The drug liquid membrane can present either hydrophilic or hydrophobic ends to the permeants (see Chapter 6). It is observed (see Chapter 6) that a change in the orientation of the drugs can alter the transport of permeants. Whichever orientation shows alterations in permeability (similar to those observed in biological cells) is of predictive value. In the majority of drugs investigated in Chapter 6 e.g. halperidol, chlorpromazine hydrochloride, reserpine, tricyclic antidepressants, local anaesthetes, hypnotics / sedatives, antihistamines, diuretics etc. it has been found that resistance to the transport of permeants is maximum when hydrophobic ends of the surface-active drugs face the approaching permeants. This implies that the receptors for these drugs are likely to be oriented in such a manner that their hydrophilic moieties are projected outwards to which hydrophilic ends of the drugs get attached. Therefore, the hydrophobic ends of the drugs project outwards to face the permeants. Such an orientation can be rationalized if one examines the nature of the receptors, in general, in relation to the lipid bilayer part of the biomembranes.

These receptors, in general, are membrane proteins and hence should be surface-active nature. Hence, they should have both hydrophilic and hydrophobic moieties in their structures. Since the exterior environment of biological cells is aqueous in nature, it is logical to expect that the hydrophobic part of these membrane proteins will be associated with the hydrophobic core of the lipid bilayers and only the hydrophilic part will face the exterior. Prediction about similar orientation of receptor proteins and also the membrane proteins, in general, has been made in the literature [6]. Thus, the studies on liquid membranes generated by drugs are capable of indicating the possible orientation of receptors responsible for interaction with the drugs.

Since the biological membrane is comprised of different types of lipids and proteins, a drug can alter transport across the membrane by one of the following mechanisms:

1. The drug itself may form a liquid membrane which can reasonably explain alteration of transport across the membrane,
2. The drug / lipid interaction may be responsible for the observed biological effect
3. The drug / protein interaction may be the causative action.

In the case where the first possibility is ruled out because an effect similar to that on biological tissues is not mimicked by the drug liquid membrane alone, interaction with the liquid membrane formed by the lipids needs to be studied. In the case of dizzepam [7], it was found that the biological actions of the drug, i.e. facilitating actions of GABA, could not be mimicked in either orientations of the drug, but interaction with lecithin-cholesterol liquid membrane showed an increase in the permeability towards GABA (Table 39 of Chapter 6).

The multiplicity of biological actions exerted by surface-active drugs can be explained well on the basis of liquid membrane hypothesis. For example, antihistamines are also known [8] to have anti-cholinergic and local anesthetic action. Chlorpromazine and a few other low-potency phenothiazines have mild anti-histamine [9] and anti-serotonin [9] activity. Such actions can be explained as a result of alteration of transport of relevant permeants because of the liquid membrane interposed between the permeant and the biomembrane.

Although multiplicity of biological actions have been discussed all along with the various drugs in the previous chapter, Srivastava etal has presented an in-depth discussion of the multiplicity of biological actions of psychotropic drugs. [10] In the light of the "liquid membrane hypothesis of drug action." A concise account of this discussion is presented below.

Clinically chlorpromazine, halperidol and reserpine are used as antipsychotics whereas imipramine is used as an antidepressant. In addition to antipsychotic and antidepressant effects, these drugs are reported to exert a variety of other biological actions as well (multiplicity of drug action). Although the values of the solute permeability of relevant permeants in the presence of drug liquid membranes are already given in Chapter 6, the normalized values of solute permeability for relevant permeants in the presence of the drug liquid membranes in case of each of the four psychotropic drugs viz. haloperidol, chlorpromazine, reserpine and imipramine as obtained in the earlier studies are recorded in Table 1. for ready reference. The data in Table 1 which are from two sets of model experiments, the one in which the permeants face hydrophilic surface of the drug liquid membrane and the other in which they face the hydrophobic surface. For experimental details the original papers should be referred to [1, 2, 11, 12].

It is well known that central regulation of the pituitary is mediated by the hypothalamus, which in turn is under the influence of neurotransmitters. It is, therefore, logical to expect that the drugs modifying the permeability of neurotransmitter molecules should display altered pituitary functions resulting from the alteration in the release of adenohypophysins and neurohypophysIns.

For the actions of dopamine the hydrophilic portions of dopamine receptors, which are located in higher brain centres, have been considered important [14, 15]. It is therefore expected that in the liquid membranes formed at the site of action by haloperilol, chlorpromazine or reserpine which are known to be dopamine antagonists, the hydrophilic parts of the drugs would be preferentially oriented towards the hydrophilic parts of the receptor and the hydrophobic parts of the drugs would be drawn outwards way from it. Therefore the agonist, dopamine molecules, would face the hydrophobic surface of the drug liquid membranes interposed between the agonist and its site of action. Thus the data (Table 1) on the transport of dopamine in the specific orientation of the drug liquid membranes with its hydrophobic surface facing the permeant, dopamine, appear relevant to its biological actions. Similar considerations apply to adrenaline and nor-adrenaline and also to serotonin, which act on pre-and post- synaptic receptor [16-18].

Neurotransmitters through, their actions on median eminence promote or inhibit the release of hypo-physiotrophic hormones, both release and release inhibiting hormones. These hypophysiotropic hormones in turn act on adenohypophysis and regulate the release of adenohypophysins [19].

Table 1. Normalized values (r) of solute permeability in the presence of liquid membranes generated by the drugs: r = ($\omega/\omega_{control}$), ω being the value of solute permeability given by the equation $\omega = (J_s / \Delta\pi)_{Jv=0}$ where J_s and J_v represents the solute flux and the volume flux per unit area of the membrane and $\Delta\pi$ is the osmotic pressure difference across the membrane and $\omega_{control}$ being the value from control experiments in which no drug was used.

Permeants	r_1	r_2
Haloperidol (Ref. 1, 13)		
Dopamine	0.766	2.938
Nor-adrenaline	0.869	3.883
Adrenaline	Undetectable	4.682
Serotonin	0.488	1.797
γ-Aminobutyric acid (GABA)	0.723	1.270
Chlorpromazine hydrochloride (Ref. 2, 13)		
Dopamine	0.340	0.524
Nor-adrenaline	0.214	0.783
Adrenaline	0.119	0.789
Serotonin	0.195	0.397
γ-Aminobutyric acid (GABA)	0.775	0.795
Reserpine (Ref. 12, 13)		
Dopamine	0.649	0.777
Nor-adrenaline	0.059	0.570
Adrenaline	0.487	0.756
Serotonin	0.293	0.488
γ-Aminobutyric acid (GABA)	0.586	1.604
Imipramine hydrochloride (Ref. 11, 13)		
Dopamine	0.394	1.632
Nor-adrenaline	0.487	1.223
Adrenaline	0.382	1.489
Serotonin	0.428	4.199

r_1 : permeants facing the hydrophobic surface of the drug liquid membrane.
r_2 : permeants facing the hydrophilic surface of the drug liquid membrane.

Dopamine, nor-adrenaline and adrenaline are reported [19-23] to inhibit the release of CRH (corticotrophin releasing hormone). This inhibition of the release of CRH in turn inhibits the release of ACTH (adrenocorticotropic hormone). The neuroleptic drugs namely haloperidol, chlorpromazine and reserpine which impede the transport of dopamine, nor-adrenaline and adrenaline (Table 1) and thereby reduce the access of these neurotransmitters to their site of action in median eminence should enhance the release of CRH and consequently of ACTH. This expectation is in agreement with literature reports that administration of these drugs viz. haloperidol, chlorpromazine and reserpine does enhance the release of ACTH [21, 22, 24]. The role of serotonin, whose permeability is reduced in the presence of imipramine (Table 1), on the release of ACTH is complicated. Both stimulatory and inhibitory actions have been reported [25, 26]. This is consistent with the fact that no significant effect of imipramine, which blocks both pre and post-synaptic receptors, on the release of adenohypohysins including ACTH have been reported [27]. Gamma aminobutyric acid (GABA) is known to have inhibitory effect on the release of CRH [21]. The impediment in the transport GABA in the presence of the three neuroleptics namely haloperidol, chloropromazine and reserpine (Table 1) in the specific orientation of the drug liquid membranes with their hydrophobic surface facing the permeants is consistent with the reported enhancement in the release of ACTH by these drugs [21, 22, 24].

It is documented that dopamine, and GABA have inhibitory effect on the release of TSH (thyroid stimulating hormone} whereas nor-adrenaline and adrenaline have stimulatory effects [21, 22, 28-30]. The impediment in the transport of nor-adrenaline and adrenaline in the presence of the chlorpromazine liquid membrane (Table 1) could be a plausible explanation for the reported inhibition of TSH release by chlorpromazine [28]. It has been reported [31], that chlorpromazine has specific affinity for accumulation in the hypothalamus area.

At hypothalamic level, the secretion of prolactin in mammals is controlled by the inhibitory hormone P-RIH (prolactin release inhibiting hormone} and possibly a ptolactin releasing hormone, P-RH, the role if any of P-RH is only of secondary importance [19]. The release of P-RIH from neuro-endocrine transducer cells, in the median eminence is controlled primarily by hypothalamic dopamine. Neuroleptic drugs which inhibit the transport of dopamine (Table 1) are, therefore, expected to enhance the prolactin secretion which indeed is substantiated by literature reports [21, 32]. It may be mentioned that there are very strong evidence to suggest that dopamine itself functions as P-RIH [21, 33].

The role of GABA in the secretion of prolactin is similar to that of dopamine at the level of median eminence and also at the hypophyseal level [34]. The impediment in the transport of GABA due to the liquid membranes generated by the neuroleptic drugs (Table 1) could also be a factor contributing to the increased prolactin secretion brought about by the neuroleptics. The role of a adrenaline, nor-adrenaline and serotonin at the level of median eminence , in prolactin secretion is stimulatory in nature, the magnitude of the stimulatory effect being much smaller in comparision to the inhibitory effect of dopamine [21, 35, 36]. The neuroleptic drugs should, therefore, bring about a decrease in prolactin release on account of impediment in the transport of serotonin, nor-adrenaline and adrenaline due to the liquid membranes generated by these drugs. In order that this effect becomes observable, one should first block the dopamine receptors and then study the effect of neuroleptics on prolactin release. Such experiments, which are called for to substanitiate this surmise, have, however, not come to our notice.

At the hypophyseal level the effects of nor-adrenaline and adrenaline on the release of prolactin are similar to that of dopamine [37, 38]. It has been shown that adding extracts of hypothalamus to the cultured antirior pituitaries decreased the quantity of prolactin released into the medium [39-41]. Danon *et al.* showed that hypothalamus obtained from rats treated with phenothiazine derivative, perphenazine, when added to the cultures of pituitaries does not inhibit prolactin secretion [42]. The impediment in the transport of nor-adrenaline, adrenaline and dopamine due to the liquid membranes of phenothiazine drugs like chlorpromazine (Table 1) could be a plausible explanation for the antagonistic effect of phenothiazine on the normally operating inhibitory influence of the hypothalamus on prolactin secretion at the level of pituitary.

The fact that dopamine plays a key role at the level of median eminence in stimulating the release of LH/FSH-RH (luteinizing hormone/follicle stimulating hormone-releasing hormone) is well established [19]. The release of dopamine at the median eminence is in turn controlled by other neurotrans mitters viz. nor-adrenaline, adrenaline, GABA, serotonin and also dopamine at the higher brain centres [43]. The observation that neuroleptics like chlorpromazine, haloperidol and reserpine block the release of LH/FSH-RH can be explained in terms of the resistance offered to the transport of the neurotransmitters by the neuroleptic drugs (Table 1), which are known to accumulate not only in the median eminence but also in the higher brain centres [22].

Dopamine and nor-adrenaline have been reported to increase growth hormone release in animals and man [44]. Neuroleptic drugs like haloperidol, reserpine, chlorpromazine etc. caused reduction in the release of growth hormone [21, 45]. The impediment in the transport of these neurotransmitters viz. dopamine and nor-adrenaline (Table 1) due to the liquid membranes generated by the neuroleptic drugs can be utilized to explain this observation – the liquid membranes reduce the access of the neurotransmitter to their action sites.

Dopamine, which acts not only on the median eminence but also on somatotrophs, has a paradoxical inhibitory effect on the release of growth hormone in acromegalics. This is attributed to somatotrophic cells themselves [21]. It may be mentioned that treating acromegaly with phenothiazines which inhibit the transport of dopamine (Table 1) has met with little success [46]. This is not unexpected in view of the paradoxical inhibitory action of dopamine in the release of growth hormone in acromegalic patients.

MSHs (melanocyte simulating hormones) secretion by the pars intermedia of pituitary gland is inhibited by catecholamines viz. nor-adrenaline, adrenaline and dopamine [47]. Neuroleptic drugs such as haloperidol, reserpine and chlorpromazine are reported to simulate MSH secretion [47]. This is an expected observation in view of the reduced permeability of adrenaline, nor-adrenaline and dopamine in the presence of the liquid membranes generated by the neuroleptic drugs reducing access of the catecholamines to their relevant sites of action.

Neurohypophysial secretions containing ADH (antidiuretic hormone), oxytocin and neruophysins are evoked by different stimuli. Acetylcholine and nicotine injected into carotid circulation cause the release of ADH, oxytocin and neruophysins while nor-adrenaline inhibits their release [48, 49]. It is therefore logical to expect that the drugs like chlorpromazine and reserpine which are likely to reduce the access of nor-adrenaline to the relevant site of action due to its reduced permeability through the liquid membranes generated by these drugs (Table 1) should reduce the inhibitory effect and thereby facilitate the release of neurohypophysial hormones like ADH, oxytocin etc. This indeed has been found to be the case [22, 50, 51].

Reduction in the concentration of serotonin at the postsynaptic receptor resulting in defective neurotransmission has been implicated in migraine [52]. Imipramine in some cases is known to act beneficially in migraine [27], whereas neuroleptics like reserpine are known to aggravate it [53]. Blockade of reuptake of serotonin due to its reduced permeability through the imipramine liquid membrane (Table 1) likely to be formed at the presynaptic receptors resulting in improved neurotransmission could also be a possible explanation for the curative action, of imipramine. Similarly blockade of serotonin due to its reduced permeability through the liquid membranes likely to be generated by the neuroleptic drugs (Table 1) at the post synaptic receptor resulting in poor neurotransmission could be a plausible explanation for the aggravation of migraine by the neuroleptic drugs like reserpine.

Neuroleptic drugs e.g. haloperidol, chlorpromazine, and antidepressant drugs like imipramine are reported to cause hypothermia [27, 54]. This effect can also be explained on the basis of the modification in the permeability of neurotransmitters due to the liquid membranes, which may be formed by these drugs in the hypothalamic region. The hormones ACTH/MSH whose secretion is inhibited by nor-adrenaline, adrenaline and dopamine at the level of median eminence [19-23, 47] have been shown to cause a fall in body temperature [54]. Since neuroleptic drugs and also the antidepressant drugs like imipramine reduce the permeability of nor-adrenaline, adrenaline and dopamine (Table 1), the presence of these drugs at the hypothalamic level may reduce the access of these neurotransmitters to the relevant sites of action causing thereby an increase in the secretion of ACTH/MSH. This in turn may be responsible for the hypothermic effect. The poikilothermic effect of chlorpromazine that is sometimes used to facilitate the induction of surgical hypothermia [27] can also be rationalized in terms of modification in the permeability of neurotransmitters due to the presence of neuroleptic drugs like chlorpromazine (Table 1). Neurotransmitters have also been implicated in central regulation of body temperature in normothermia [54]. While most neurons are temperature insensitive, warm sensitive neurons \and cold sensitive neurons located in the preoptic area and in the anterior hypothalamic area have been implicated in thermoregulation in mammals. The firing rates of warm sensitive neurons increase with warming or decrease with cooling while reverse are the case in cold sensitive neurons. In mammals intra cerebroventricular injection of serotonin produces a rise in body temperature [55]. It is therefore logical to guess that serotonin acts on cold sensitive neurons and nor-adrenaline on warm sensitive neurons. It is documented that cold sensitive neurons loose their thermo sensitivity during synaptic blockade while the warm sensitive neurons do not [56, 57]. Chlorpromazine during synaptic blockade would therefore impair the thermo sensitivity of cold sensitive neurons and leave the thermo sensitivity of warm sensitive neurons unaffected leading to poikilothermia.

Parkinson's disease is known to be due to deficiency of neurotransmitters like dopamine in basal ganglia [58]. Neuroleptic drugs e.g. haloperidol, chlorpromazine and reserpine are known to cause Parkinson's disease [32, 58]. Reduced permeability of dopamine (Table 1) in the presence of the liquid membranes likely to be generated by these drugs in the region of basal ganglia could be one of the contributing factors for this side effect i.e. drugs induced Parkinsonism. The extra pyramidal effects of antipsychotic drugs like haloperidol are reported to be resistant to levodopa therapy [59]. Since reduced concentration of serotonin in cerebrospinal fluid has also been linked with the a defect of extra-pyramidal function [60, 61] the reduced permeability of serotonin in the presence of antipsychotic drugs like haloperidol (Table 1) offers a clue to the causation of extra pyramidal symptoms.

Most neuroleptic drugs have marked protective action against nausea and emesis inducing effects of dopamine agonists, which can interact with the central dopaminergic receptors in the chemoreceptor trigger zone of the medulla [27]. This effect can also be explained by the reduction in the permeability of dopamine due to the liquid membranes generated by the neuroleptic drugs (Table 1). It is reported that neuroleptics and also tricyclic antidepressants e.g. imipramine cause orthostatic hypotension [27]. For example, in normal man intravenous administration of chlorpromazine causes orthostatic hypotension due to a combination of central action and peripheral α-adrenergic blockade [62]. Although the actions of these drugs on cardiovascular system are complex because these drugs produce direct effects on the heart and blood vessels and also indirect ones through actions on central nervous system and autonomic reflexes, reduced permeability of noradrenaline in the presence of these drugs (Table 1) could also contribute to the causation of orthostatic hypotension.

Neuroleptic drugs particularly chlorpromazine and reserpine, during coition, are known to impair ejaculation without interfering with erection [27, 63]. Attribution of this effect to adrenergic blockade though logical remains unsubstantiated [27]. Reduction in the permeability of nor-adrenaline due to the liquid membranes generated by the neuroleptic drugs (Table 1) is consistent with the conjecture that impairment of ejaculation may be due to adrenergic blockade.

Thus it appears that modification in the permeability of relevant neurotransmitters due to the liquid membranes generated by the psychotropic drugs may be one of the causal factors for the multiple actions of these drugs.

7.2 The Liquid Membrane Hypothesis vis-à-vis Existing Theories of Drug Actions

The liquid membrane hypothesis for drug action proposes that in a series of structurally-related drugs, which are congeners of a common chemical moiety and which act by reducing permeability of hydrophilic substances, any structural variation which increases hydrophobicity of the compound will increase resistance towards transport of the hydrophilic permeant. In other words, any deviation in structure leading to increase in hydrophobicity will reduce the CMC of the drug, make it more potent and increase resistance towards a hydrophilic permeant. However, this sequence of events will continue so long as the hydrophilic group of the drug, responsible for interaction with the biomembranes is unaltered. Any alteration of the hydrophilic moieties of the drug may alter specificity towards the membrane and, therefore, may alter the nature of response towards the permeants, e.g., after alteration of the hydrophilic structure, the drug may inhibit transport of another permeant more specifically that the earlier permeant. This finding offers a clue towards the structure-activity relationship. An increase in hydrophobicity will alter the drug action quantitatively, i.e., it will increase the potency. The studies on 5-fluorouracil and its derivatives reported in previous chapter substantiate this conclusion; of course, in several other cases also this conclusion can be seen to be valid. The change in hydrophilicity, however, may alter the nature of action qualitatively, i.e., the specificity of the resistance towards different permeants may change. Similar comments have been made by Burger [64] in connection with structure activity relationship.

This conclusion that keeping the hydrophilic group unaltered, increase in hydrophobicity will result in the increase in potency of the drug can be further understood in the light of the alternative view of enzymatic reactions put forth by Dewar and Storch [65].

The Michaelis-Menten mechanism for enzyme reactions, in its simplest form, is written as

$$E + S \rightleftharpoons ES \rightarrow P$$

where E stands for enzyme, S for substrate, ES for enzyme-substrate complex and P for products. Dewar and Storch [65] have presented a critique of Michaelis-Menten mechanism, which is hard to ignore. These authors have pointed out that adsorption of the substrate in the active site of an enzyme can occur only if all solvent is squeezed out from them. This means that the first step, which should precede the formation of enzyme-substrate complex, is desolvation or dehydration (if water is the solvent, which it is in most cases) of the active site. Since this step is not considered in the Michaelis-Menten mechanism, the interpretations of enzyme reactions in terms of solution chemistry according to Dewar and Storch are misguided.

The mechanism of drug-receptor interaction is considered analogous to Michaelis-Mentenn mechanism: one has to replace E by the receptor R, substrate by the drug D, enzyme-substrate complex ES by the drug-receptor complex RD and product by response, i.e.

$$R + D \rightleftharpoons RD \rightarrow \text{response.}$$

Ideas of Dewar and Storch can, therefore, be extended [66] to the case of drug receptor interaction provided the receptor is located in the aqueous phase. As a first step, preceding the drug-receptor interaction, water should be squeezed out of the active site of the receptor. It, therefore, follows that more hydrophobic is the drug more capacity it should have to dehydrate the active site of the receptor and hence more potent it should be. This is consistent with the occurrence of the hydrophobic term in the Hansch's [67] structure-activity relationship. Thus this particular observation on the increase in potency of the drug with the increase in the hydrophobicity may also have a rationale in the ideas of Dewar and Storch [65], i.e. without tampering with the hydrophilic moiety, the increase in hydrophobicity would result in increase in the capacity of the drug molecule to dehydrate the active receptor site leading to a more effective binding of drug with receptor in the drug receptor complex.

According to Paton's rate theory [68, 69], the dissociation rate constant can be a good indicator of the nature of action shown by a drug. An antagonist is expected to have a low dissociation constant (K_2) as compared to the agonist. Consequently, in a series of antagonists, as the dissociation constant continues to decrease, the potency of the compound as an antagonist increase. In a series of monoquaternary salts, it is indicated that $K_2=0.0038\times2.65^{12-n}$, where n is the number of carbon atoms in the alkyl chain [69]. In other words, K_2 falls by a constant factor of 2.65 for each methylene group added. In the case of monoquaternary salts, Paton [68] has commented: "The association of these mono-quaternaries is dictated by long range ionic forces, that is, by the cationic head, but once the molecule is bound, its dissociation is more or less hampered by Van der Walls binding of the molecule to the receptor surface. The association rate would then be similar for all the compounds, but the dissociation constant would be sensitive to the length of the alkyl chain to a degree comparable with the manner in which the surface tension of alkyl carboxylic acids varies with alkyl chain length". These comments can also be understood in terms of the liquid membrane hypothesis. The addition of each methylene group in an antagonist will increase its hydrophobicity, resulting in reduction of its CMC. Lowering of the CMC may be linked to an increase in potency of the compound as discussed earlier. Besides reducing the CMC, an increase in the methylene groups will strengthen the hydrophobic core of the drug liquid membrane and may offer more resistance to the transport of hydrophilic permeants. The CMC of a drug, therefore, appears to provide the same information, which the dissociation constant provides in case of rate theory [68].

If the dose-response curve of an agonist is compared with the dose-response curve of mixture of an agonist and an antagonist, there is a flattening of the dose-response curve in the latter case [68-71]. This change leads to a parallel right shift in the case of competitive antagonists. The proposition that an agonist replaces an antagonist is ruled out [68]. This effect is further substantiated by low dissociation constants in the case of antagonists [68]. The observations related to the dose-response curves can also be explained on the basis of the liquid membrane hypothesis. A liquid membrane generated by a surface-active 'antagonist' drug is interposed between receptors in the biomembrane and the agonist. As a consequence, transport of agonist is likely to be reduced, resulting in lesser amount of agonist reaching the receptor. Hence, to achieve the same quantum of response, a higher amount of agonist will be needed. This effect will result in the shifting of dose-response curves to the right. The nature of the liquid membrane and the extent of the resistance offered to the agonist will determine the nature and extent of the shift in the dose-response curve.

One experimental observation in relation to the dose-response curve of the agonist-antagonist mixture has necessitated the hypothesis of "spare receptor". It is observed [69], that a mixture of an agonist and an antagonist elicits the same maximum response as in the case of an agonist alone, but at a comparatively higher concentration. The dilemma is: if the antagonist occludes the receptors, how is it possible to obtain parallel dose-response curves with and without an antagonist? Or, in spite of a sizeable section of receptors being occupied by an antagonist, how can a maximal response be obtained? The dilemma has been resolved by proposing [72] the existence of "spare receptors"; i.e., those receptors without combining with which the agonist alone was capable of eliciting maximum response. However, there is a criticism about this hypothesis. A direct experimental demonstration for "spare receptors" is still awaited [73]. Efforts have also been made [74], to demonstrate experimentally that there are no "spare receptors". Paton has commented [69] "for occupancy theory, existence of spare receptors merely seems a puzzling extravagance". In the liquid membrane hypothesis for drug action, the existence of spare receptors is not necessary. The rate of transport of an agonist across the liquid membrane of an antagonist is dependent on the concentration gradient of the agonist across the liquid membrane. As the concentration of agonist is increased, the rate of flow of the agonist across the liquid membrane generated by the antagonist will also increase and at a certain higher concentration of the agonist it will elicit the same quantum of response as in the absence of the antagonist. Thus, rather than the existence of "hypothetical additional receptors" the resistance offered by the liquid membrane generated by the antagonist to the flow of agonist is likely to decide the strength of the biological response. An indication of this proposition is available in the literature. According to the "potentialsevergiftung theory" [75], the "action of the agonist was related to its flux across the cell membrane, which in turn was related to the driving force". The driving force is the concentration gradient.

While commenting [73] on the rate theory, it is mentioned that, in general, for the rate of action of the drug, any one of the following four steps may be the rate-determining step:

1) Access to the receptor
2) Conversion of the drug from an inactive to an active form
3) Rate of combination with the receptor
4) Rate of production of the response

Among these steps, access to the receptor seems to be the most common rate-limiting step [73]. Hence, any event which is likely to reduce access of the agonist to the receptor should have profound influence on the nature and sequence of the agonist-receptor interaction and, hence, the consequent biological response. Generation of a liquid membrane having the ability to reduce access of the agonist to the receptor is one such step. As a result, it is likely to affect the agonist-receptor interaction in a notable manner.

To explain the kinetics of reversible antagonism in aortic strips, a biphasic model was proposed [76, 77]. According to this hypothesis, it was suggested that receptors are situated in a biphasic separated from the extra cellular space by an interfacial barrier through which agonists (but not antagonists) penetrate quickly; penetration of this barrier is considered as the rate-limiting step dictating the kinetics of antagonism. However, the existence of such a barrier in the case of antagonists has been ruled out experimentally [69]. Another prediction of the biphasic hypothesis, i.e., the dose-ratio (the ratio by which the agonist dose must be increased in order to restore a standard response in the presence of antagonist) should rise/fall exponentially when an antagonist is added/removed, is also not true [69]. It is occupancy and not the dose ratio that is observed change exponentially. The liquid membrane hypothesis resolves this problem. Though there is no barrier for the antagonist to reach the receptor, a liquid membrane generated by an antagonist can act as a barrier to the flow of the agonist.

A general comment regarding the validity of the liquid membrane hypothesis for drug action needs special mention. It is known that the majority of transport processes in biological systems (especially those of neurotransmitters) are "active" in nature. Hence, only after showing that a drug-liquid membrane also impedes the active process would the role of the liquid membrane phenomenon in the action of antagonistic drugs become acceptable. For any process of active transport, the rate of access of the permeant to the active site is an important factor. If this itself is impeded, because of the resistance to its transport, even the active transport will be reduced. This reduction can result in antagonism. This is especially true in the case of drug/receptor interaction because access to the receptor has been considered [73] to be a rate limiting process in the whole sequence of drug action.

Thus, the liquid membrane hypothesis of drug action points towards a new facet of drug action. This aspect of drug action has hitherto gone unnoticed. The hypothesis provides a physical basis for the action of the drugs, which are surface-active in nature. The liquid membrane hypothesis, however, needs further experimentation. Characterization of the drug liquid membrane in terms of thickness, dielectric properties and studies on the specificity of the drug liquid membrane provide a few such areas.

REFERENCES

[1] S.B. Bhise, P.R. Marwadi, S.S. Mathur and R.C. Srivastava, J. Pharm. Sci., 71 (1982)526.

[2] S.B. Bhise, P.R. Marwadi, S.S. Mathur and R.C. Srivastava, Biophys.Chem., 17 (1983)187.

[3] S.H. Snyder, Am. J. Psychiatry, 133(1976)197.

[4] R.C. Srivastava, R.K. Sharma, R. Srinivasan and S.B. Bhise, J. Colloid Interface Sci., 94(1983)456.

[5] A.G. Lee, Nature, 262(1976)545.

[6] I.D. Kuntz, In M.E. Wolf (ed.) Burger's Medicinal chemistry, Part I, John Wiley & Sons, New York, 1980, p.291.

[7] R.C. Srivastava, R.K. Sharma and S.B. Bhise, J. Colloid Interface Sci., 93(1982)72.

[8] W.W. Douglos, In A.G. Gilman, L.S. Goodman and A. Gilman, (eds.) The Pharmacological Basic of Therapeutics, Macmillan New York, 1980, pp. 623-625.

[9] R.J. Baldessarini, In A.G. Gilman, L.S. Goodman and A. Gilman (eds.) The Pharmacological Basis of Therapatics Macmillan, New York, 1980, pp.404.

[10] R.C. Srivastava, A.N. Nagappa, D.B. Raju and A.K. Das, Indian J. Chem., 31A(1992)533.

[11] R.C. Srivastava, R.P.S. Jakhar and S.B. Bhise, J. Colloid Interface Sci., 87(1982)56.

[12] S.B. Bhise, P.R. Marwadi, S.S. Mathur and R.C. Srivastava, J. Pharm. Sci., 72(1983)599.

[13] R.C. Srivastava, S.B. Bhise and S.S. Mathur, Adv. Colloid Interface Sci., 20(1984)131.

[14] P.Seeman, Pharmacol Rev., 32(1980)229.

[15] W.C. Bowman and M.J. Rand, Text Book of Pharmacology, Blackwell, Oxford 1980, p.11, 27.

[16] N. Weiner, In A.G. Gilman, L.S. Goodman and A. Gilman (eds.) The pharmacological Basis of Therapeutics, Macmillan, New York, 1980, pp. 138-175.

[17] F.E. Bloom, In AG Gilman L.S. Good man and A. Gilman (eds.) Macmillan, New York, 1980, pp. 235-257.

[18] W.R. Fuller, Annu. Rev. Pharmacol, Toxicol, 20(1980)111.

[19] W.C. Bowman and M.J. Rand, Text Book of Pharmacology, Blackwell, Oxford, 1980, pp. 19.6

[20] E.E. Adams, M.J. Ashby, S.M. Brown, M.C. Whiteand K, Mashiter, elin. Endocrinol., 15(1981)479.

[21] J. Tuomisto and P. Mannisto, Pharmacol, Rev., 37(1985)249.

[22] D. Deweid, Pharmacol, Rev., 19(1967)251.

[23] R.P. Maickal and R.R. Martel, Pharmacol. Biochem. Bchav., 19(1983)321.

[24] L. Debreceni, T.L. Csonka and B. Esete., Acta, Physiol. Acad. Sci., Hung, 36(1969)425.

[25] D.T. Krieger, Mt. Siana, J. Med., 40(1973)302.

[26] D.T. Krieger, Ann. N.Y. Acad. Sci., 297(1977)257.

[27] R.J. Baldessarini, In A.G. Gilman, L.S. Goodman and A. Gilman (eds.), The pharmacological Basis of Therapeutics, Macmillan, New York, 1980, pp. 391-447.

[28] K. Anderson, Brain Catecholamines and Pituitary Thyroid Axis; A Quantitative Histofluorimetrical and functional Analysis in the male Rat, Academic Dissertation, Karolinska Institute, Stockholm, 1984

[29] J. Tuomisto, T. Ranta, A. Sarinen, P. Mannisto and J.Leppaluoto, Lanect, 2(1973)510.

[30] E. Vijayan and S.M. McCann, Endocrinology, 103(1978)1888.

[31] A.W. Wase, J. Christensen and E. Polly, Arch. Neurol. Psychiat, Chicago, 75(1956)54.

[32] W.C. Bowman and M.J. Rand, Text Book of Pharmacology, Blackwell, Oxford, 1980, pp.15.31.

[33] N. Ben Jonathan, C. Oliver, H.J. Weiner, R.S. Mical ad J.C. Porter, Endocrinology, 100(1977)452.

[34] A.V. Schally, T.W. Redding, A. Arimura, A. Dupont and G.L. Linthicum, Endocrinology, 100(1977)681.

[35] C. Denef and M. Baes, Endocrinology, 111(1982)356.

[36] E.E. Muller, G. Nistico and U. Scapagnini, Neurotransmitters and Anterior Pituitary Function, Academic Press, New York, 1977.

[37] R.I, Weiner and W.F. Ganong, Physiol. Rev., 58(1978)905.

[38] J. Nicoletu, P. Filipponi, M. Strappini, L. Fedli, S. Petrelli, G. Gregorini, F. Santeusanio and P. Brunetti, Horn. Res., (Basel) 19(1984)158.

[39] J.L. Pasteels, C.R. Acad Sci., (Paris), 253(1961)3074.

[40] J.L. Pasteels, C.R. Acad Sci., (Paris), 254(1962)2664.

[41] J. Meites, C.S. Nicoll and P.K. Talwalker, In A.V. Nalbandov(Ed.), Advances in Neuroendocrinatology, University of Illinois Press, Urbana, 1963 pp. 238-277.

[42] A Dannon, S. Dickestein and F.G. Sulman, Proc. Soc., Exp. Biol. Med. 114(1963)366.

[43] S.M. McCann, Annu. Rev. Pharmacol, Toxicol., 22(1982)491.

[44] J. Meites, J. Simpkins, J. Bruni and J. Advis, IRCS, J. Med. Sci., 5(1977)1.

[45] J. Meites and W.E. Sonntag, Annu. Rev. Pharmacol, Toxicol, 21(1981)295.

[46] R.C. Dimond, S.R. Brammav, R.L. Atkinson, W.G. Howard and J.M. Earll, J. Clin. Endocrinol. Metab, 36(1973)1189.

[47] A.J. Thody, In A B Simonds (ed.), Advances in Drug Research, Vol. 11, Academic Press New York, 1977, pp. 23-74.

[48] W.C. Bowman and M.J. Rand, Text Book of Pharmacology, Blackwell, Oxford, 1980, pp. 19.1-19.7.

[49] S.M. Sief and A.G. Robinson, Annu. Rev. Physiol., 40(1978)345.

[50] R.R. Chaudhary, Brit. J. Pharmacol, 17(1961)297.

[51] R. Gaunt, J.J. Chart and A.A. Renzi, Annu. Rev. Pharmacol., 3(1963)109.

[52] N.H. Raskin, Annu. Rev. Pharmcol. Toxicol., 21(1981)463.

[53] W.C. Bowman and M.J. Rand, Text Book of Pharmacology, Blackwell, Oxford, 1980 pp. 16.40.

[54] J.M. Lipton and W.J. Clark, Annu, Rev. Physiol., 48(1986)613.

[55] W.C. Bowman and M.J. Raud, TextBook of Pharmacology, Blackwell, Oxford, 1980 pp 31.9

[56] J.A. Boulant and J.B. Dean, Annu. Rev., Physiol, 48(1986)639.

[57] S.R. Kalso and J.A. Boulant, Am. J. Physiol, 243(1982)480.

[58] R.M. Pinder, In G.P. Ellis and G.B. West (Eds.), Progress in Medicinal Chemistry, North Holland, New York, 1973 pp. 191-274.

[59] R. Byck, In A.G. Gilman, L.S. Goodman and A. Gilamn (eds.) The Pharmacological Basis of Therapeutics, Macmillan New York, 1975, p.152-200.

[60] T.N. Chase and D.L. Murphy, Annu. Rev. Pharmacology, 13(1973) 181.

[61] O. Hornykiewics, Pharmacol, Rev., 18(1966) 925.

[62] G. Sakalis, S.H. Curry, G.P. Mould and M.H. Lader, Clin. Pharmacol, Ther., 13(1972)931.

[63] W.C. Bowman and M.J. Rand, Text Book of Pharmacology, Blackwell, Oxford, 1980, pp. 20.24.

[64] A. Burger, In B. Uvnas, K.J. Burnings and P. Lindren (eds.), Modern Concepts in the Relationship between Structure and Pharmacological Activity, Vol.7, Pergamon Press, Oxford, 1963, pp.35-74.

[65] M.J.S. Dewar and D.M. Storch, Proc. Natl. Acad. Sci., USA, 82(1985)2225.

[66] A.N. Nagappa, P.V. Pandi and R.C. Srivastava, J. Surf. Sci., Technol, 18(2002)191.

[67] E. Coats, W.R. Glave and C. Hansch, J. Med. Chem., 13(1970)913.

[68] W.D.M. Paton, Proc. Roy. Soc., B154(1961) 21.

[69] W.D.M. Paton and H.P. Rang, Adv. Drug. Res., 3(1966)57.

[70] A.J. Clark, The mode of action of drugs on cells, E. Arnold Co., London 1933.

[71] J.H. Gaddum, Pharmacol Rev. 9(1957)211.

[72] R.P. Stephenson, Brit. J. Pharmacol., 11(1956)379.

[73] D.R. Waud, Pharmacol. Rev., 20(1968)49.

[74] M.May, J.F. Moran, H.Kinelberg and D.J. Triggle, Mol. Pharmacol, 3(1967) 28.

[75] W. Straub, Arch. Gesamte Physiol, 119(1907)127.

[76] R.F. Furchgott, Pharmacol Rev., 7(1955)193.

[77] R.F. Furchgott, Annu. Rev. Pharmacol, 4(1964)21.

Chapter 8

Application of surface activity in therapeutics

Chapters 1 to 7 are devoted to the development of a "liquid membrane hypothesis of drug action" which we believe is a new aspect of the role of surface activity in drug action. The account presented in chapters I-7 centers on the efforts made by our research group and is in a way an autobiographical account.

In this chapter a general discussion on the application of surface activity in therapeutics is presented. Some of the areas we intend to cover in this chapter are drug absorption, stability, targeting and novel drug delivery systems etc. This we believe will give a sense of completeness to the monograph, which entitles "surface activity in drug action".

Surface activity is of ubiquitous presence in living system. Lung surfactants, bile acids and salts are well known for their physiological role are surface.

Surfactant and polymer systems play an important role in modern drug delivery, where they may allow control of the drug release rate, enhance effective drug solubility, minimize drug degradation, contribute to reduced toxicity, and facilitate control of drug uptake. In all, they contribute significantly to therapeutic efficiency.

The exploration of principles of surface activity for therapeutics has proved to be quite interesting and rewarding. Beginning with simple suspensions and emulsions the principles of surface activity are making advancement in to sophisticated drug delivery systems like Microemulsion (ME) liposome, noisome, polymeric micelles and self-emulsifying drug delivery systems (sedds). The role of surface activity is also explored to improve and alter pharmacokinetics and pharmacodynamic properties of drug molecules. The therapeutic benefits achieved relate to improved bioavialability and reduced toxicity.

Many pharmacologically active compounds are amphiphilic by nature, and hence behave like surfactants, which tend to associate as micelles in aqueous medium. Micelles are highly dynamic structures, with a tendency to aggregate by stepwise mechanism as that of bile salts [1,2]. The surface-active drugs of quite different chemical structures are reported to self associate and bind to membranes causing disruptions and solubilization, in a detergent like manner.

Surfactants are commonly used to supplement delivery systems. A relatively small concentration of surfactant incorporated into a drug can drastically increase the effectiveness of drug delivery systems. Surfactants perform by increasing absorption of drug into cell membrane, solubility of drug into carrier and stability of drug in a delivery system. Therefore, surfactants can complement other drug delivery strategies.

By acting as a wetting agent, surfactants facilitate the absorption of a drug into the cell wall and membrane. They increase the contact area between the drug and the cells by encouraging interfacial contact between the drug and the target. Surfactants raise the permeability of membranes, which allows for easier absorption. However, interactions of surfactants with membranes can cause disruption of biological membranes.

The solubility of poorly soluble drugs can be enhanced depending on the type of carrier and the amount of drug that must be dissolved. Surfactants can greatly enhance the ability of a solvent to dissolve a drug. Introducing a surfactant into a solution lowers the surface tension and increases the solubility of the organic material into the solvent.

Micelles too can dissolve many substances in the body. For instance, anionic micelles display significant interaction with certain proteins. A high concentration of surfactants over a long period of time may disturb some metabolic processes.

Surfactants are often used to stabilize ME, emulsion, and suspension systems into which drugs are dissolved. Surfactants decrease the surface tension of multicomponent systems. Thus, the measurement of surface tension is a means to determine the properties of micelles and to understand their effect on solubility, fluid flow, and membrane transport.

Although micelles offer a wide range of applications as drug carriers, they are not yet as practical as many methods. The main reason for this is the equilibrium, which exists between the dissolved drug and the micelle's surroundings. Also affecting the use of micelles is their potential toxicity. A great deal of research is being done on these issues, but because of the previously mentioned complications, the use of surfactants as primary delivery mechanisms is limited.

Topical and transdermal drug delivery systems comprises of iontophoresis, electroporation, mucosal (buccal, rectal, vegianal, ocular and nasal), oral delivery, injection technologies, inhalation delivery, targeted delivery, liposome, micro encapsulation, colloidal chemistry, polymers, nanotechnology, micro-electro-mechanical system, and microchip are some of the key areas of emerging technologies where surface activity is explored. Dendimers, niosome, micro emulsions, lipid emulsions, pluronic micelles, sedds, along with gene delivery and protein delivery are other emerging areas of delivery where the application of surface activity is extensively explored.

8.1 Drug Absorption

There is a good deal of literature on the use surface-active agents as absorption enhancers. In this section, the most important absorption enhancers for topical, transdermal and mucosal drug delivery are reviewed.

Absorption describes the rate at which a drug leaves its site of administration and the extent to which this occurs. Absorption is governed by factors such as surface area of the absorption, blood flow to the site of absorption, the physical state of the drug (solution suspension or solid dosage form), its water solubility and concentration at the site of absorption. Many surface-active agents are known to alter absorption of drugs, e.g. bile salts.

8.1.1. Topical and transdermal absorption enhancers.

It is well known that surfactants have effects on the permeability characteristics of several biological membranes, including skin [3,4] and for this reason they can enhance the skin penetration of other compounds present in the formulation. Therefore, in recent years they have been employed to enhance the permeation rates of several drugs [5,6].

The effect of surface-active agents on the skin barrier function depends on the surfactant chemical structure. In general, anionic surfactants tend to be more effective than cationic ones, whereas nonionic surfactants are considerably less effective. Most anionic surfactants can induce swelling of the *stratum corneum*, as well as uncoiling and stretching of α-keratin helices, thereby opening up the protein controlled polar pathways [7].

Pergolide mesylate is a potent dopamine receptor agonist. *In vitro* transport of pergolide from surfactant-based elastic vesicles through human skin were studied in order to elucidate the possible mechanisms of action and to establish the optimal conditions for surfactant based elastic vesicles. The findings show a strong correlation between the drug incorporation to saturated levels and the drug transport, both of which were influenced by the pH of the drug-vesicular system [8].

Skin penetration and mechanisms of action in the delivery of the D2-agonist rotigotine from surfactant-based elastic vesicle formulations was reported by Honeywell-Nguyen et al. They concluded elastic vesicles are not only penetration enhancers but also promising vehicles for transdermal drug delivery [9].

For the controlling transdermal delivery of triprolidine, the application of ethylene vinyl acetate membrane containing permeation enhancer could be useful in the development of transdermal drug delivery system [10].

The percutaneous absorption of p-chlorometaxylenol (PCMX), a topical antiseptic, was investigated using pigskin. *In vitro* diffusion studies showed that the diffusion of PCMX through pigskin followed a steady-state flux model [11].

Release of salicylic acid, diclofenac acid and diclofenac acid salts from isotropic and anisotropic nonionic surfactant systems across rat skin was studied by Gabboun et al. Results indicated that the rate-determining step in the transport process was the release of the drug from the specified donor system [12].

Lorazepam is an anxiolytic, antidepressant agent, having suitable feature for transdermal delivery. The enhancing effects of various surfactants like SLS, CTAB and benzalkonium chloride or Tween 80 with different concentrations on the permeation of lorazepam were evaluated using Franz diffusion cells fitted with rat skins. The *in vitro* permeation experiments with rat skin revealed that the surfactant enhancers varied in their ability to enhance the flux of lorazepam. The increase in flux at low concentrations is normally attributed to the ability of the surfactant molecules to penetrate the skin and increase its permeability. Reduction in the rate of transport of the drug present in enhancer systems beyond 1% w/w is attributed to the ability of the surfactant molecules to form micelles and entrapment of drug in micelles [13].

The potential application of highly biocompatible o/w MEs as topical drug carrier systems for the percutaneous delivery of anti-inflammatory drug ketoprofen was investigated. Ketoprofen loaded MEs showed an enhanced permeation through human skin with respect to conventional formulations [14].

Jimenez et al. studied the *in vitro* transdermal permeation and skin accumulation of octyl methoxy cinnamate (OMC) an ultraviolet absorber--through pigskin and to determine the quantity of OMC in the skin surface and different pigskin layers. Data showed that OMC accumulation enhanced when formulated as emulsions [15].

Evidence for lymphatic transport of insulin by topically applied biphasic vesicles lipid-based delivery system, which was previously shown to deliver sustained physiological levels of basal insulin in a pain-free manner across the skin, was evaluated in a diabetic rat model. Transdermal patches containing insulin in biphasic vesicles were applied to the streptozotocin-induced diabetic rats showed a lymphatic transport of a protein [16].

The transdermal penetration of a model lipophilic drug (estradiol) through human epidermis from phosphatidylcholine (PC) based liposomes with cholate-containing ultradeformable (Transfersomes) and membrane-stabilized (cholesterol-containing) vesicles were used. Iontophoretic studies revealed the superiority of ultradeformable vesicles regarding drug skin penetration and deposition compared to traditional liposomes [17].

Colloids in an aqueous suspension form can cross skin barrier only through hydrophilic pathways. Various colloids have different ability to do this by penetrating narrow pores in the barriers modeling the skin. Amphipats that acceptably weaken the membrane (surfactants, solvents such as certain alcohols etc.) consequently facilitate controlled local bilayers destabilization and increase lipid bilayer flexibility. Drug delivery by means of highly adaptable drug carriers, more over allows highly efficient and well-tolerated drug targeting into the skin proper. Sustained drug release through the skin into systemic blood circulation is another field of ultra deformable drug carrier application [18].

Ethosomes are soft vesicles carriers, which are mainly composed of phospholipids PC, ethanol at relatively high concentrations, and water. It found that ethosomes can penetrate the skin and allow enhanced delivery of various compounds to the deep strata of the skin or to the systemic circulation [19,20,] Recently, it was found that ethosomes carriers, phospholipids vesicular systems containing relatively high concentrations of alcohol, were very effective at enhancing dermal and transdermal delivery of both lipophilic and hydrophilic molecules. Fluorescent probes delivered from ethosomes systems reached the deep strata of the skin. Delivery of minoxidil to the pilo sebaceous units from ethosomes was much greater compared to delivery from classic liposomes [21]. In addition, clinical studies with acyclovir showed that ethosomal formulations were superior to the currently available topical therapy at treating recurrent herpes labialis [22]. *In vivo* skin permeation of testosterone from patches containing ethosomal drug were more effective at delivering testosterone through rabbit pinna skin than commercially available Testoderm patches [23]. Results using trihexyphenidyl hydrochloride ethosomes indicated that this system has the potential to be further developed into an antiparkinsonian patch [24]. The transdermal delivery of insulin from an ethosomal carrier resulted in lower blood glucose levels in normal and diabetic rats *in vivo*, with a plateau effect [25, 26]. Ethosomes enable cannabidiol's skin permeation and its accumulation in a depot at levels that demonstrate the potential of transdermal cannabidiol to be used as an anti-inflammatory treatment [27].

Efficient delivery of antibiotics to deep skin strata from ethosomal applications could be highly beneficial, reducing possible side effects and other drawbacks associated with systemic treatment. Furthermore, ethosomal delivery systems could be considered for the treatment of a number of dermal infections, requiring intracellular delivery of antibiotics, whereby the drug must bypass two barriers: the subcutaneous and the cell membrane [28].

Preliminary studies with plasmids and insulin revealed that the ethosomal carrier might be used for enhanced delivery of these agents. In further work, the ethosomal technology was broadened to introduce agents into cultured cells and microorganisms. Enhanced delivery of bioactive molecules through the skin and cellular membranes by means of an ethosomal carrier opens numerous challenges and opportunities for the research and future development of novel improved therapies [29].

A single photomechanical wave was applied to the skin of rats in the presence of SLS as novel technique of transdermal drug absorption. It was concluded that the combination of photomechanical waves and surfactants could enhance transdermal drug delivery [30].

Phloretin is the aglycone of phlorizin, a polyphenolic substance occurring in the root bark of apple trees. The potential use of phloretin, a polyphenolic compound, as a penetration enhancer in the transdermal delivery of lignocaine hydrochloride (L-HCl) has been investigated. The results suggest the potential use of phloretin as penetration enhancer in the delivery of L-HCl through skin [31].

The synergistic effect of ultrasound and surfactants on transdermal drug delivery has been demonstrated. Low-frequency ultrasound (20 kHz) and surfactants have been individually shown to enhance transdermal drug transport. The data shows that ultrasound and surfactants synergistically enhance skin permeability. Two mechanisms are shown to play a role in this synergistic effect. First, ultrasound enhances surfactant delivery (enhanced delivery) into the skin and, second, ultrasound disperses surfactant (enhanced dispersion) within the skin [32].

Gabiga et al reported effect of penetration enhancers on isosorbide dinitrate [ISDN] penetration through rat skin from a transdermal therapeutic system. The increased skin-penetration enhancing effect of oleic acid and propylene glycol in comparison to PEG400 and isopropyl myristate on percutaneous permeation of ISDN was shown [33].

The effect of a series of polyoxyethylene non-ionic surfactants on the membrane transport of barbiturates in goldfish has been assessed using the overturn time technique. The results are comparable with data obtained with the same surfactants using other epithelial membranes [34].

A transdermal ME system containing ketoprofen was developed using, oleic acid as the oil phase, as it showed a good solublizing capacity and excellent skin permeation rate of the drug [35]. In a study, ketoprofen-loaded MEs showed an enhanced permeation through human skin with respect to conventional formulations. No significant percutaneous enhancer effect was observed for ketoprofen-loaded oleic acid–lecithin MEs. The human skin tolerability of various ME formulations was evaluated on human volunteers [36].

Fluorinated lipids and fluorinated surfactants can be used to elaborate and stabilize various colloidal systems, including different types of emulsions, vesicles and tubules, that also show promise for controlled release drug delivery. A large variety of light and heavy, linear and cyclic fluorocarbons were gelified. Several of these gels are stable enough to withstand heat-sterilization conditions. These fluorocarbon gels have potential for topical applications. They could serve to protect a wound while remaining permeable to gases. Water-soluble drugs (antibiotics), nutrients, and other substances of therapeutic value could be incorporated the aqueous film. Preliminary experiments showed irritation when such fluorocarbon gels were spread on scarified or non-scarified rabbit skin [37].

Topical liposome or niosomes may serve as solubilization matrix, as a local depot for sustained release of dermally active compounds, as penetration enhancers, or as rate-limiting membrane barrier for the modulation of systemic absorption of drugs. Investigators have mostly focused on dermal corticosteroid liposome products. However, localized effects of liposome-associated proteins such as superoxide dismutase, tissue growth factors and interferons appear to be enhanced. The delivery of liposome-encapsulated proteins and enzymes into deeper skin layers has been reported, although the mechanism of delivery remains to be elucidated [38].

Gelation of aqueous solutions of poly (oxyethylene)-poly (oxypropylene)-poly (oxyethylene) block co-polymeric non-ionic Surfactants (Pluronics or poloxamers) by the action of γ-rays is reported. The gels obtained from poloxamers with ethylene oxide have a very high water uptake capacity and can be used as sustained release drug delivery systems [39].

A proniosomal based transdermal drug delivery system of levonorgestrel was developed and extensively characterized both *in vitro* and *in vivo*. The proniosomal structure was liquid crystalline-compact niosomes hybrid, which could be converted into niosomes upon hydration. The study demonstrated the utility of proniosomal transdermal patch bearing levonorgestrel for effective contraception [40].

The incorporation of small amounts of surfactants into a polymeric dispersion can dramatically alter the polymer conformation and the viscosity of the dispersion. Aggregation processes can appear as a consequence of hydrophobic interactions between the non-polar surfactant tail and the polymer backbone, the electrostatic interactions between the polar heads of the surfactant and the charged groups of the polymer, or both [41,42]. Surfactant-can also greatly modify the responsiveness of temperature-sensitive polymers. These effects were reported for cellulose ethers [43], poloxamer-co-acrylic acid [44], or poly (*N*-isopropylacrylamide) [45].

Barreiro-Iglesias et al reported the analysis of the interactions and the effects exerted by the addition of small amounts of surfactants of different nature and low toxicities [46] on the macro and microrheological properties and pH-responsiveness of Carbopol 934 gels. The results show that by choosing the appropriate proportion of the most suitable surfactant, it is possible to modulate the flow behavior, elastic properties, and diffusion microenvironment of carbopol gels, without losing the pH-dependent gelling ability, which could improve the suitability of carbopol gels for drug delivery through different routes [47].

Transdermal delivery of cyclosporin A solubilized in mixed micelles through mice skin was assessed by an *in vitro* permeation technique and *in vivo* study was also undertaken. Mixed micelles were shown to be efficient carrier for the transdermal delivery of the lipophilic polypeptide when kept in solution during the application process [48].

A transdermal preparation containing ketoprofen was developed using O/W ME system with Labrasol/Cremophor RH 40 surfactant and cosurfactant. The optimum formulation of the ME consisted of 3 percent ketoprofen, 6% oleic acid, and 30% Labrasol/Cremophor RH 40 (1:1) and water [49].

ME systems allow for the microscopic co-incorporation of aqueous and organic phase liquids. Novel ME systems with Tween 80 and transdermal enhancers such as n-methyl pyrrolidone oleyl alcohol were incorporated in to all system with out disruption of the stable emulsion. The novel ME system in this study potentially offers many beneficial characteristics for transdermal drug delivery [50].

Iontophoresis and enhancers were performed to enhance percutaneous absorption of enoxacin so as to compare the enhancement between these two enhancing methods. The cationic surfactant of benzalkonium chloride showed the highest enhancing activity for enoxacin for all pH values of buffer vehicles. The combination of benzalkonium chloride and iontophoresis exerted a synergistic effect for anionic enoxacin in pH 10.0, which was possibly due to the shielding of negative charge in skin and the water molecules carried by chloride [51].

Triprolidine-containing matrix was fabricated with ethylene–vinyl acetate (EVA) copolymer to control the release of the drug. The permeation rate of triprolidine in the stripped skin was greatly larger than that in the whole skin. The permeability of triprolidine was markedly increased with stripping of the mouse skin to remove the stratum corneum that acts as a barrier of skin permeation. For the controlling transdermal delivery of triprolidine, the application of EVA membrane containing permeation enhancer could be useful in the development of transdermal drug delivery system [52].

The *in vitro* transdermal absorption of apomorphine from MEs was studied using the skin of the hairless mouse as a membrane.. Forming apomorphine–octanoic acid ion-pairs, increased the lipophilicity of the drug. Apomorphine in MEs, protected from light with antioxidants, showed no degradation for up to 6 months [53].

Matrix-type transdermal delivery systems of testosterone (TS) were formulated with three different pressure-sensitive adhesives (PSA). The effects of PSA, skin permeation enhancers, and solubilizers on the rat skin permeation rate of TS were systematically investigated. This study led to the development of a non-scrotal matrix-type transdermal delivery system of testosterone [54].

Cannabidiol (CBD) is a new drug candidate for treatment of rheumatic diseases. However, its oral administration is associated with a number of drawbacks. Ethosomes enable CBD's skin permeation and its accumulation in a depot at levels that demonstrate the potential of transdermal CBD to be used as an anti-inflammatory treatment [55].

In vitro transdermal permeation and skin accumulation of ultraviolet (UV) absorberoctyl methoxycinnamate (OMC) through pigskin was evaluated in oil-in-water, water-in-oil emulsions and nanocapsules. Data showed that UV absorber exhibited increases in skin accumulation when is formulated in emulsions in free form. Study demonstrate that the inclusion of OMC-encapsulated in sunscreen formulations decreases the skin accumulation of the cinnamate since the *in vitro* release mechanism of OMC-nanocapsules is governed by hydrophobicity and crystallinity of the polymer and by the high lipophilicity of the drug. The crystallinity of the polymer has the ability of reflecting and scattering UV radiation on their own thus leading to photoprotection without the need for molecular sunscreens [56].

Highly deformable hydrophilic lipid vesicles have been studied for transdermal delivery of therapeutically active small molecules and proteins. Ultradeformable liposomes (UL) with sodium cholate, sodium deoxycholate, and Tween 80 were tested as transdermal gene delivery. These data suggest that UL might be of use as a transdermal delivery system of plasmid DNA, and that the choice of edge activators may play an important role in the transdermal delivery of plasmid DNA via UL [57].

Triptolide (TP) has been shown to have anti-inflammatory, immunosuppressive, anti-fertility and anti-neoplastic activities. However, its clinical use is restricted to some content due to its poor water solubility and some toxic effects. The transdermal delivery capacity and anti-inflammatory activity were evaluated. The results indicated that these solid lipid nanoparticles dispersions and MEs could serve as efficient promoters for the TP penetrating into skin [58].

8.1.2. Oral and mucosal absorption enhancers

Oral drug administration is the most preferred and widely used route for small molecule drugs because of the ease of formulation, patient convenience and compliance, and usually good absorption in the intestine. However, therapeutic proteins are exposed to high levels of proteases, such as pepsin and are readily inactivated [59]. Mucus membranes of the oral cavities, rectal, vaginal, ocular and nasal are explored as novel routes of administration of drugs.

The idea of the intestinal mucosa being a mere physical barrier to absorption of drugs, which by-passed only by aqueous pores or lipid channels in the membrane, had now been discarded. Modern concepts suggested that the mucosa also provided barriers in the form of metabolic enzymes, particularly the cytochrome P450 family, and transporting proteins such as p-glycoprotein. This prevented permeation of drugs whose physicochemical characteristics otherwise suggested facile transport by passive processes. Common excipients, previously considered innocuous in the biopharmaceutical sense, might have an effect on these enzymes and transporters and hence might affect the bioavailability of a drug in unexpected ways. Typically, studies were carried out *in vitro* by noting changes in drug concentration on either side of model membranes in diffusion cells.

The transport of drugs across membranes in the presence and absence of non-ionic surfactants (Cremophor EL a Polyethoxylated castor oil, Tween 80 and PEG 300) had been studied by Borchardt and his colleagues [60,61]. They had found that these excipients could affect the activity of p-glycoprotein in the intestinal mucosa, thus affecting the protein's ability to limit the transmucosal transport of drugs such as Taxol. For example, inclusion of 20 per cent PEG 300 in the transport buffer typically used in Caco-2 (a cell culture model of the intestinal mucosa) cell transport experiments totally inhibited polarised efflux of [3H]

Taxol. The attention has, more recently, been focused on the mechanism of this inhibition. Using appropriate probes, it could be shown that PEG 300 changed membrane fluidity by its effect on either lipid side chains or polar head groups. Similar results were found using a Madin-Darby canine kidney cell line transfected with the gene for p-glycoprotein [62].

Albendazole (ABZ) is a widely used broad-spectrum benzimidazole (BZD) anthelmentic. Low hydro solubility and poor/erratic gastrointestinal (GI) absorption play against the systemic availability and resultant clinical efficacy of BZD compounds. Effect of amphiphilic surfactant agents on the gastrointestinal absorption of albendazole in cattle was studied by Virkel et al. SLS-mediated enhanced dissolution and absorption of ABZ accounted for the observed increased systemic availability of the active ABZSO metabolite in cattle. These results should be considered among strategies to improve the use of BZD anthelmentic [63].

ME formulations, which can be used to improve the bioavailability of poorly soluble drugs, were designed using only pharmaceutical excipients. Several types of oils and surfactants were tested and it was found that propyleneglycol monoalkyl ester and glycerol monoalkyl ester were solubilized easily in an aqueous medium by various types of surfactants. These ME formulations can be administered as a form of water-in-oil ME or surfactant-oil mixture, and are expected to convert to oil-in-water ME in the small intestine [64].

Eiamtrakarn et al studied gastrointestinal mucoadhesive patch system (GI-MAPS) for oral administration of granulocyte colony stimulating factor (G-CSF), as a model protein. The system consists of four layered films, both drug and pharmaceutical additives including an organic acid, citric acid, and a non-ionic surfactant, polyoxyethylated castor oil derivative (HCO-60), were formulated in the middle layer. The results suggest the usefulness of GI-MAPS for the oral administration of proteins [65].

Amoxicillin-loaded polyethylcyanoacrylate nanoparticles were studied for influence of PEG coating on the particle size, drug release rate and phagocytic uptake. Experimental data suggest that nanoparticles could be useful for treatment of disease by *per oral* administration [66].

The feasibility of incorporating significant quantities of the anionic surfactant, SLS, into an immediate release tablet formulation of a poorly water-soluble immunosuppressive agent was investigated. Employing the Korsmeyer–Peppas model of Fickian and non-Fickian drug release, it was further shown that release of the drug from the dosage form was governed largely by surface erosion of the surfactant-enriched tablet matrix [67].

Labrasol and Gelucire 14 are well-defined mixtures of mono-, di- and triglycerides and mono- and di-fatty esters of polyethylene glycol, which contain predominant fatty acids composed of caprylate/caprate and laurate, respectively. These commercially available surfactants Labrasol and Gelucire have been shown to greatly improve the bioavailability of poorly absorbable drugs such as gentamicin [68], insulin [69] vancomycin [70] albendazole [71] and luekotriene B4 inhibitors [72].

Different strategies are under investigation to improve intestinal absorption of low molecular weight heparins (LMWHs). The carrier compound, sodium *N*- [10-(2-hydroxybenzoyl) amino] deaconate (SNAD), was found to increase the entral absorption of LMWH in experimental animals [73] .The saturated fatty acids having the carbon C10—C16, especially C10 and C12, exhibited enhancement of the absorption of LMWH from intestine [74]. Entral absorptions enhancement of LMWH have been reported by the use of absorption enhancers, *e.g.* rectal administration with sodium cholate in rats and human subjects [75], duodenal administration with Carbopol 934P in rats and pigs [76], intestinal administration with chitosan derivatives or mono-*N*-carboxymethyl chitosan in rats [77].

Liposomes are thought to improve intestinal drug absorption, and to be a good candidate as an oral dosage form for drugs that are extremely poorly absorbed. A recently developed type of liposome, designated here as a ''double liposome'' (DL), was confirmed to consist of smaller liposomes enveloped in lipid bilayers [78]. Ebato et al evaluated hypocalcemic effects of DL prepared by the modified glass-filter method as an oral dosage form of salmon calcitonin [79].

To improve the oral absorption of ceftriaxone (CTX) a broad-spectrum third generation cephalosporin antibiotic, the CTX-carragenan gel complex was prepared, and several kinds of compounds were screened as potential oral enhancers in rats. The results suggest that Capmul MCM C10 is a promising carrier, having a good balance between bioavailability enhancing activity and safety, for the oral delivery of CTX when it is co administered with complex [80].

For nitrendipine a poorly soluble model drug, employing the ME formulations enhanced its absorption significantly. The absorption behavior also varied with the type of surfactant [81].

Small nanospheres prepared by spontaneous polymer-protein self-assembling are an attractive concept, since the use of solvents and surfactants can be avoided. Spontaneous formation of polymer conjugates with variety of proteins, such as tetanus toxoid, recombinant nerve growth factor and insulin were investigated. It has been suggested that above polymers could be an effective mucosal protein delivery system [82].

Octreotide acetate is a somatostatin analogue used for the control of endocrine tumors of the gastrointestinal (GI) tract and the treatment of acromegalics. The oral absorption of octreotide is limited because of the limited permeation across the intestinal epithelium. Both chitosan hydrochloride and *N*-trimethyl chitosan chloride (TMC), a quaternized chitosan derivative, are nonabsorbable and nontoxic polymers that have been proven to effectively increase the permeation of hydrophilic macromolecules across mucosal epithelia by opening the tight junctions .The intestinal absorption of octreotide when it is co administered with the polycationic absorption enhancer TMC in Caco-2 cell monolayers were significantly increased the absorption of the peptide analogue, resulting in a 5-fold increase of octreotide bioavailability compared with the controls (octreotide alone)[83].

The first line of attempts to improve peptide absorption is to increase drug solubility, especially for very lipophilic compounds, such as cyclosporin A, by use of new drug formulations on the basis of bile [84,85] olive oil [86], water-in-oil [87] or oil-in-water MEs

[88]. The approaches appropriate for passively absorbed peptides are the use of absorption enhancers [89 94]. Ionic and non-ionic ten sides [88,95-99] as well as various fatty acids or salicylate [100, 101] have been used. Tensides are surface-active substances, which by reducing the surface energy of water mediate the wetting of surfaces or promote that substances insoluble in water can be emulgated or dispersed in it. These enhancers improve absorption by different partly unknown mechanisms.

FK224 a tachykinin natural killer (NK)-1 NK-2 receptor antagonist is a cyclo peptide with a low aqueous solubility. Following oral administration to rats, poor absorption was observed due to proteolysis in the gastrointestinal tract. Nakate et al studied the effect of the pulmonary route on the systemic absorption and the bioavailability of FK224. From absorption studies on the PEG400 solution given by various routes (intranasal, subcutaneous, intra tracheal and intravenous as reference), it was shown that pulmonary administration was a potentially attractive route for FK224 [102].

Bagger et al studied nasal bioavailability of peptide T (synthetically produced protein derived from gp-120 the protein on the surface of HIV which binds to CD4 molecule) from aqueous formulations containing sodium glycocholate, an absorption enhancer with known effect on epithelial tight junctions, and/or glycofurol in a crossover study in rabbits. It was shown that the bioavailability of peptide T was significantly enhanced when glycofurol or sodium glycocholate was added to a nasal formulation [103].

Suitable microspheres formulations were designed in order to provide the absorption of a high polar drug through nasal mucosa. For this purpose, gentamicin sulfate (GS) was chosen as a model drug and used at different drug/polymer ratios in the microspheres formulations. The microspheres were prepared by spray drying technique. Hydroxypropyl methylcellulose was used as a mucoadhesive polymer in the formulations to increase the residence time of the microspheres on the mucosa. Sodium cholate was added into the formulations for increasing the absorption of GS through nasal mucosa [104].

Extensive research efforts have been directed towards the systemic administration of therapeutic proteins and poorly absorbed macromolecules via various nontraditional, injection-free administration sites such as the lung. As a portal for noninvasive delivery, pulmonary administration possesses several attractive features including a large surface area for drug absorption. Nevertheless, achieving substantial bioavailability of proteins and macromolecules by this route has remained a challenge, chiefly due to poor absorption across the epithelium. The lungs are relatively impermeable to most drugs when formulated without an absorption enhancer/promoter. In an attempt to circumvent this problem, many novel absorption promoters have been tested for enhancing the systemic availability of drugs from the lungs. Various protease inhibitors, surfactants, lipids, polymers and agents from other classes have been tested for their efficacy in improving the systemic availability of protein and macromolecular drugs after pulmonary administration [105].

Chitosan is a non-toxic, biocompatible polymer that has found a number of applications in drug delivery including that of absorption enhancer of hydrophilic macromolecular drugs. Chitosan, protonated (pH<6.5), is able to increase the Para cellular permeability of peptide drugs across mucosal epithelia [106].

The nonionic, water-soluble surfactant Myrj 51(ICI surfactant), which is well tolerated physiologically, was tested in a preformulation study and bioavailability of sodium valproate suppositories. The Myrj series of surfactants are polyoxyethylene derivatives of stearic acid. Bioavailability achieved was comparable to that of oral administration, which establishes the surfactant Myrj 51 is an appropriate excipients in formulations designed for rectal administration [107].

Cefpirom is a new semi-synthetic amino-2-thiazolyl-methoxyimino cephalosporin that has been substituted in position 3 with a cyclopenteno-pyridinium group in order to create a zwitterionic compound. It exhibits highly hydrophilic properties, as shown from its extremely low partition coefficient, and therefore its lipophilicity was increased using bile salts. The effect of this on the partition coefficients determined in the n-octanol/buffer system was confirmed using an in-vitro transport model with artificial and biological membranes [108]. Piroxicam is a non-steroidal anti-inflammatory drug that is characterized by low solubility and high permeability. *In vitro* and *in vivo* evaluation indicated enhanced bioavailability of piroxicam using Gelucire (GL) 44/14 and Labrasol surfactant. The results of the *in vivo* study revealed that the GL dosage form would be advantageous with regards to rapid onset of action, especially in various painful conditions where an acute analgesic effect is desired [109].

Regional absorption of inhaled insulin together with an enhancer (sodium di-octyl-sulfosuccinate [DOSS]) in the rabbit airways and lung was studied by Dahlback et al. Insulin was administered with or without DOSS aerosol inhalation, intratracheal infusion, intranasally, sublingually, and without DOSS intravenously. Inhaled insulin together with the absorption enhancer DOSS decreased the blood glucose level more effectively than insulin given intratracheally, intranasally, or sublingually. The effect on blood glucose reflected the difference in plasma insulin concentration for the different routes of administration [110].

A self-emulsifying system (SES), a mixture of oil and a surfactant, which forms an oil-in-water emulsion, is expected to improve the *in vitro* drug dissolution and enhance the *in vivo* drug absorption. Kim and Ku studied, a poorly water-soluble drug, indomethacin bioavailability in SES rectal administration. Gelatin hollow type suppositories, filled with the SES containing indomethacin and indomethacin powder physically mixed with the SES, was administered to rats rectally and results indicate an increased absorption of indomethacin significantly by the SES [111].

Gentamicin is a polarized water-soluble compound having very poor intestinal membrane permeability resulting in low oral bioavailability. Gentamicin formulations containing labrasol were evaluated in beagle dogs. Labrasol was found to improve the intestinal absorption of gentamicin in rats [112].

Vancomycin hydrochloride (VCM) is a glycopeptide antibiotic used for the treatment of infections caused by methicillin-resistant staphylococci. It is water soluble, having a high molecular weight, and poorly absorbed from the gastrointestinal tract. Mixtures of VCM with Labrasol and α-tocopheryl PEG1000 succinate (TPGS) were prepared to improve oral absorption of VCM. The results of the study indicate that formulations containing Labrasol and TPGS improve intestinal absorption of hydrophilic macromolecular drug, VCM [113].

8.2 Solublizing agents

The range of solid-state forms that a drug substance exhibits (i.e., amorphous, polymorphs, solvates, or co-crystals) is a key issue in the development of a pharmaceutical product because the form can affect bioavailability and stability in significant ways [114,115]. Studies in the pharmaceutical literature have focused on the dissolution process and the effect of surfactants, with little attention given to how surfactants affect the crystallization steps involved in solution-mediated transformations [119-118]. It would seem that surfactants, given their interfacial properties and their ability to form a variety of supramolecular structures (micelles, hemi micelles, etc.) in aqueous solution, could have profound effect on crystallization in addition to their effects on dissolution. Surfactants have been shown to promote or inhibit crystallization steps nucleation, crystal growth in organic and inorganic crystals [119-124] and direct the crystallization of solid phases that otherwise are not formed [125,126].

Solubilization of amphotericin B (AMB) by dioctadecyldimethylammonium bromide (DODAB) bilayer fragments inspired this evaluation of its *in vivo* activity from survival and tissue burden experiments against systemic candidacies in a mouse model with the traditional deoxycholate/AMB formulation (DOC/AMB). DODAB/AMB was as effective as DOC/AMB for treating systemic candidacies in a mouse model [127]. Conductometric and spectrofluorimetric characterization of the mixed micelles constituted by dodecyltrimethylammonium bromide and a tricyclic antidepressant drug amitryptaline in aqueous solution have been analyzed. The solubilization of the drug in the mixed micelles has been also studied through the mass action model, by determining the association constant between the micelles and the drugs of amitryptaline family [128].

The association of hydrophobic, cationic drugs with lung surfactant was determined to assess the pharmacokinetic implications on drug disposition and retention in the lung. Cationic drugs have very favorable distributions from an aqueous solution to the lipid phase of lung surfactant [129].

The cosolubilization phenomenon of three non-polar drugs (hydrocortisone, beta-estradiol, and ethynylestradiol) in polysorbate 80 solutions was studied. It was found that the solubility of any drug decreased in the presence of other steroidal compounds. In an attempt to understand the observation, a model has been proposed to describe and to predict the drug solubility in the presence of other non-polar drugs in a non-ionic surfactant [130].

Surfactants had a noticeable effect in increasing the solubility of cyclosporin A (CsA), a neutral undecapeptide. Twenty percent solutions of Tween 20, Tween 80, and Cremophor EL increased the solubility by 60 to 160 fold. Cyclodextrins can increase the CsA solubility, but alpha cyclodextrin was more effective than Hydroxypropyl beta cyclodextrin. Cosolvents on the other hand did not increase the solubility of CsA as much as expected from the LOGP (logarithm of water-octanol partition coefficient) value of CsA [131].

Based on an investigation on furbiprofen solubilization in polysorbate 80 solutions at different pH, an equilibrium-based model was proposed to characterize the drug–surfactant interactions in pH-controlled system [132].

Nagarajan has reviewed the solubilization capacity of block copolymer aggregates in aqueous solutions for hydrophobic molecules [133].

Alkanoyl-6-O-ascorbic acid esters are easily obtained from vitamin C, and produce self-assembled aggregates in water solutions, with an inner hydrophobic pool surrounded by an external hydrophilic shell. Compared to ascorbic acid, their solubility in oils and fats is greatly enhanced, while the peculiar antioxidant activity is retained in the polar head groups of such surfactants. In virtue of their amphiphilic nature, ascorbic acid-based supramolecular systems can dissolve relevant amounts of hydrophobic, poorly water soluble chemicals such as drugs, vitamins, and so on, and at the same time they provide a suitable shield against oxidative deterioration of valuable materials [134].

Lutein, a naturally occurring carotenoid, is widely distributed in fruits and vegetables and is particularly concentrated in the *Tagetes erecta* flower. Epidemiological studies suggest that a high lutein intake increases serum levels that are associated with a lower risk of cataract and age-related macular degeneration. Lutein can either be free or esterified (myristate, palmitate, or stearate). Both are practically insoluble in aqueous systems, and their solubility in food grade solvents (oils) is very limited, resulting is low bioavailability. ME, reversed micelles have been utilized to solubilize both luteins for improved bioavailability [135].

Anthocyanin is one of the most widely distributed water-soluble pigments responsible for pink, red, and purple color in fruit and vegetables. Despite attractive color and pharmaceutical potential, low stability of anthocyanins limited their application as a food colorant. Various factors such as pH, temperature, oxygen, ascorbic acid, light, metals, and their combinations affect the stability of anthocyanins. Recently, anthocyanins are regarded as biologically active substances because of their strong antioxidant activities, which are able to reduce the risk of cancer and heart disease BY anthocyanins solubilized in reverse micelle [RM]. The color stability of anthocyanins in RM formed Aerosol OT in hexane was higher than that of buffered anthocyanins during the storage [136].

The combined effects of sodium n-octyl sulfate (SOS) and CTAB on the solubilization of (+)-limonene in aqueous solution were studied using a headspace gas chromatography technique. The findings showed the mixing of SOS and CTAB resulted in positive synergistic effects on the solubility of (+)-limonene. The positive synergistic effects are explained from the perspective of the phase behavior of this mixed surfactant system [137].

Diphenylmethane derivatives as anti-histamine agents [138] phenothiazine derivatives as psychotropic agents [139,140] were solubilized due to intrinsic self-association in aqueous solution. These drugs had common characteristic that the molecules possessed amphiphilic structures comprising both hydrophilic and hydrophobic moieties. Formation of the counter-ion salt characteristic of higher dissociation constant is one of the promising procedures extensively employed to improve the aqueous solubility of ionic poorly water-soluble drugs [141,142].

N-[2-(3,5-Di-*tert*-butyl-4-hydroxyphenethyl)-4, 6-difluorophenyl] -*N*-[4-(*N*-benzypip-eridyl)] urea (N-4472) is a newly developed drug having a lipid-lowering effect.

To improve the aqueous solubility of a N-4472, organic acid/N-4472 evaporates were prepared by using succinic acid, L-tartaric acid, citric acid and vitamin C (VC). Among these evaporates, the VC/N-4472 evaporate at a molar ratio of more than 2 (VC/N-4472) formed stable colloidal particles in aqueous solution. Further more, it was evidenced that all of these complexes were composed of amphiphilic structures comprising both hydrophobic N-4472 moiety and hydrophilic VC moiety, thereby properly accounting for surface-active property of these complexes [143].

Duval-Terrie et al developed a new surfactants for membrane protein solubilization, from a natural, biodegradable polymer: the polysaccharide pullulan. A set of amphiphilic pullulans (HMCMPs), differing in hydrophobic modification ratio, charge ratio, and the nature of the hydrophobic chains introduced, were synthesized and tested in solubilization experiments with outer membranes of *Pseudomonas fluorescens*. The membrane proteins were precipitated, and then resolubilized with various HMCMPs. The decyl alkyl chain (C10) was the hydrophobic graft that gave the highest level of solubilization. Thus, HMCMPs appear to constitute a promising new class of polymeric surfactants for membrane protein studies [144].

Complexation and micellization are two effective ways of solubilizing drugs. The combined effect of surfactant and complexant on the solubilization of a poorly water-soluble compound a tubulin inhibitor (NSC-639829, a benzoylphenylurea) is investigated. With increasing concentration of SLS in solutions of fixed concentration of sulfobutylether- β-cyclodextrin, cyclodextrin, the total solubility of the drug decreases linearly, reaches a minimum and then increases linearly. The above observation is attributed to the fact that the surfactant molecule competes with the drug to "fit" in the non-polar cyclodextrin cavity. The combined use of two solubilizing agents, a surfactant and a complexant, results in a much lower solubility than when either one is used alone at the same concentration. The surfactant molecule acts as a competitive inhibitor in the solubilization of the drug by the complexant. Similarly the complexant "pulls" the surfactant out of solution, making it unavailable for solubilizing the drug [145].

Triclosan, an antimicrobial, although widely incorporated into many skin care products, toothpastes, and liquid soaps, presents formulation difficulties because it is practically insoluble in water. The order of solubilizing performance of different solubilizers was evaluated [146].

The effect of a plant sterol, β-sitosterol (SI), and a plant stanol, sitostanol (SS), on the solubilization of cholesterol (CH) by model dietary mixed micelles was examined under *in vitro* conditions. Free SI and SS were shown to reduce the concentration of CH in dietary mixed micelles via a dynamic competition mechanism that is clearly beneficial for the reduction of the intestinal uptake of cholesterol [147].

The combined effects of sodium n-octyl sulfate (SOS) and CTAB on the solubilization of (+)-limonene in aqueous solution were studied. The findings showed the mixing of SOS and CTAB resulted in positive synergistic effects on the solubility of (+)-limonene. The positive synergistic effects are explained from the perspective of the phase behavior of this mixed surfactant system [148].

The effect of alkanols and cyclodextrins on the phase behavior of an isopropyl myristate ME system was studied and the effect on solubility of model drugs progesterone and indomethacin were determined. Isopropyl myristate-based ME systems alone could increase the solubility values of progesterone and indomethacin up to 3300-fold and 500-fold, respectively, compared to those in water. But the two types of cyclodextrins studied affected isopropyl myristate-based ME systems negatively [149].

The micelles formed by di- and tri-block copolymers have been widely examined for their ability to solubilize poorly water-soluble drugs [150,151]. In most systems the hydrophilic block has been poly (oxyethylene), which serves as a stabilizing interface between the hydrophobic core and the external aqueous medium and also in some cases as a locus for the solubilization of more hydrophilic solutes. Micelles formed from copolymers with a wide range of hydrophobic blocks have been explored for their solubilizing potential for poorly soluble drugs. Diblock copolymers of poly (DL-lactide) and PEG have been reported as effective solubilizers of taxol [152].

The solubilization capacities of micellar solutions of a series of block copolymers composed of hydrophilic poly (oxyethylene) (E) and hydrophobic poly (oxypropylene) (P), poly (oxybutylene) (B) or poly (oxyphenylethylene) (S, from styrene oxide) have been compared using the poorly water-soluble drug griseofulvin as a model solubilizate [153].

The phase solubility of several different compounds, i.e., cholesterol, ibuprofen, diflunisal, alprazolam, 17b-estradiol and diethylstilbestrol, and various charged and uncharged cyclodextrins were investigated. The results indicate that drug/cyclodextrin complexes can self-associate to form water-soluble aggregates, which then can further solubilize the drug through non-inclusion complexation [154].

Aqueous miconazole (MCZ) aggregates were solubilized and/or colloidally stabilized by bilayer-forming synthetic amphiphiles such as dioctadecyldimethylammonium bromide (DODAB) or sodium dihexadecylphosphate (DHP) dispersions. Inexpensive, synthetic bilayer fragments offered a large area of hydrophobic nanosurfaces dispersed and electrostatically stabilized in water opening new prospects for drug solubilization and colloid stabilization of insoluble drug particles [155].

The usefulness of sugar surfactants as solubilizing agents was assessed and compared to commercial polyoxyethylene-based surfactants. The sugar surfactants examined comprised of monosaccharides or disaccharides with alkyl chains ranging from C_8 to C_{12}. Each surfactant was investigated with respect to solubilization capacity for felodipine and haemolytic activity the haemolytic activity was determined using a static method in which surfactant solutions were added to fresh dog blood. The polyoxyethylene-based surfactants were found to be more suitable as solubilizing agents than the sugar surfactants due to better solubilization capacities combined with lower haemolytic activities [156].

The effect of two surfactants SLS and polysorbate 80, and two hydrophilic polymers: PEG 6000 and polyvinylpyrrolidone (PVP) on the aqueous solubility of indomethacin was studied. It was found that all the above solubilizing agents increase the solubility of the drug in the following order: SLS > polysorbate 80 > PEG 6000 > PVP [157].

8.3 Dissolution

The range of solid-state forms that a drug substance exhibits (i.e., amorphous, polymorphs, solvates, or co-crystals) is a key issue in the development of a pharmaceutical product because the form can affect bioavailability and stability in significant ways [158,159]. Studies in the pharmaceutical literature have focused on the dissolution process and the effect of surfactants, [160-162], with little attention given to how surfactants affect the crystallization steps involved in solution-mediated transformations. It would seem that surfactants, given their interfacial properties and their ability to form a variety of supramolecular structures (micelles, hemimicelles, etc.) in aqueous solution, could have a profound effect on crystallization in addition to their effects on dissolution. Surfactants have been shown to promote or inhibit crystallization steps (nucleation, crystal growth) in organic and inorganic crystals [161-168] and direct the crystallization of solid phases that otherwise are not formed. [169-170].

The influence of two structurally different anionic surfactants on the anhydrous-to-dihydrate transformation of carbamazepine (CBZ) was investigated. The surfactants studied were SLS, a surfactant commonly used in compendial dissolution methods, and sodium taurocholate (STC), an important surfactant in the solubilization and absorption of drugs and lipids in the gastrointestinal tract. Results show that both surfactants promoted the crystallization of CBZ dihydrate [CBZ (D)] during dissolution of the anhydrous monoclinic polymorph [CBZ (A)]). Examination of crystal surfaces showed that SLS facilitated the surface-mediated nucleation of CBZ (D) on CBZ (A) crystals at surfactant concentrations below the CMC [171].

The dissolution mechanism of the neutral drug danazol into solutions of the ionic surfactant SDS has been studied [172]. The effect of counter ion concentration on drug dissolution was also studied by controlling the solution ionic strength. The model-predicted dissolution rates were compared with the experimental data [172].

It has been revealed that the surfactant molecular complexes (SMCs) derived from long-alkyl-chain surfactants are sufficiently soluble in water through the path of micellar dispersion above the CMC of the complex surfactants, whereas the short-chain homologues cannot dissolve in water but dissociate the complexes, resulting in a heterogeneous phase made up of the liberated additives. The fact agrees perfectly with the familiar aspects of solubilization by surfactant; i.e., the longer is the alkyl chain of the surfactant the more effective it is for solubilization. Based on these results, it has been deduced that the possibility for any pair of surfactant and solubilizate (additive) to realize solubilization simply depends on the relative importance of equilibrium of dissociation or association of the SMC species in aqueous medium [173].

New grades of ultra-fine microcrystalline cellulose, without or with variable percentage SLS, were prepared by an ultrasonic homogenisation process from Avicel PH-101, prior to recovery by spray drying. Both new grade types were found to be inferior compared with grade C in a tableting application for paracetamol, resulting largely from poor flow of the feed material. However, both new grades proved superior to grade C in an aqueous extrusion/spheronisation application for the preparation of indomethacin pellets, producing smoother pellets in greater yield [174].

The α tocopheryl PEG1000 succinate (TPGS) was used to increase the aqueous solubility and dissolution rate of furosemide. The infrared spectroscopic analysis showed that an interaction, in the solid dispersion, such as an association between the functional groups of furosemide and TPGS might occur in the molecular level. The infrared spectroscopy and differential thermal analysis showed the physicochemical modifications of the furosemide from the solid dispersion. The solid dispersion technique with TPGS provides a promising way to increase the solubility and dissolution rate of poorly soluble drugs [175].

Equilibrium solubility, the profile and the rate of ibuprofen dissolution in solutions of SDS, Tween 60 and Brij 35 were studied. It was found that the solubility of ibuprofen was enhanced by the formed micellar system. The size and shape of formed micelles depended on the used particular surfactant [176].

8.4 Drug stabilization

Surfactants are known to affect drug stability. A brief account of surfactants influencing drug stability is given below:

Solid lipid nanoparticles show different degradation velocities by the lipolytic enzyme, pancreatic lipase, as a function of their composition i.e., lipid matrix, surfactant. The influence of surfactants can be both accelerating and degrading (e.g. cholic acid sodium salt) or a slowing down effect on due to steric stabilization (e.g. Poloxamer 407). As a second steric stabilizer, Tween 80 has been used and the results showed a less pronounced effect on hindering the degradation process than for Poloxamer 407[177].

In order to be used as drug carriers, pluronic micelles require stabilization to prevent degradation caused by significant dilution accompanying intravenous injection. At low pluronic concentrations, when the drugs were located in the hydrophilic environment, drug uptake was increased. In contrast, when the drugs were encapsulated in the hydrophobic cores of pluronic micelles, drug uptake by the cells was substantially decreased [178].

The cooperative nature of interaction of cationic surfactants with short oligonucleotides leading to eventual stabilization of DNA duplexes is demonstrated. Understanding the cooperative binding of the cationic surfactants to the DNA may have implications for rational design of DNA binding drugs and DNA delivery systems [179].

A series of glycolipid surfactants derived from Tris (hydroxymethyl) acrylamidomethane (THAM) and bearing hydrocarbon or perfluorocarbon tails and an acryloyl group attached to their polar head were prepared to explore the aqueous behavior of the supramolecular systems they form. To stabilize these vesicles, polymerization by ultra violet irradiation was carried out. During this reaction, a precipitation in water was observed for the hydrocarbon surfactants, whereas fluorocarbon structures provide stable vesicles without any alteration of their size [180].

Current and potential application possibilities of block copolymer colloidal assemblies as stabilizers, flocculants, nano reservoir, among others, controlled delivery of bioactive agents, catalysis, latex agglomeration and stabilization of non-aqueous emulsion are discussed by Riess [181].

Protein aggregation and inactivation are major problems associated with the encapsulation of pharmaceutical proteins in biodegradable micro spheres. The use of PEG as the emulsifying agent in the aqueous and organic phase prevented gamma-chymotrypsin inactivation and aggregation during encapsulation. The stabilization approach was also worked for the model protein *Horseradish peroxidase* and thus is of a general nature [182].

The mean aggregate number of the antipsychotic drug chlorpromazine hydrochloride (CPZ) nanostructure was investigated by fluorescence quenching using 9-methylanthracene as the quencher. An increase in the aggregate size with increasing drug concentration confirmed the stepwise aggregation theory of CPZ micelle formation. Differential scanning calorimetry was used to examine the effects of concentration on the thermodynamics of micellization. The enthalpy of demicellization increased with increasing CPZ concentration (5–12 mM), suggesting a greater stability of the aggregates at higher concentrations [183].

The solubility and stability of the chemically unstable drug tetrazepam, which has poor water solubility, have been studied in bile salts–PC-mixed micelles (BS–PC-MM). Sodium glycocholate–Soya PC-MM interfered with the degradation kinetics and displayed better stabilizing effects under both aerobic and anaerobic conditions [184].

Lipid emulsion particles containing 10% of medium chain triglycerides were prepared using 2% w/w for use as drug carrier. The emulsion stability did not notably change in the presence of a model destabilizing drug indomethacin. The use of a second hydrophilic surfactant to adjust the packing properties of the lecithin at the oil–water interface provided an increase in the stability of lipid emulsions, and this may be of importance in the formulation of drug delivery systems [185].

Gelatin-containing, electrically conducting, rigid water-in-oil ME-based organogels, both with and without the presence of a model drug, have been prepared using pharmaceutically acceptable oils and surfactants. Aerosol-OT was found to be the single most effective surfactant for stabilizing the w/o MEs [186].

Pacheco and Carmona-Ribeiro has reported the effects of synthetic lipids on solubilization and colloid on stability of hydrophobic drug [187]. Aqueous miconazole aggregates were solubilized and/or colloidally stabilized by bilayer-forming synthetic amphiphiles such as dioctadecyldimethylammonium bromide or sodium dihexadecylphosphate dispersions. Inexpensive, synthetic bilayer fragments offered a large area of hydrophobic nanosurfaces dispersed and electrostatically stabilized in water, opening new prospects for drug solubilization and colloid stabilization of insoluble drug particles [187].

The stability of fluorinated phospholipid-based vesicles in terms of detergent-induced release of encapsulated carboxyfluorescein has been evaluated. It has been shown that fluorinated liposomes have a promising potential as drug carrier and delivery systems for oral administration [188].

The preservation of biological activity of protein drug in formulations is still a major challenge for successful drug delivery. The enzyme L-asparaginase, which exhibits a short *in vivo* half-life and is only active against leukaemia in its tetrameric form, was encapsulated in poly (D-L-lactide-co-glycolide) nanospheres by the (w/o)/w-emulsion solvent evaporation technique in presence of various potential stabilisers. The preparation of *l asparaginase* nanospheres with trehalose, PEG 400, and glycerol as components of the inner aqueous phase yielded colloidal formulations with increased biological activity [189].

Barreiro-Iglesias et al has studied the solubilization and stabilization of camptothecin in micellar solutions of pluronic-*g*-poly (acrylic acid) copolymers. The capability of a family of copolymers comprising pluronic surfactants covalently conjugated with poly (acrylic acid) to enhance the aqueous solubility and stability of the lactone form of camptothecin (CPT) was studied. The loading of CPT into the poly (acrylic acid) micelles was able to prevent the hydrolysis of the lactone group of the drug for 2 h at pH 8 in water [190].

The pulmonary drug delivery of proteins presents an alternative to parenteral and oral administration. Nebulization of aqueous protein solutions is an ideal method for pulmonary application of therapeutic proteins. The surfactants like poloxamer 188, PEG 8000 solutol HS15 and octanoyl-*N*-methyl-glucamide on nebulized aviscumine has indicated stabilization of formulation [191].

8.5 Surfactants in drug targeting

The drug targeting involves application of new chemical and biological approaches to the preparation of targeted biopharmaceuticals, including ribonuclease–antibody chimeras, biospecific antibodies, enzyme–antibody conjugates, and folate bearing conjugates. Applications include the targeting of myocardial infarctions using liposomal systems, the use of monoclonal antibodies to target malaria and HIV-infected cells, the targeting of fusion proteins to leukemias and lymphomas, and the use of antioxidant enzyme–antibody conjugates to target the pulmonary endothelium. Polymeric micelles demonstrate a series of attractive properties as drug carriers, such as high stability both *in vitro* and *in vivo* and good biocompatibility. Polymeric micelles have been shown to incorporate well poorly soluble pharmaceuticals and efficiently deliver them into pathological areas with compromised vasculature (tumors, infarcts) via the enhanced permeability and retention effect.

Current literature suggests that bio-distribution of therapeutic particles following intravenous administration is modulated by the size of the particle and surface charge. A hydrophilic and uncharged surface on the particle reduces their uptake by the phagocytic cells of reticular endothelial system and aggregation, significantly prolonging circulation time. For targeting DNA complexes to cancer cells, particles should be small enough to pass through fenestrations in the tumor vasculature. The particle size and surface hydrophilicity can easily be modulated by PEGylated DNA particles. [192].

Yang etal studied the distribution of camptothecin solid lipid nanoparticles (SLN) and targeting effect on brain. The results indicate that SLN are a promising sustained release and drug targeting system for lipophilic antitumour drugs, and may also allow a reduction in dosage and a decrease in systemic toxicity [193].

Carboplatin was incorporated into nonphospholipid vesicles by the method of membrane dispersion and studied for its lung targeting. Antitumor effect was increased in comparison with the original drug of the same dose [194].

Drugs can be delivered to brain with the aid of poly (butylcyanoacrylate) (PBCA) nanoparticles coated with polysorbate 80. These carriers can penetrate BBB and deliver drugs of various structures, including peptides, hydrophilic compounds, and lipophilic compounds eliminated from brain with P-glycoprotein [195].

Bae et al has studied the metalloprotease-specific PEG methyl ether-peptide-doxorubicin (mPEG-GPLGV-DOX) conjugate for targeting anticancer drug delivery based on angiogenesis. The mPEG-GPLGV-DOX conjugate formed a micelle structure in aqueous solution, with a CMC of about 0.25 mg/ml. An increase was observed of 20% chemotherapeutic activity compared with free doxorubicin [196].

Low-density lipoprotein (LDL) has been found suitable as a targeting carrier for cytotoxic drug. However, higher drug loading into LDL particles without disrupting their native integrity remains a major obstacle. It is indicated that physicochemical factors significantly affect drug-loading efficiency and may need to be considered to optimize drug incorporation into LDL particles [197].

Caliceti et al has studied PEG avidin bioconjugates as suitable candidates for tumor pretargeting. Interestingly, all conjugates accumulated significantly in the tumor mass [198].

Peritoneal dissemination in gastric cancer is a common fatal clinical condition with few effective therapies available. Iinuma et al studied the therapeutic effect of a tumor-targeting drug delivery system that uses cisplatin-encapsulated and transferin (Tf)-conjugated PEG liposomes (Tf-PEG liposomes) in nude mice with peritoneal dissemination of human gastric cancer cells. The results suggest that cisplatin-encapsulated Tf-PEGliposomes may be useful as a new intracellular targeting carrier for treatment of gastric cancer with peritoneal dissemination [199].

Gijsens et al studied targeting of the photocytotoxic compound AlPcS4 to Hela cells by Tf-conjugated PEG-liposomes. The images of intracellular accumulation and reactive oxdative species production matched the accumulation and photocytotoxicity profile of the different photoactive compounds. The photodynamic activity of the Tf-Lip-AlPcS4 conjugate on HeLa cells is much more potent than free AlPcS4 as a result of selective transferrin receptor mediated uptake [200].

Lipoproteins are endogenous particles that transport lipids through the blood to various cell types, where they are recognized and taken up via specific receptors. These particles are, therefore, excellent candidates for the targeted delivery of drugs like wide range of lipophilic, amphiphilic, and polyanionic compounds to hepatocytes and tumour cells [201].

Hong etal studied the prolonged blood circulation of methotrexate (MTX) by modulation of liposomal composition. Various compositions of liposomes were prepared with 2:1 of PC and cholesterol (CH) with or without distearoylphosphatidyl-ethanolamine-N-PEG 2000 (DSPE-PEG). Results indicated that DPPC/CH/DSPE-PEG liposomes most effectively prolonged the blood circulation, and reduced hepatosplenic and kidney uptake of MTX. DPPC/CH/DSPE-PEG liposomes may have potential as an efficient delivery system for MTX [202].

In aqueous media, certain PEG/ phosphatidylethanolamine (PEG–PE) conjugates form very stable micelles. The PEG-based corona makes these micelles long circulating, whereas the lipid hydrophobic core may be used as a cargo space for poorly soluble compounds, including many anti- cancer drugs [203]. Micelles prepared from PEG of various lengths conjugated with phosphatidylethanolamine (PEG–PE) were loaded with various poorly soluble anticancer agents. PEG–PE micelles selectively accumulated in Lewis lung carcinoma (LLC) tumors implanted in mice. Modification of the micelles with tumor specific antibodies further enhanced the efficiency of tumor accumulation [204].

8.6 Surfactants as wetting agents

The potential effect of the common pharmaceutical wetting agent SLS, on the transport of the hydrophilic bisphosphonate, tiludronate, was investigated. It was revealed that SLS increased tiludronate Para cellular transport through its specific and transient effect on the permeability of the intercellular space [205].

The cationic water soluble drugs (chlorpheniramine maleate, pseudoephedrine HCI and propranolol HCl) were bound to a cation-exchange resin (Amberlite IRP 69) and microencapsulated using an aqueous solvent evaporation method, whereby the resin particles were dispersed in an organic polymer solution [ethyl cellulose, poly (methyl methacrylate), Eudragit RS 100] followed by emulsification into an external aqueous phase. A key variable for the successful encapsulation was the preferred wettability of the resin particles by the polymer phase. High encapsulation efficiencies were obtained, at high drug loading capacity of the resin, with drugs with high binding affinity and with a wetting agent. Phosphatidyl choline was the preferred wetting agent in order to avoid the partitioning of the resin into the external phase [206].

The process of non-aqueous spheronization was investigated for the *in situ* formation of an enteric co precipitate of nifedipine with Hydroxypropyl methylcellulose phthalate (HP-55) in spherical pellets. The final product formed containing SLS 2% as wetting agent and sodium starch glycolate 5% as disintegrant processed with optimum solvent level, gave a high yield of acceptable spheres, which showed poor drug release at low pH and enhanced release at the pH of the upper small intestine [207].

A number of poly (ether-ester) azo polymers consisting of various concentrations of 4-{2-[2-(2-hydroxyethoxy) ethoxy] ethoxy} benzoic acid (HEEEBA), 4-{2-[2-(2-hydroxyethoxy) ethoxy] ethoxy} phenylazobenzoic acid (HEEEPABA) and either 12-hydroxydodecanoic or 16-hydroxyhexadecanoic acid were synthesized. When 1% polysorbate 80 (a wetting agent) was added to the polymers in the degradation medium, all polymers showed excellent degradation. Azo polymers with HEEEBA have the potential to be used as colon-specific drug release materials [208].

Dissolution profiles of oxodipine and griseofulvin were obtained from binary and ternary systems prepared with PEG 6000 and PEG 6000/Tween 20, respectively. The dissolution of drugs (oxodipine or griseofulvin) from ternary systems, drug/PEG 6000/Tween 20, is better than that obtained from the respective binary systems, since in the former the wetting action of surfactant agent and the solubilizing effect of PEG 6000 over drug are additive [209].

Kapsi and Ayres studied the bioavailability of solutions of itraconazole, a water insoluble antifungal drug. Initial treatment of itraconazole with the wetting agent/cosolvent glycerol prior to making itraconazole into a solid solution improved drug dissolution, and also reduced the PEG amount required to dissolve drug to form solid solution [210].

8.7 Synergistic effects

A synergistic effect is one in which the combined effect of two or more chemicals is greater than the sum of the effects of individual drug effect.

Zhu and Feng reported synergistic solubilization of polycyclic aromatic hydrocarbons by mixed anionic-nonionic surfactants. The noted synergism for the mixed surfactants is attributed to the formation of mixed micelles, the lower CMC of the mixed-surfactant solutions, and the increase of the solute's molar solubilization ratio or micellar partition coefficients (K_{mc}) because of the lower polarity of the mixed micelles. Solubility enhancement efficiencies of surfactants above the CMC follow the order of Triton X-100>Brij 35> Triton X305 >SDS. [211].

Limpens and Scheper reported the synergistic effects of locally administered cytostatic drugs and a surfactant on the development of delayed-type hypersensitivity (DTH) to keyhole limpet haemocyanin in mice. The immunomodulating effects of cyclophosphamide-derivative Z 7557(mafosfamide) and the plant alkaloid VP-16 (Oral Etopodise), were compared with the effects of systemically administered cyclophosphamide and several established adjuvants: Freund's complete adjuvant, dextran sulphate, and dimethyl dioctadecyl ammonium bromide (DDA). All agents tested promoted the development of DTH. The results show that DDA and sub optimal amounts of locally administered cytostatic drugs act synergistically on DTH [212].

Potent synergistic *in vitro* interaction between non antimicrobial membrane-active compounds and itraconazole (ITZ) against clinical isolates of *Aspergillus fumigatus* resistant to ITZ were studied by Afeltra et al. [213], to develop new approaches for the treatment of invasive infections caused by *Aspergillus fumigatus*.

The mechanisms of herpes simplex virus (HSV) inactivation by SLS and n-lauroylsarcosine (LS), two anionic surfactants with protein denaturant potency, have been evaluated in cultured cells. The pretreatment of HSV-2 strain 333 with specific combinations of SLS and LS concentrations inhibited the viral infectivity in a synergistic manner and resulted in only a small increase in their toxicities for exponentially growing Vero cells compared with that caused by each compound alone [214].

Permeability enhancing effect of the surfactants, 1-palmitoyl-2-lyso-sn-gycero-3-phosphocholine (lysoPPC) and palmitic acid (PA), on lipid membranes was studied by Davidsen et al. when lysoPPC and PA were added simultaneously in equimolar concentrations. A dramatic synergistic permeability-enhancing effect was observed [215].

Synergistic effects in mixed wormlike micelles of dimeric and single-chain cationic surfactant at high ionic strength has been investigated. The variation of the cmc of the mixtures, measured by surface tension experiments, with composition revealed synergism in micelle formation. These effects are ascribed to a progressive intermicellar cross linking resulting from a continuous increase of the end-cap energy with in the mixture [216].

8.8 Prodrugs

Micellar prodrugs are prepared in order to improve the solubility of a poorly soluble drug, which on administration are converted to active drugs by metabolic process [217].

Hydrophilic and lipophilic polymers are widely used excipients to control the release rate of drugs from matrices. Surfactants were investigated to determine the release rate of highly soluble drug (propranolol HCl) from hydroxy propyl methylcellulose Eudragit matrices. Surfactants like SLS as an anionic surfactant, CTAB as a cationic surfactant, Tween 65 and Arlacel 60 as non-ionic surfactants were selected. The results indicated that, the type and ionization of surfactant, hydrophilicity and lipophilicity of surface-active agent and various ratios of surfactants are important factors in controlling the release rate of propranolol [218].

Amantadine was modified by direct acylation with succinic and glutaric anhydrides and the covalent bond with α,β -poly(*N*-hydroxyethyl)-DL-aspartamide (PHEA) was realized. The amount of amantadine in the obtained copolymers was evaluated by hydrolysis of the conjugates. The enthalpic effect due to the interaction between the two synthesized polymeric prodrugs and SDS micelles as a simple system mimicking the cellular membrane was measured by the calorimetric technique and compared with that of PHEA. Binding of PHEA-succinylamantadine (PHEA-S-AMA) to surfactant micelles appeared to be stronger than that shown by PHEA-glutarylamantadine (PHEA-G-AMA) [219].

The viral agents, acyclovir and valaciclovir, were coupled with activated PEG. The ability of the macromolecular conjugate to release the free drug was also evaluated in plasma, in which PEG-valacyclovir was proved to be the most effective prodrug [220].

Very strong intramolecular bonds and very weak intermolecular interactions uniquely characterize fluorocarbons and fluorocarbon moieties. This results in a combination of exceptional thermal, chemical and biological inertness, low surface tension, high fluidity, excellent spreading characteristics, low solubility in water, and high gas dissolving capacities, which are the basis for innovative applications in the field of prodrugs [221].

Cholesteryl ester prodrugs of ibuprofen and flufenamate were synthesized and solubilized in MEs [222]. Similarly, PEG derived prodrugs of amino group-containing compounds were synthesized and were found to have extended plasma circulation with apparent accumulation of the drug in the tumor [223].

8.9 Surfactants and drug delivery

Drug delivery is of great importance in relation to therapeutic efficiency. Surfactant systems and polymers contribute to drug delivery in many important ways. A better understanding of this field would allow the rational design of more advanced drug delivery formulations, by establishing structure activity performance relationships.

Application of principles of surface activity and surfactant behaviors is much explored area and has resulted in effective products like novel drug delivery systems. The novel drug delivery systems in which surface activity is directly or indirectly involved will be discussed in the fallowing sections.

8.9.1 Dendrimers

The branched polyamines, including polyamidoamide dendrimers, polypropyleneimine, and polyethyleneimine, are able to purge prion protien (Prp), the protease-resistant isoform of the Prp, from scrapie-infected neuroblastoma cells in culture. The studies suggest that branched polyamines might be useful therapeutic agents for treatment of prion diseases and perhaps a variety of other degenerative disorders [224]. The anti-prion activity of new cationic phosphorus-containing dendrimers (P-dendrimers) with tertiary amine end-groups was tested. These molecules had a strong anti-prion activity, decreasing both PrP and infectivity in scrapie-infected cells at non-cytotoxic doses. They can bind PrP and decrease the amount of pre-existing PrP from several prion strains, including the BSE strain. More importantly, when tested in a murine scrapie model, the dendrimers were able to decrease PrP accumulation in the spleen by more than 80 percent. These molecules have a high bioavailability and therefore exhibit relevant potential for prion therapeutics for at least post-exposure prophylaxis [225].

Joester et al studied amphiphilic dendrimers are as novel self-assembling vectors for efficient gene delivery [226]. Percec et al reported a library of amphiphilic dendritic dipeptides that self-assemble in solution and in bulk through a complex recognition process into helical pores. The molecular recognition and self-assembly process is sufficiently robust to tolerate a range of modifications to the amphiphilic structure, while preliminary proton transport measurements establish that the pores are functional. This class of self-assembling dendrimers will allow the design of a variety of biologically inspired systems with functional properties arising from their porous structure [227]. The PEGylated dendrimers based on melamine were evaluated as candidate vehicles for drug delivery by Chen et al Cationic dendrimers were found to be more cytotoxic and hemolytic than anionic or PEGylated dendrimers [228].

8.9.2 Gene delivery systems

Extra cellular and intracellular factors influencing gene transfection mediated by 1,4-dihydropyridine (DHP) amphiphilic was reported by Hyvonen et al. Some extra cellular glycosaminoglycans and serum interfere with 1,4-DHP-amphiphile-mediated transfection by destabilizing the amphiplexes [229].

Novel, double-chained pyridinium compounds have been developed that display highly efficient DNA transfection properties. Most importantly, the pyridinium compounds were found to be essentially nontoxic toward cells. By systematically modifying the structure of the pyridinium amphiphile, i.e., by changing either the head group structure or the alkyl chains, some insight was obtained that may lead to unraveling the mechanism of amphiphile-mediated transfection, and thus to protocols that further optimize the carrier properties of the amphiphile [230]. Gebhart et al studied design and formulation of polyplexes based on pluronic-polyethyleneimine conjugates for gene transfer [231]. Intracellular delivery of nanometric DNA particles via the folate receptor was achieved using a designed cationic thiol-detergent, tetradecyl-cysteinyl-ornithine [232].

For two series of polyethylenimine-*graft*-poly (ethylene glycol) (PEI-*g*-PEG) block copolymers, the influence of copolymer structure on DNA complexation was investigated and physicochemical properties of these complexes were compared with the results of blood compatibility, cytotoxicity, and transfection activity assays. The degree of PEGylation and the MW of PEG were found to strongly influence DNA condensation of polyethylenimine and therefore also affect the biological activity of the PEI-*g*-PEG/ DNA complexes. These results provide a basis for the rational design of block copolymer gene delivery systems [233]. Self-assembly of polyamine-poly (ethylene glycol) copolymers with phosphorothioate oligonucleotides was studied by Vinogradov et al. The simplicity of preparation and long shelf life make these systems attractive as potential pharmaceutical formulations for oligonucleotides [234].

Glutathione (GSH)-sensitive stabilization of polyion complex micelles entrapping antisense oligo deoxy nucleotides (ODN) was achieved by the reversible cross-linking of the core through disulfide bonds, aiming at the development of a novel DNA carrier system for antisense therapy. The stability of the ODN in the core cross-linked micelles against nuclease was appreciably increased compared to that of free ODN and that in the micelles without cross-linking. On the other hand, the micelles dissociated to release ODN in the presence of GSH at a concentration comparable to the intracellular environment, featuring the potential ability of this system for intracellular ODN delivery [235].

A new class of PEG derivative, Chol-PEG-A, having both cholesteryl- and amino-pendant groups was synthesized. This amphiphilic PEG derivative forms a cationic polymer assembly in water showed higher gene expression [236]. Polystyrene nanoparticles were prepared by surfactant-free emulsion polymerization using water-soluble cationic initiators to induce a positive surface charge. ODN release can be induced by the addition of anionic surfactants or by increasing the pH of the dispersion medium [237].

The liposomes are microscopic lipid membrane vesicles that provide a strategy for gene delivery. Raghavachari and Fahl has used and found that nonionic liposomes are best suited for transdermal delivery of gene [238].

PEG has been coupled to many cationic polymers such as polyethylenimine to improve the stability and transfection efficiency [239]. However, the longer PEG side chains also interrupt the gene delivery to the cells due to the more efficient steric hindrance by longer PEG side chains, and thus the transfection efficiency diminishes as PEG side chains [239].

8.9.3 Lipid emulsions

The hemolytic activity of sodium oleate, a high lytic agent, was investigated in different surfactant solutions and lipid emulsion formulations. A new explanation of the protective function of these systems is proposed. It was found that the hemolytic activity of the lytic agent was greatly decreased in solutions and/or dispersions with the surfactant Cremophor EL (Polyethoxylated castor oil), Solutol H16 and phospholipids, which can usually build a micellar or liposomal structure. In the case of Pluracare F 68 (a block copolymer of ethylene glycol and propylene glycols), where the micelle formation is still controversial, the hemolytic activity of the lytic agent was practically not affected and complete hemolysis was observed. In contrast to this, all emulsion formulations, independent of the emulsifier type, showed a stable erythrocyte behavior. Additionally, in the case of lipid emulsions only, a larger amount of the lytic agent could be added without any remarkable increase in the hemolytic activity. As an explanation for these effects it is proposed that the lytic agent is either incorporated into the lipophilic core or intercalates between the emulsifier molecules at the interface. This decreases the direct contact of the lytic agent with the erythrocyte membrane. As a result, the erythrocytes will effectively be protected from hemolytic damage, which can otherwise be induced by such substances [240]

Long-circulating sub micron lipid emulsions, stabilized with PEG-modified phosphatidylethanolamine (PEG-PE), are promising drug carriers with substantial capacity for solubilization of lipophilic anticancer agents. Lundberg et al has studied the conjugation of an anti B cell lymphoma monoclonal antibody to the long-circulating drug-carrier lipid emulsions. The results indicate that emulsion-antibody complexes might be a useful drug-carrier system for more specific delivery of anticancer drugs to B cell malignancy [241].

Paclitaxel is an anticancer agent with poor solubility in water and requires a suitable formulation for intravenous administration. Paclitaxel is formulated for clinical use in ethanol and Cremophor EL, a solvent system associated with severe adverse effects. Paclitaxel was entrapped in lipid emulsion droplets with triolein as oil core and dipalmitoyl PC as the principal emulsifier. The emulsion was further stabilized with polysorbate 80 and PEG-dipalmitoyl phosphatidylethanolamine. The result indicates that long-circulating sub micron lipid emulsions may prove useful, not only for replacement of the more toxic Cremophor EL vehicle, but also by improving the distribution of the drug to the tumors [242].

Tamilvanan and Benita have reviewed the potentials of lipid emulsions for ocular delivery of lipophilic drugs. For nearly a decade, oil-in-water lipid emulsions containing either anionic or cationic droplets have been recognized as an interesting and promising ocular topical delivery vehicle for lipophilic drugs. The review covers an update on the state of the art of incorporating the lipophilic drugs, a brief description concerning the components and the classification of lipid emulsions [243].

Injectable lipid emulsion is an important component of parenteral nutrition. ClinOleic is a lipid emulsion composed of olive oil (80%) and soybean oil (20%). A study on the efficacy and safety of ClinOleic in adults already receiving parenteral nutrition, comparing it to their usual lipid (soybean-oil-based) was reported by Thomas-Gibson et al. The results indicate ClinOleic may be used as a safe alternative to standard soybean-oil-based lipid emulsions [244].

Lipid-coated micro bubbles represent a new class of agents with both diagnostic and therapeutic applications. Micro bubbles have low density. Stabilization of micro bubbles by lipid coatings creates low-density particles with unusual properties for diagnostic imaging and drug delivery [245].

To assess the prolongation of epidural bupivacaine by a novel lipid formulation, a physically stabilized bupivacaine containing dry emulsion was prepared by spray drying. The onset time of epidural anesthesia was similar for both formulations of bupivacaine used, while a significant blockade prolongation was observed with the emulsion compared to the solution, suggesting a controlled release of bupivacaine. Dry emulsions could be promising dosage forms to optimize the disposition of epidurally administered local anesthetics [246].

In an attempt to develop an artificial lipoprotein-like particle, lipid nano-sphere (LNS), incorporating dexamethasone palmitate (DMP) was formulated using soybean oil and egg lecithin. Potential drug carriers were compared with a conventional fat emulsion for intravenous nutrition, lipid microspheres (LM), which is already used clinically. LM easily entered reticuloendothelial systems, such as the liver, and was rapidly cleared from the circulation. However, LNS showed much higher plasma levels of DMP after intravenous administration to rats and recovered more than 80 percent of the injected dose in the perfusate in single-pass rat liver perfusion. Nanometer-sized lipid emulsion particles, LNS, seem to be a promising carrier system for passive drug targeting of lipophilic drugs [247].

8.9.4 Liposomes

Liposomes have gained considerable interest as drug delivery tool for therapy of cancer and infectious diseases. They have therapeutic advantages, such as the ability to deliver large amounts of drugs to specific sites, spare healthy tissue from toxic effects, and increase the systemic circulation time of the drug [248]. However, one problem with liposomes is their poor stability in the circulatory system. To improve this, poly (ethylene glycol), PEG, is frequently used as a component in the lipid bilayer. PEG stabilizes the liposomes sterically and also effectively protects the liposomes from degradation by the reticuloendothelial system [249].

Kullberg, et al investigated conjugation of epidermal growth factor (EGF), to liposomes using the micelle-transfer method. The conjugate was shown to have EGF-receptor-specific cellular binding in cultured human glioma cells [250].

Di- octadecyldimethylammonium bromide (DODAB) liposomes were tested as effective inducers of delayed hypersensitivity (DH) towards The 18 kDa antigenic protein from *Mycobacterium leprae (P)* or its N-acyl derivative. Immunization tests DH indicate an enhancement of cell- mediated immunological response towards P / DODAB complexes that is not obtained for the isolated protein [251].

Encapsulation of technetium-99m sestamibi ((99m) Tc-MIBI) in polyethyleneglycol-liposomes could extend the duration of its circulation in blood and alter its biodistribution, enabling its concentration in tumors to be increased. The (99m) Tc-MIBI-PEG-liposomes demonstrated a longer blood circulation time, enabled distinction between chemotherapy-sensitive and resistant cells and improved tumor to background contrast *in vivo* imaging. (99m) Tc-MIBI-PEG-liposomes therefore show promising potential for tumor imaging [252].

The design of targeted oral liposomes is anticipated to improve the systemic delivery of poorly absorbed agents, such as proteins and peptides. A poly (ethylene oxide) (PEO)-folic acid (FA) derivative was prepared and evaluated for improving liposome transport across a model gastrointestinal cell line FA-PEO-cholesterol derivatives were synthesized and adsorbed at (Caco-2). Liposome surfaces encapsulating Texas Red Dextran 3000 (TR-Dex)

TR- Dex, is a poorly absorbed, neutral, hydrophilic, large molecular weight marker. An increase in intracellular accumulation of TR-dex associated with FA-PEO-coated liposomes, but not other formulations, was evidence of the potential of FA-targeted liposomes in the oral delivery of poorly absorbed, large agents like proteins and peptides [253].

Recent advances in liposome technology have shown promise relative to the introduction of chemotherapeutic agents with reduced toxicity, extended longevity, and potential for cell-specific targeting. A liposomal delivery of doxorubicin was targeted specifically to C6 glioma *in vitro* by coupling transferrin to the distal ends of liposomal PEG chains. Competitive binding assays support the receptor-mediated mechanism of targeting [254].

A major obstacle in the development of red cell substitutes has been overcoming their short circulation persistence. Distearoyl phosphoethanolamine PEG5000 (PEG-PE) was added to the formulation of liposome-encapsulated hemoglobin (LEH) to decrease reticuloendothelial system uptake and prolong LEH circulation persistence. The addition of PEG-PE to the LEH formulation greatly prolongs the circulation persistence of LEH and represents a significant step in the development of red cell substitutes with prolonged oxygen delivery [255].

Rapid and pH-sensitive release of a highly water-soluble fluorescent aqueous content marker, pyranine, from egg PC liposomes following incorporation of N- isopropylacrylamide (NIPA) copolymers in liposomal membranes was reported by Zignani et al. The results indicate the pH-sensitive copolymers of NIPA represents promising strategy for improving liposomal drug delivery [256].

Allen et al describe a new technique for the formation of ligand-targeted liposomes that can be used with whole antibodies, antibody fragments, peptides or other ligands. The ligands are coupled to PEG micelles and then transferred in a simple incubation step from the micelles into the outer monolayers of pre-formed, drug-loaded liposomes. This versatile method allows a combinatorial approach to the design of targeted liposomes that minimizes manufacturing complexities, allowing a variety of ligands to be inserted into a variety of pre-formed liposomes containing a variety of drugs. This allows the ligand-targeted therapeutics to be tailored to the needs of individual patients [257].

Kozubek et al studied amphiphile-based natural dihydroxyphenols and their semi synthetic derivatives as vesicular drug carriers. The presence of these compounds in lipid composition enhances liposomal drug encapsulation, reduces the amount of the lipid carrier necessary for efficient entrapment of anthracycline drugs, stabilizes liposomal formulation [258].

To improve epidural pain treatment, a long acting, single-dose gel injection is being developed in using liposomal systems to control the release and dural permeation of ibuprofen. The liposomal gel controlled ibuprofen release and dural permeation *in vitro* showed a permeation pattern favorable for maintaining constant drug levels. The liposomal poloxamer gel represents a new formulation approach to increase the local epidural availability of ibuprofen [259].

The adhesion of bacteria to medical implants and the subsequent development of a biofilm frequently result in the infection of surrounding tissue and may require removal of the device. DiTizio etal have developed a liposomal hydrogel system that significantly reduces bacterial adhesion to silicone catheter material. The system consists of a PEG-gelatin hydrogel in which liposomes containing the antibiotic ciprofloxacin are sequestered. This new antimicrobial coating shows promise as a prophylactic and/or treatment for catheter-related infection [260].

Angiogenesis is a key process in the growth and metastasis of a tumor. Disrupting this process is considered a promising treatment strategy. Therefore, a drug delivery system specifically aiming at angiogenic tumor endothelial cells was developed. For this, peptides with affinity for this integrin were coupled to the distal end of PEG coated long-circulating liposomes (LCL). Peptide–LCL containing doxorubicin showed superior efficacy over non-targeted LCL in inhibiting C 26 doxorubicin-insensitive tumor out growth *in vitro* and *in vivo* [261].

Targeted delivery of radionuclides and therapeutic agents to specific biomarkers of breast cancer has important implications for the diagnosis and therapy of breast cancer. Vasoactive intestinal peptide receptors (VIP-R) are approximately five times more expressed in human breast cancer, compared to normal breast tissue. Dagar et al have used VIP, a 28 amino acid mammalian neuropeptide, as a breast cancer-targeting moiety for targeted imaging of breast cancer. Collectively, these data showed that Tc99m-HMPAO (a radio nucleotide) encapsulating VIP–SSL can be successfully used for the targeted imaging of breast cancer [262].

Antibody or ligand-mediated targeting of liposomal anticancer drugs to antigens expressed selectively or over-expressed on tumor cells is increasingly being recognized as an effective strategy for increasing the therapeutic indices of anticancer drugs. Sapra and Allen have reviewed some recent advances in the field of ligand-targeted liposomes for the delivery of anticancer drugs [263].

Immunostimulating complexes (ISCOMs) are spherical, micellar assemblies of about 40 nm made of the saponin mixture Quil A, cholesterol and phospholipids. Liposomes and ISCOMs have a long history as vehicles for antigen delivery. Liposomes can carry both membrane-associated antigens as well as water soluble molecules. In recent years, considerable progress has been achieved with respect to the use of better-defined saponin. Kersten and Crommelin have reported the progress in clinical trials with ISCOMs [264].

It was recently reported that the partitioning of unsaturated cosurfactant molecules between aqueous and lipid environments in detergent-based lamellar phase systems could be determined using avoided-level-crossing muon-spin resonance (mSR) [265].

Based on a recent report on an avoided-level-crossing mSR study of cosurfactant partitioning in lamellar phase systems, it is proposed that the same technique can be used to determine the partitioning of drug molecules between aqueous environments such as cell fluids and lipid-like environments of a cell membrane, or of polymer or model-lipid based liposomes. The use mSR study as a novel probe for such investigations was reported [266].

Cardiovascular disease processes such as atherosclerosis, restenosis, and inflammation are typically localized to discrete regions of the vasculature, affording great opportunity for targeted pharmacological treatment. Liposomes are potentially advantageous targeted drug carriers for such intravascular applications. To facilitate their use as drug delivery vehicles, Lestini et al have considered three components of liposome design: (i) identification of candidate cell surface receptors for targeting; (ii) identification of ligands that maintain binding specificity and affinity; and (iii) prevention of rapid nonspecific clearance of liposomes into the reticuloendothelial organs. Together these studies demonstrate the feasibility of using peptides to guide liposomes to desired receptors, and illustrate the influence of vesicle stability on liposome interactions *in vivo*. Furthermore, they underscore the importance of simultaneously considering both targeting specificity and vesicle longevity in the design of effective targeted drug delivery systems [267].

Kim etal has reported enhancement of PEG modified cationic liposome-mediated gene deliveries and their effects on serum stability and transfection efficiency. The results suggest that the PEG-added transfection complexes could be a useful non-viral vector because of their simplicity in preparation, enhanced stability and prolonged circulation compared with the conventional transfection complexes [268].

Hong et al has reported pH-sensitive, serum-stable and long-circulating liposomes as a new drug delivery system with dioleoylphosphatidylethanolamine (DOPE) and oleic acid (DOPE/oleic acid liposome) or DOPE and 1,2-dipalmitoylsuccinylglycerol (DOPE/DPSG liposome). The inclusion of DSPE-PEG enhanced the serum stability of both DOPE/oleic acid and DOPE/DPSG liposomes, but also shifted the pH-response curve of pH-sensitive liposomes to more acidic regions and reduced the maximum leakage percentage the liposomes at tumor tissues suggests that the liposomes might be useful for the targeted delivery of drugs such as anticancer agents [269].

To achieve a sustained and targeted delivery of liposomes to liver parenchymal cells Chittima Managit et al modified distearoyl-PC)/cholesterol (60:40) liposomes with a galactosylated cholesterol derivative and polysorbate (Tween) 20 or 1,2-distearoyl-*sn*-glycero-3-phosphoethanolamine-*N*-polyethylene glycol. After intravenous injection, liposomes were rapidly eliminated from the blood circulation and mostly recovered in the liver. The blood elimination of modified liposomes was slightly reduced as compared to galactosylated liposomes. In contrast, a significant reduction in the blood elimination was observed modified liposomes. The results suggest that modified liposomes can control the delivery rate drug to liver PC without losing its targeting capability [270].

PEG-immunoliposome are newly developed pendant-type PEG-immunoliposomes, carrying monoclonal antibodies or their fragments (Fab') at the distal ends of the PEG chains. In terms of target binding of Type C, two different anatomical compartments are considered viz., mouse lung endothelium and the implanted solid tumor. Fab'-Type C showed low RES uptake and a long circulation time, and enhanced accumulation of the liposomes in the solid tumor was seen. [271].

8.9.5.Microemulsions(ME)

New ME vehicle facilitates percutaneous penetration *in vitro* and cutaneous drug bioavailability *in vivo* was investigated in quaternary systems composed of glyceryl oleate polyoxyl 40 fatty acid derivatives (surfactants)/tetra glycol (cosurfactant)/isopropyl palmitate/water. It has been suggested that these MEs loaded with lidocaine would provide adequate analgesia in relatively shorter periods of time [272].

An oral heparin would avoid the inconvenience of subcutaneous injections and adverse events associated with warfarin. A mild chitosan/PEG/calcium alginate micro encapsulation process, as applied to encapsulation of biological macromolecules such as heparin and low molecular weight heparin (LMWH) was investigated. Heparin and LMWH entrapped alginate beads were further surface/enteric coated with chitosan and cellulose acetate phthalate via carbodiimide functionalities These results established the feasibility of modifying the formulation in order to obtain the desired controlled release of bioactive agent for a convenient pH dependent delivery system [273].

Jain et al. has described a novel *in situ* method for the preparation of injectable biodegradable poly (lactide-co-glycolide) microspheres for the controlled delivery of drugs .The drug release was retarded with increase in the molecular weight of the encapsulated drug. Substitution of triacetin by triethyl citrate and miglyol 812 by soybean oil resulted in variation in the release of the drug from the in situ formed microspheres [274].

The administration of a sparingly soluble drug is always problematic, especially when the drug has to be released from the degradable matrix of a polymeric drug delivery system. Attempts were made to achieve the complete release of 1-[2-(fluorobenzoyl) amino ethyl]-4-(7-methoxynaphtyl) piperazine, a potential anxiolytic and antidepressor hydrophobic compound, from racemic poly (lactic acid) (PLA50)-based microparticles, 100 percent release was required at a low rate in order to allow monthly repeated S.C. or I.M. injections of this potent compound [275].

Stable oil-in-water (o/w) MEs used as vehicles for dermal drug delivery have been developed using lidocaine and prilocaine in oil form (eutectic mixture). These MEs were able to solubilize up to 20 percent eutectic mixture of lidocaine and prilocaine without phase separation [276].

Park and Kim have evaluated the flurbiprofen-loaded ME for parenteral delivery. The MEs of flurbiprofen prepared with ethyl oleate and Tween 20 can be used as a parenteral drug carrier for this and other poorly water-soluble drugs, provided that physical stability can be properly addressed [277].

An ethyl laurate-based ME system with Tween 80 as surfactant, propylene glycol and ethanol as cosolvents was developed for intranasal delivery of diazepam. Diazepam, a practically water-insoluble drug, displayed a high solubility and a fairly rapid nasal absorption from ME within 2-3 min, and enhanced bioavailability of 50 percent when compared with intravenous injection. These results suggest that this ethyl laurate-based ME may be a useful approach for the rapid-onset delivery of diazepam during the emergency treatment of status epileptics [278].

A novel lecithin-based ME containing Amphotericin B (AmB) was developed to reduce the toxic effects of the drug, comparing it with the commercial formulation Fungizone1.Different studies are described in this work to prove the stability of these new dosage forms. Acute toxicity results, determined by a graphic method, the probit binary model and the Reed and Muench method showed that lethal dose 50 for AmB MEs was of 2.9 mg/kg compared to 1.4 mg/kg for the commercial deoxycholate suspension, Fungizone1. The overall results indicate that treatment with AmB MEs was less toxic than Fungizone1, suggesting a potential therapeutic application [279].

Eye drops are the most commonly used dosage form by ocular route, in spite of low bioavailability and the pulsed release of the drug. However, due to their intrinsic properties and specific structures, the MEs are a promising dosage form for the natural defense of the eye. The *in vivo* results and preliminary studies on healthy volunteers have shown a delayed effect and an increase in the bioavailability of the drug. The proposed mechanism is based on the adsorption of the nanodroplets representing the internal phase of the ME, which constitutes a reservoir of the drug on the cornea and should then limit their drainage [280].

Kantaria et al has studied the gelatin-containing, electrically conducting, rigid water-in-oil (w/o) ME-based organogels (MBG), both with and without the presence of a model drug, have been prepared using pharmaceutically acceptable oils and surfactants. ME formation was observed upon its combination with a variety of non-ionic surfactants. Upon addition of gelatin to the w/o ME, MBG could be formed when using AOT as stabilizer with most of the oils investigated [281].

Biodegradable carriers containing gentamicin for local treatment of bone infections were developed. The gentamicin sulphate loading of the microspheres, after a methylene chloride-water extraction procedure, exceeded 90% of the theoretical value. Experimental results showed that the biodegradable poly-L-lactic acid: PEGgentamicin delivery system had a potential for prophylaxis of post-operative infection [282].

Multilamellar vesicles called spherulites have recently been discovered and are being developed for encapsulation applications. It appeared that the dilution of a complex dispersion formulated with no external aqueous phase containing a hydrophilic surfactant provided the slowest release of encapsulated ions. Furthermore, this formulation maintained a difference of pH between the internal and external aqueous phases for a few hours. These new systems of Spherulites known as complex dispersions show great potential for pharmaceutical applications such as controlled release and protection of encapsulated substances [283].

Poly (D, L-lactic-co-glycolic acid)(PLGA) multiphase microspheres were prepared by the multiple emulsion solvent evaporation method using acetonitrile as the polymer solvent and mineral oil as the evaporation medium. This anhydrous emulsion system employing the solid suspension core and containing a dispersion of TNF-α (Tissue Necroting Factor α) enabled the encapsulation of this protein without loss of activity. It was concluded that the anhydrous emulsion system is a suitable approach to prepare multiple microspheres as an alternative to the W/O emulsion system, especially when solvent sensitive proteins are incorporated into the microspheres [284].

Paclitaxel is an anticancer agent with low aqueous solubility. More extensive clinical use of this drug is somewhat delayed due to lack of appropriate delivery vehicles. The average life span of ascitic-tumor-bearing mice was prolonged significantly by the treatment of paclitaxel-emulsion .The formulated emulsion is a promising carrier for paclitaxel and other lipophilic drugs [285].

Lipid vehicles and surface-active agents have been successfully used to increase oral absorption and availability of free and encapsulated proteins. In order to investigate if these vehicles could also enhance the serum IgG responses elicited after the oral administration of protein antigens, free bovine serum albumin was orally administered to mice in different vehicles: a 0.3 percent sodium bicarbonate aqueous solution, and ethyl oleate/0.3 percent sodium bicarbonate o/w emulsion or ethyl oleate containing the previously described surface active agents. The immune response elicited by the free antigen was enhanced by the use of these substances, especially when the free protein was administered as an oil suspension containing the surface-active agents [286].

Honeywell-Nguyen and Bouwstra reported the *in vitro* transport of pergolide from surfactant-based elastic vesicles through human skin. Several aspects of vesicular delivery were studied in order to elucidate the possible mechanisms of action and to establish the optimal conditions and drug candidates for usage with elastic vesicles. The findings show that there was a strong correlation between the drug incorporation to saturated levels and the drug transport, both of which were influenced by the pH of the drug-vesicular system [287].

8.9.6 Nanoparticles

Amphiphilic β cyclodextrin (β CD) modified on the primary face were synthesized, and evaluated of their potential as novel excipients in the preparation of nanocapsules. The results indicated that derivatives with six carbon aliphatic chains on the primary face proved to be the most efficient among the amphiphilic β CD in this study [288].

Miyazaki et al prepared poly n-butylcyanoacrylate (PNBCA) nanocapsules loaded with indomethacin and to evaluated the carrier system to deliver the drug systemically after its topical application. The presented data show that indomethacin loaded PNBCA nanocapsules can improve the transdermal delivery of indomethacin compared to a conventional gel formulation using Pluronic F-127. This might be due to their ultra fine particle size and their hydrophilic and hydrophobic surface characteristics [289].

Rolipram containing nanoparticles were prepared and tested as a potential drug carrier and targeting system for the treatment of inflammatory bowel disease. A more controlled release was obtained after 2 days of dissolution with the pressure homogenization-emulsification method, no further significant drug release was observed with the spontaneous emulsification solvent diffusion method [290].

Nanoparticles of griseofulvin, a model drug with poor solubility and low bioavailability, were prepared from water dilutable MEs by the solvent diffusion technique. Solvent-in-water ME formulations containing water, butyl lactate, lecithin, taurodeoxycholate sodium salt or dipotassium glycyrrhizinate (KG), 1,2-propanediol or ethanol were used. Nanoparticles size using KG, were below 100 nm with low polydispersity and an increased dissolution rate [291].

Araujo et al reported influence of the surfactant concentration on the body distribution of nanoparticles. The rapid uptake of nanoparticles (after i.v. inj.) especially by the liver reticuloendothelial system can be reduced and coating them with non-ionic surfactants can alter the body distribution. The nanoparticles were coated with poloxamine 908 and polysorbate 80, and the influence of different surfactant concentrations on the body distribution was investigated. The results indicate that the type of interaction and the strength of the adsorptive binding to the nanoparticles are different with different surfactants [292].

A novel non-competitive NMDA receptor antagonist MRZ 2/576 is a potent but rather short-acting (5-15 min) anticonvulsant following intravenous administration to mice as estimated by the prevention of maximal electroshock induced convulsions. This is most probably due to a rapid elimination of the drug from the central nervous system by transport processes that are sensitive to probenecid. Intravenous administration of the drug bound to poly (butylcyanoacrylate) nanoparticles coated with polysorbate 80 prolongs the duration of the anticonvulsive activity in mice up to 210 min and after probenecid pre-treatment up to 270 min compared to 150 min with probenecid and MRZ 2/576 alone. The results of this study demonstrate that polysorbate 80 coated poly (butylcyanoacrylate) nanoparticles used so far as a delivery system to the brain for drugs that do not freely penetrate the blood brain barrier can also be used as a parenteral controlled release system to prolong the CNS availability of drugs that have a short duration of action [293].

Calvo et al studied the long-circulating PEGylated polycyanoacrylate nanoparticles as new drug carrier for brain delivery. Biodistribution profiles and brain concentrations of [14C]-radio labeled PEG-PHDCA, polysorbate 80 or poloxamine 908-coated PHDCA nanoparticles, and uncoated PHDCA nanoparticles were determined by radioactivity counting after intravenous administration in mice and rats. This study highlights the requirements such as long-circulating properties of the carrier, and appropriate surface characteristics for targeting the drug to the brain [294].

The utility of nanoparticles of polyalkylcyanoacrylate as a targeted delivery system for nifurtimox against *Trypanosoma cruzi*, responsible for Chagas' disease was studied by Gonzalez-Martin. An emulsion was prepared by polymerization process. Ethylcyanoacrylate nanoparticles and formulations containing different concentrations of nifurtimox, polyethylcyanoacrylates and surfactants were investigated and analysed for size and drug content. It was concluded that the nanoparticles loaded with nifurtimox constitutes a good carrier of the drug against *T. cruzi*. The loaded-nanoparticles significantly increase trypanocidal activity [295].

The blood--brain barrier represents an insurmountable obstacle for a large number of drugs, including antibiotics, antineoplastic agents, and a variety of central nervous system active drugs, especially in neuropathies. One of the possibilities to overcome this barrier is a drug delivery to the brain using nanoparticles. The nanoparticles may be especially helpful for the treatment of the disseminated and very aggressive brain tumors The most likely mechanism is endocytosis by the endothelial cells lining the brain blood capillaries. Nanoparticles-mediated drug transport to the brain depends on the over coating of the particles with polysorbates, especially polysorbate 80. Over coating with these materials seems to lead to the adsorption of apolipoprotein E from blood plasma onto the nanoparticles surface. The particles then seem to mimic low-density lipoprotein (LDL) particles and could interact with the LDL receptor leading to their uptake by the endothelial cells. After this the drug may be released in these cells and diffuse into the brain interior or the particles may be transcytosed. Other processes such as tight junction modulation or P-glycoprotein inhibition also may occur. Moreover, these mechanisms may run in parallel or may be cooperative thus enabling a drug delivery to the brain [296].

Mitra and Lin have studied the effect of surfactant on fabrication and characterization of paclitaxel-loaded polybutylcyanoacrylate (PBCA) nanoparticulate delivery systems. The feasibility of applying biodegradable PBCA nanoparticulate delivery systems (NDSs) for the controlled release of paclitaxel was investigated. Paclitaxel-loaded and unloaded PBCA-NDSs containing various surfactants (dextran 70, cholesterol, polyvinyl alcohol and lecithin) were prepared by anionic polymerization. PBCA-NDSs were smaller in particle size, higher zeta potential and better drug entrapment efficiency, and better controlled release of paclitaxel [297].

Tobio have studied stealth PLA-PEG nanoparticles as protein carriers for nasal administration. A model protein antigen, tetanus toxoid (TT), within hydrophobic (PLA) and surface hydrophilic (PLA-PEG) nanoparticles and was evaluated for transport of proteins through the nasal mucosa. The transport of the radio labeled protein through the rat nasal mucosa was highly affected by the surface properties of the nanoparticles. Thus a novel nanoparticulate system has been developed with excellent characteristics for the transport of proteins through the nasal mucosa [298].

Barbault-Foucher et al has designed poly-epsilon-caprolactone nanospheres (PECNS) coated with bioadhesive hyaluronic acid (HA) for ocular delivery. The strategies to attach HA on PECNS surface were achieved by coating the core by chain entanglement with HA or coating PECNS by HA adsorption and also by coating PECNS by electrostatic interactions between negatively charged HA and a cationic surfactant (stearylamine or benzalkonium chloride). The HA is strongly attached on PECNS which is positively charged by cationic surfactant [299].

Prolonged circulation of anticancer agent in blood is expected to decrease the host toxicity and enhance the anticancer activity. The prolonged and sustained release of anticancer agent adriamycin using biodegradable poly (gamma-benzyl-L- glutamate)/poly (ethylene oxide) (PBLG/PEO) polymer nanoparticles was reported by Oh et al. PBLG/PEO polymer is a hydrophilic/hydrophobic block copolymer and forms a micelle-like structure in solution These results suggest usefulness of PBLG/PEO nanoparticles as a sustained and prolonged release carrier for adriamycin [300].

Recent studies have shown that drugs that are normally unable to cross the blood-brain barrier (BBB) following intravenous injection can be transported across this barrier by binding to poly (butyl cyanoacrylate) nanoparticles and coating with polysorbate 80. However, the mechanism of this transport so far was not known. Kreuter et al investigated the possible involvement of apolipoproteins in the transport of nanoparticles-bound drugs into the brain. Polysorbate 80-coated nanoparticles adsorb these apolipoproteins from the blood after injection and thus seem to mimic lipoprotein particles that could be taken up by the brain capillary endothelial cells via receptor-mediated endocytosis. Bound drugs then may be further transported into the brain by diffusion following release within the endothelial cells or, alternatively, by transcytosis [301].

8.9.7 Niosomes

Niosomes are vesicles formed by self-assembly of non-ionic surfactants. They are vesicular delivery systems, which are formed via aqueous dispersion of non-ionic surfactant films. Non-ionic surface active agents based discoidal vesicles (discomes) bearing timolol maleate were prepared. The prepared system could produce or sustain a suitable activity profile upon administration into the ocular cavity; however, systemic absorption was minimized to a negliable level. The discomes were found to be promising and of potential for controlled ocular administration of water-soluble drugs [302].

Pillai and Salim has studied the inhibition of platelet aggregation *in vitro* by niosome-encapsulated indomethacin .The objective was to study the effect of niosomal-encapsulated indomethacin on platelet function such as inhibition of aggregation and ATP release induced by a variety of agonists (adenosine 5'-diphosphate (ADP), epinephrine, arachidonic acid, ristocetine) and to explore the feasibility of carrier-mediated drug delivery to the platelets. At equimolar doses, the niosomal drug proved to be more efficient in inhibiting platelet aggregation than the free drug, probably due to greater quantity of the drug reaching the specific site of inhibition in the interior of the platelets and acting directly on the cyclo-oxygenase system to prevent thromboxane formation [303].

Arunothayanun et al has reported the extrusion of niosomes from capillaries resembles a pulsed delivery device. The expulsion of single or groups of intact polystyrene microspheres or tetradecyl-beta-D-maltoside niosomes with sizes smaller than the exit diameter can be achieved readily. The stepwise release profile of luteinizing hormone releasing hormone (LHRH) obtained after pulsatile expulsion of groups of niosomes entrapping LHRH indicates the feasibility of this system for pulsatile delivery of vesicles [304].

The feasibility to develop a per oral vaccine delivery system based on niosomes was evaluated using BALB/c mice. Ovalbumin was encapsulated in various lyophilized niosome preparations consisting of sucrose esters, cholesterol and dicetyl phosphate. Only encapsulation of ovalbumin into Wasag7 (sucrose esters) niosomes resulted in a significant increase in antibody titres [305].

Sihorkar and Vyas prepared and appended with a polysaccharide cap using hydrophobic anchors for oral drug delivery. Hydrophobized polysaccharides, O-palmitoyl pullulan and cholesteroyl pullulan were anchored on to propranolol HCL containing preformed niosomes. Furthermore, the exceptional shelf stability of the coated vesicles establishes the potential of polysaccharide-coated niosomes as an oral delivery system for water-soluble agents [306].

The niosomes were prepared with hydrated mixture of various non-ionic surfactants and cholesterol was studied. The entrapment efficiencies of the vesicles and micro viscosities of the vesicular membrane depended on alkyl chain length of non-ionic surfactants and amount of cholesterol used to prepare vesicles. Cholesterol was used to complete the hydrophobic moiety of single alkyl chain nonionic surfactants for vesicle formation. Niosome prepared with Tween 61 bearing a long alkyl chain and a large hydrophilic moiety in the combination with cholesterol at 1:1 molar ratio was found to have the highest entrapment efficiency of water-soluble substances [307].

Optimizations of drug delivery through human skin are important in modern therapy. Drug–vehicle interactions (drug or prodrug selection, chemical potential control, ion pairs, coacervates and eutectic systems) and the role of vesicles and particles like liposomes, transfersomes, ethosomes, niosomes play a significant role in drug delivery. One can modify the stratum corneum by hydration and chemical enhancers, or bypass or remove this tissue via micro needles, ablation and follicular delivery. Electrically assisted methods (ultrasound, iontophoresis, electroporation, magnetophoresis, photomechanical waves) show considerable promise. Of particular interest is the synergy between chemical enhancers, ultrasound, iontophoresis and electroporation [308].

The skin permeation and partitioning of a fluorinated quinolone antibacterial agent, enoxacin, in liposomes and niosomes, after topical application, were elucidated in the present study. The enhanced delivery across the skin of liposome and niosome encapsulated enoxacin had been observed after selecting the appropriate formulations. The optimized formulations could also reserve a large amount of enoxacin in the skin. The ability of liposomes and niosomes to modulate drug delivery without significant toxicity makes the two vesicles useful to formulate topical enoxacin [309].

A procedure is described for producing a dry product, which may be hydrated immediately before use to yield aqueous niosome dispersions similar to those produced by more cumbersome conventional methods. Proniosomes minimize problems of niosome physical stability such as aggregation, fusion and leaking, and provide additional convenience in transportation, distribution, storage, and dosing. In all comparisons, proniosomes-derived niosomes are as good or better than conventional niosomes [310].

To prepare niosomes that have high encapsulation capacity for soluble drugs, starting from Span 60 and cholesterol, an improved method, evaporation-sonication method, was proposed. To obtain the highest encapsulation efficiency, several factors including the structure of surfactant, level of lipid, content of drug and cholesterol were investigated and optimized. The results indicate that the Span 60 is the most ideal surfactant among four kinds of Span. Furthermore, the release studies of colchicine and 5-fluorouracil *in vitro* from niosomes exhibited a prolonged release profile as studied over a period of 24 h. The results demonstrated that niosomes prepared in this way not only have high encapsulation capacity but also is expected that side effects of drugs may be reduced [311].

Nebulization of niosomal all-trans-retinoic acid (ATRA) is an inexpensive niosomes delivering system for inhaled aerosol was designed. Niosomes may provide a means to reduce the toxicity of ATRA and pharmacokinetics in a manner similar to liposomes. In addition, the low cost of the surfactants used for niosomes and their greater stability compared with liposomes makes them an attractive alternative. The results are very encouraging and alternative approach to the respiratory delivery of ATRA by aerosolization [312].

8.9.8 Pluronic and polymeric micelles

Amphiphilic block copolymers (ABCs) have been used extensively in pharmaceutical applications ranging from sustained-release technologies to gene delivery. The utility of ABCs for delivery of therapeutic agents results from their unique chemical composition, which is characterized by a hydrophilic block that is chemically tethered to a hydrophobic block. Although the Pluronics, composed of poly (ethylene oxide)- blockpoly (propylene oxide)-block-poly (ethylene oxide), are the most widely studied ABC system, copolymers containing poly (L-amino acid) and poly (ester) hydrophobic blocks have also shown great promise in delivery applications. ABCs have been used for numerous pharmaceutical applications including drug solubilization/stabilization, alteration of the pharmacokinetic profile of encapsulated substances, and suppression of multidrug resistance [313].

Amphiphilic graft copolymers comprising monomeric units of methoxy PEG (mPEG)-acrylate, 2- hydroxyethyl methacrylate (HEMA)–cholesterol conjugates and HEMA were synthesized and their properties characterized. The loading capacity of pyrene decreases with increasing CMC [314].

A novel approach in the field of polymeric drug delivery systems was introduced by the formation of polymeric micelles and subsequently by functionalized polymeric micelles functionalized polymeric micelles are expected to find a wide application in the fields of drug delivery and diagnosis since the possibility of coupling to bioactive substances is provided. A large number of densely packed functional groups on the outer shell of the micelle allow immobilization of biologically active substances at a high concentration. This is a great advantage for utilizing this particular type of nanosphere in the biomedical field [315].

The Methoxy PEG/polylactic acid block copolymer with a polymerizable methacryloyl end was prepared by anionic ring-opening polymerization followed by end capping with methacrylic anhydride. A stable core shell type, spherical micelle with a number-averaged diameter of ca. 30nm was obtained by a dialysis method, and the resulting aqueous micelle solution was core-polymerized chemically and photo chemically to produce more stable nanoparticles, which was evidenced by spectroscopic and light-scattering techniques. Taxol-incorporated micelles were prepared to entrap Taxol into block copolymer micelles by the O/W emulsion method [316].

To estimate the feasibility of novel containers for drugs, poly- (ethylene oxide) poly (,-benzyl L-aspartate) micelles were prepared by dialysis against water using different solvents. The solvent selected is very important because it drastically affects the stability of polymeric micelles. The CMC of the prepared micelles in distilled water was determined by a fluorescence probe technique using pyrene. Indomethacin (IMC) as a model drug was incorporated into the micelles by dialysis and an oil/water emulsion method. The release rate of IMC from the micelles was increased by increasing the pH of the medium and indicated that the release rate of IMC from the micelles are controlled by the partition coefficient of IMC based on the pH of the medium and interaction between IMC and the hydrophobic portion of the micelles [317].

Amphiphilic AB-type diblock copolymers composed of hydrophobic poly (lactide) (PLA) segments and hydrophilic poly (glycolic acid lysine) segments with amino side-chain groups self-associated to form PLA-based polymeric micelles with amino surfaces in an aqueous solution were prepared. The PLA-based polymeric micelle as bioabsorbable vehicles for hydrophobic drugs, the entrapment of drugs micelles and their release with doxorubicin as a hydrophobic drug was studied [318].

Pluronic block copolymers are found to be an efficient drug delivery system with multiple effects. The micelles formed by incorporation of drugs results in increased solubility, metabolic stability and circulation time. The interactions of the pluronic unimers with multidrug-resistant cancer cells result in sensitization of these cells with respect to various anticancer agents. Furthermore, the single molecular chains of copolymer, unimers, inhibit drug efflux transporters in both the blood–brain barrier and in the small intestine, which provides for the enhanced transport select drugs to the brain and increases oral bioavailability. These and other applications of Pluronic block copolymers in various drug delivery and gene delivery systems are considered [319, 320].

Marin et al studied drug delivery in pluronic micelles and the effect of high-frequency ultrasound on drug release from micelles and intracellular uptake. The effect of high-frequency ultrasound on doxorubicin (DOX) release from pluronic micelles and intracellular DOX uptake was studied for promyelocytic leukemia HL-60 cells, ovarian carcinoma drug-sensitive and multidrug-resistant cells and breast cancer MCF-7 cells. Ultrasound significantly enhanced the intracellular DOX uptake from Pluronic micelles [321].

In vitro and *in vivo* evaluation of pluronic F127-based timolol maleate [TM] and PF127 containing 3 percent methylcellulose ocular delivery systems and their ocular bioavailability were studied by El-Kamel. *In vivo* study showed that 2.5 and 2.4 fold increased the ocular bioavailability of TM, measured in albino rabbits, respectively, compared with 0.5% TM aqueous solution [322].

A new class of pluronic based injectable controlled release depots was developed and evaluated for the protein release kinetics by DesNoyer et al. These consist of blends of poly (ethylene oxide) (PEO)/poly (propylene oxide) (PPO)/poly (ethylene oxide) (PEO) triblock copolymers (Pluronics) with poly (D, L-lactide) (PDLA)/1-methyl-2-pyrrolidinone (NMP) solutions. Increasing the pluronic concentration beyond a critical point resulted in a transition from a burst-type profile to an extended-release profile [323].

Nifedipine (N) and nifedipine pluronic F-68 solid dispersion (SD) pellets were developed and characterized for drug release mechanisms from a multi-unit erosion matrix system for controlled release. Due to increased solubility of N in SD, the drug release mechanism from the multi-unit erosion matrix changed from pure surface erosion to an erosion/diffusion mechanism, thereby altering the release rate and kinetics [324].

Altered organ accumulation of oligonucleotides using polyethyleneimine (PEI) grafted with poly (ethylene oxide) or pluronic as carriers was studied by Ochietti et al. PEI is a potent non-viral system that has been known to deliver efficiently both plasmids and oligo deoxy nucleotides (ODNs) *in vitro*. However, *in vivo* systemic administration of DNA/PEI complexes has encountered significant difficulties because these complexes are toxic and have low biodistribution in target tissues. Evaluation of PEI grafted with poly (ethylene oxide) (PEO (8K)-g-PEI (2K)) and PEI grafted with non-ionic amphiphilic block copolymer, pluronic P85 (P85-g-PEI (2K)) as carriers for systemic delivery of ODNs was carried out. The results of this study suggest that formulating ODN with PEO (8K)-g-PEI (2K) and P85-g-PEI (2K) carriers allows targeting of the ODN to the liver or kidneys, respectively. The variation in the tissue distribution of ODN observed with the two carriers is probably due to the different hydrophilic-lipophilic balance of the polyether chains grafted to PEI in these molecules. Therefore, polyether-grafted PEI carriers provide a simple way to enhance ODN accumulation in a desired compartment without the need of a specific targeting moiety [325].

Pluronic block copolymers for overcoming drug resistance in cancer was reported by Kabnov et al. Pluronic block copolymers have been used extensively in a variety of pharmaceutical formulations including delivery of low molecular mass drugs and polypeptides. Novel applications of Pluronic block copolymers in the treatment of drug-resistant tumors discovered. Pluronic block copolymers interact with multidrug-resistant cancer (MDR) tumors resulting in drastic sensitization of these tumors with respect to various anticancer agents, particularly, anthracycline antibiotics [326].

Effects of pluronic block copolymers on drug absorption in Caco-2 cell monolayers suggest that pluronic P85 unimers increase accumulation of a P glycoprotein (Pgp) dependent drug in Caco-2 monolayers through inhibition of the Pgp efflux system. The mechanism of the micelle effect is not known, however, it is very similar to the micelle effects involving vesicular transport of the micelle-incorporated drug. The study suggests that Pluronic copolymers can be useful in increasing oral absorption of select drugs [327].

Pluronic block copolymers are recognized pharmaceutical excipients listed in the USP and British Pharmacopoeia. They have been used extensively in a variety of pharmaceutical formulations including delivery of low molecular mass drugs and polypeptides. The applications of pluronic block copolymers in gene therapy. In particular, these molecules can modify the biological response during gene therapy in the skeletal muscle, resulting in an enhancement of the transgenic expression as well as an enhancement of the therapeutic effect of the transgene. Furthermore, pluronic block copolymers are versatile molecules that can be used as structural elements of the polycationic-based gene delivery systems (polyplexes) [328].

Kataoka etal reported the formulation of doxorubicin (DOX)-loaded poly (ethylene glycol)-poly (beta-benzyl-L-aspartate) copolymer micelles (PEG-PBLA) and evaluated their pharmaceutical characteristics and biological significance. DOX loaded in the micelle showed a considerably higher antitumor activity compared to free DOX against mouse C26 tumor by i.v. injection, indicating a promising feature for PEG-PBLA micelle as a long-circulating carrier system [329].

Inhalation is gaining increasing acceptance as a convenient, reproducible, and non-invasive method of drug delivery to the lung tissue and/or the systemic circulation. However, sustained drug release following inhalation remains elusive, due in part to the lack of appropriate materials designed specifically for use in the lungs to control the release of bioactive compounds. Fu etal synthesized a new family of ether-anhydride copolymers composed entirely of FDA-approved monomers, including PEG. Sebacic acid, a hydrophobic monomer, was co polymerized with PEG in order to produce water-insoluble polymers capable of providing continuous drug release kinetics following immersion in an aqueous environment. The preparation of large porous particles with these new polymers was systematically approached, utilizing central composite design, to develop improved particle physical properties for deep lung delivery. Microparticles containing model drugs were made with sizes suitable for deposition in various regions of the lung following inhalation as a dry powder. These new systems may also find application as "stealth" carriers for therapeutic compounds following intravenous injection [330].

A new drug delivery modality was developed based on drug encapsulation in polymeric micelles followed by a controlled release at the tumor site triggered by ultrasound focused on the tumor. Ultrasound not only released drug from micelles but also enhanced the local uptake of both free and encapsulated drug by tumor cells, thus providing effective drug targeting [331].

The physicochemical properties of polyion complex (PIC) micelles formed from antisense-oligodeoxynucleotides (antisense-ODN) and poly (ethylene glycol)-poly (L-lysine) block copolymers (PEG-PLL) were investigated to utilize them as a novel formulation for antisense-ODN delivery. The stability of antisense-ODN against deoxyribonuclease I (DNase I) attack was evaluated using capillary gel electrophoresis, revealing that the complexation of antisense-ODN with PEG-PLL effectively prohibited Dnase I attack. These characteristics of the PIC micelle system highlight its promising feature as ODN carrier used in the field of targeting therapy [332].

Solutions of surface-active triblock copolymer pluronic F127 in the vicinity of the CMC were prepared with or without pilocarpine (either as the hydrochloride salt or the free base) in water and phosphate buffer. Additionally, it was found that the pilocarpine solutions without F127 in water exhibit a certain surface activity. The best results were obtained for the micellar pilocarpine base solution, which exhibits significant prolongation of miotic activity and an increase of area under curve [333].

Polymeric micelles have recently emerged as a novel promising colloidal carrier for the targeting of poorly water-soluble and amphiphilic drugs. Polymeric micelles are considerably more stable than surfactant micelles and can solubilize substantial amounts of hydrophobic compounds in their inner core. Due to their hydrophilic shell and small size they sometimes exhibit prolonged circulation times *in vivo* and can accumulate in tumoral tissues. Potential medical applications, especially in cancer chemotherapy, are described and discussed [334].

Vancomycin is a glycopeptide antibiotic that is the drug of choice for the treatment of Gram-positive infections caused by methicillin resistant *Staphylococcus aureus*. It is also used in the treatment of bacterial infections in patients allergic to β lactum antibiotics. The facile reaction of vancomycin with various PEG linkers, at the Vancomycin 3 position, has been selectively accomplished by using an excess of base in DMF. All PEG-vancomycin transport forms show significant antibacterial activity that is on the same order of native vancomycin. Significant increases in the AUC were observed for all PEG-vancomycin conjugates thus making them potential single dose therapies [335].

8.9.9 Protein delivery systems

A class of new biodegradable hydrogels based on poly (ethylene glycol) methacrylate-graft-poly (glutamic acid) and poly (ethylene glycol) dimethacrylate was synthesized by photo-induced polymerization. Because all the polymeric constituents were highly hydrophilic, cross-linking could be performed in aqueous solutions. This type of cross-linked hydrogel was prepared by modifying a select number of acidic side-groups on poly (glutamic acid) with poly (ethylene glycol) methacrylate. Drug release rates from these hydrogels were found to be proportional to the protein molecular weight and the cross linker density, increasing at lower protein molecular weight or cross linker density. The biodegradable hydrogels could be an attractive avenue for drug delivery applications. The specific photo induced cross linking chemistry used would permit hydrogels to be synthesized in existence of the entrapped macromolecular drugs including peptides, proteins, and cells. In addition, the rapid feature of this polymerization procedure along with the ability to perform hydrogel synthesis and drug loading in an aqueous environment would offer great advantages in retaining drug activity during hydrogel synthesis [336].

Hyaluronic acids grafted with poly (ethylene glycol) were synthesized. The materials characterization, enzymatic degradability and peptide (insulin) release from solutions of the copolymers were examined. Distribution of bioactive peptides within the polymer chain is well known for combinations of PEG and polysaccharides as aqueous polymer two-phase systems. Such a heterogeneous-structured polymeric solution may be advantageous as an injectable therapeutic formulation for ophthalmic or arthritis treatment [337].

Formulations for natural and peptide nucleic acid based on polymeric sub micron particles by Cortesi et al. The cationic sub micron particles containing different surfactants like di iso butyl phenoxy ethyl- dimaethy benzyl ammonium bromide, di octodecyl di methyl ammonium bromides were used for DNA/DNA and DNA/Peptide nucleic acids were studied [338].

Modeling studies with beta-endorphin have clearly demonstrated that an amphiphilic secondary structural segment is a salient feature of the biologically active conformation of this 31-residue opioid peptide hormone. The synthesis of peptide models using unnatural building blocks by designing a beta-endorphin analogue (peptide 6) in which the hydrophilic linker region between the NH2-terminal enkephalin (residues 1-5) and the COOH-terminal helix was carried out. These studies clearly illustrate that one can use unusual building blocks to construct structural regions of synthetic analogues and still preserve the biological activity of peptide hormones [339].

A new microparticulate delivery system composed of a stabilizing gelatin/poloxamer microcore surrounded by a PLGA coat was designed to improve the stability of tetanus toxoid (TT) encapsulated in PLGA microspheres. The administration of a mixture of encapsulated and adsorbed TT led to significant higher and more prolonged neutralizing antibodies levels than those measured for the adsorbed toxoid [340].

Viscoelastic properties of whey protein isolate-stabilized emulsions have been investigated by determining storage and loss moduli of both fresh emulsions and heat-set emulsion gels. Gel strength increases with the increase of protein concentration in the system. The flocculated protein-covered oil droplets behave as active fillers and hence dramatically enhance the gel strength. The presence of water-soluble surfactant Tween 20 induces a dramatic reduction in emulsion gel strength, which is attributable to protein displacement from the oil-water interface. Oil droplets that are fully covered with Tween 20 do not adhere to protein gel matrix and do not contribute to gel strength. The presence of oil-soluble monopalmitin increases the viscous character of fresh emulsions and substantially reduces the modulus of heat-set emulsion gels [341].

Lipoamino acid and liposaccharide conjugates of somatostatin analogue TT-232 were synthesized to modify the physicochemical properties of the parent peptide. The relative position, the number, and the nature of the lipid and/or saccharide moieties were varied. Experiments *in vitro* clearly showed that many compounds modified at the N- and/or C-terminus with lipid or sugar moieties retained the biological activity of the parent compound. An interesting construct was synthesized containing lipid and sugar units at opposite ends of the somatostatin analogue, so that the entire molecule could be considered as an amphipathic surfactant [342].

Torres-Lugo etal has reported physicochemical behavior and cytotoxic effects of p (methacrylic acid-g-ethylene glycol) (P (MAA-g-EG)) nanospheres for oral delivery of proteins. The challenges faced to orally deliver therapeutic agents with unfavorable physicochemical properties, such as proteins, have been the primary motivation for the design and development of novel oral delivery systems that could circumvent biological barriers. The examination of the physicochemical interactions of the P (MAA-g-EG) nanosphere system with Caco-2 cell monolayers revealed that these systems possessed low cytotoxicity and were capable of opening the tight junctions between epithelial cells, therefore significantly reducing the transepithelial electrical resistance [343].

Qiu etal has prepared a hydrogel by in situ cross-linking of a thiol-containing poly (ethylene glycol)-based copolymer for protein drug delivery. The new poly (ethylene glycol)-based copolymer containing multiple thiol (-SH) groups was cross-linked in situ to form a polymer hydrogel under mild conditions. No organic solvent, elevated temperature, or harsh pH is required in the formulation or patient administration processes, making it particularly useful for delivery of fragile therapeutics, such as proteins [344].

Nonionic triblock copolymers are relatively nontoxic adjuvants that induce high-titer, long-lasting antibody responses. These facilitate the delivery of exogenous proteins into the MHC class I and class II processing pathways. These copolymers are thought to modulate hydrophobic adhesive interactions between antigens (Ag) and lymphoid cells. However, copolymers did not enhance binding of peptides to the MHC molecules on APC, presentation of endogenously synthesized Ag, or presentation of exogenous Ag delivered by electroporation. These results provide additional evidence that these nonionic triblock copolymers can serve as powerful adjuvants for augmenting both humeral and cell-mediated immunity to protein Ag [345].

Mandal and Bostanian carried out the effect of varying drug load and concentration of a surfactant, SLS on the release characteristics of a model peptide bovine serum albumin. The investigation provided a mechanistic basis toward the development of a hydrogel formulation by altering the concentration of two fundamental components, i.e., drug and surfactant, within the delivery system [346].

Bouillot etal has studied the protein encapsulation in biodegradable amphiphilic microspheres. MPOE-PLA microspheres containing bovine serum albumin (BSA) were prepared by the double emulsion method with high encapsulation efficiency. This coating improves the performance of the release system compared with PLA microspheres. Studies of the diffusion of 1% rhodamine aqueous solution into the microspheres by means of confocal microscopy showed a fast diffusion of water through the matrices containing high molecular weight MPOE chains and could explain the fast release of BSA from these microspheres [347].

Li etal has studied polyethylene glycol-coated liposomes for oral delivery of recombinant human epidermal growth factor. PEG-coated liposomes containing rhEGF was prepared and evaluated for their stability and permeability in Caco-2 cells. The gastric ulcer healing effect was significantly increased in DPPC liposome compared with PC liposome and the solution. The enhanced curative ratio of rhEGF encapsulated into dipalmitoyl PC liposome may be due to the resistance to enzyme degradation, higher permeability and increased plasma AUC. Therefore, PEG-coated liposomes containing rhEGF could be used as an oral delivery formulation with enhanced encapsulation efficiency [348].

Several tripartate releasable PEG linkers (rPEG) that can provide anchimeric assistance to hydrolysis (cyclization prodrugs) were prepared and, after conjugation to lysozyme demonstrated rapid cleavage in rat plasma compared to nonassisted, permanently bound PEG. Varying the chemical structure and adding steric hindrance can adjust the half-life of the protein conjugates from slow to very fast. The pharmacokinetics in mice was also determined for rPEG-Interleukin 2 conjugates *in vivo* using an ELISA assay. The employment of releasable PEG polymers substantially broadens the applications of PEGylation drug delivery technology by introducing the benefits of controlled release of native protein therapeutics [349].

To reduce the injection frequency and toxicity of intravenously administered protein drugs, it is necessary to develop safe and sustained injectable delivery systems. Kim et al evaluated liposomes as safe and sustained injectable delivery systems of proteins. The liposomes coated with PEG showed 3-fold higher efficiency of insulin incorporation than did the liposomes without PEG. Moreover, among the liposomes coated with PEG, dipalmitoylphosphocholine liposomes showed higher incorporation efficiency than did dimyristoylphosphocholine liposomes. These results indicate that PEG-coated liposomes could be developed as a relatively safe and sustained injectable delivery system for insulin with improved incorporation efficiency [350].

Adjuvants aimed at increasing the immunogenicity of recombinant antigens remain a focus in vaccine development. Worldwide, there is currently considerable care for the development of biodegradable microspheres as controlled release of vaccines, since the major disadvantage of several currently available vaccines is the need for repeated administration. Microspheres prepared from the biodegradable and biocompatible polymers, the polylactide or polylactide-co-glycolide have been shown to be effective adjuvants for a number of antigens. The polylactide-co-poly (ethylene glycol) microspheres have shown great potential as a next generation adjuvant to replace or complement existing aluminum salts for vaccine potential [351].

8.9.10 Self emulsifying drug delivery systems

The oral delivery of hydrophobic drugs presents a major challenge because of the low aqueous solubility of such compounds. Self-emulsifying drug delivery systems (SEDDS), which are isotropic mixtures of oils, surfactants, solvents and co-solvents/surfactants, can be used for the design of formulations in order to improve the oral absorption of highly lipophilic drug compounds. SEDDS can be orally administered in soft or hard gelatin capsules and form fine relatively stable o/w emulsions upon aqueous dilution owing to the gentle agitation of the gastrointestinal fluids. The efficiency of oral absorption of the drug compound from the SEDDS depends on much formulation related parameters, such as surfactant concentration, oil/surfactant ratio, and polarity of the emulsion, droplet size and charge, all of which in essence determine the self-emulsification ability. Thus, only very specific pharmaceutical excipients combinations will lead to efficient self-emulsifying systems. Significant improvement in the oral bioavailability of these drug compounds has been demonstrated for each case. The fact that almost 40% of the new drug compounds are hydrophobic in nature implies that studies with SEDDS will continue, and more drug compounds formulated as SEDDS will reach the pharmaceutical market in the future [352].

Traditionally intestinal lymphatic delivery has been expressed as a percentage of the dose transported in the lymph. Using this parameter results obtained to date, with lipid-based vehicles, are somewhat disappointing maximizing at approximately 20–30%, for highly lipophilic compounds including DDT and halofantrine (Hf). Recent data, monitoring Hf, in a fed versus fasted dog study, have shown that a higher degree of lymphatic transport is possible (50% dose) in the postprandial state, this study should result in stimulating renewed interest in the potential of achieving significant levels of lymphatic targeting lipid-based formulations like seeds for intestinal lymphatic delivery [353].

Lipid formulations for oral administration of drugs generally consist of a drug dissolved in a blend of two or more excipients, which may be triglycerides oils, partial glyceride, surfactants or co-surfactants. The primary mechanism of action, which leads to improved bioavailability is usually avoidance, or partial avoidance, of the slow dissolution process which limits the bioavailability of hydrophobic drugs from solid dosage forms. Recent years several successful oral pharmaceutical products have been marketed as lipid systems, notably cyclosporin and the two HIV protease inhibitors, ritonavir and aquinavir. Consequently, there is now considerable interest in the potential of lipid formulations for oral administration [354].

The self-emulsifying formulations of coenzyme Q_{10} (Co Q_{10}) were prepared using two oils (Myvacet 9-45 and Captex-200), two emulsifiers (Labrafac CM-10 and Labrasol) and a cosurfactant (lauroglycol). In all the formulations, the level of Co Q_{10} was fixed at 5.66% w/w of the vehicle. A two-fold increase in the bioavailability was observed for the self-emulsifying system compared to a powder formulation. SEDDS have improved the bioavailability of CoQ_{10} significantly. The data suggest the potential use of SEDDS to provide an efficient way of improving oral absorption of lipophilic drugs [355].

Krishna and Sheth investigated a novel self-emulsifying parenteral drug delivery system. The application of three polyhydroxy alcohols for improving parenteral emulsion formulations was investigated. It was found that anhydrous mixtures of oil, surfactants and 30% or higher concentration of glycerol formed self-emulsifying isotropic liquids, suitable for preparing parenteral SEEDS [356].

Gershanik and Benita reviewed the self-dispersing lipid formulations for improving oral absorption of lipophilic drugs recently. Numerous bioavailability studies carried out in animals and humans, reviewed suggest that hydrophobic drugs are better absorbed when administered in self-dispersing lipid formulations [357].

Kim et al prepared and studied *in vitro* evaluation of self-micro emulsifying drug delivery systems containing idebenone. A new self-micro emulsifying drug delivery system (SMEDDS) was developed to increase the dissolution rate, solubility, and, ultimately, bioavailability of a poorly water-soluble drug, idebenone. The developed SMEDDS formulation can be used as a possible alternative to traditional oral formulations of idebenone to improve its bioavailability [358].

Paclitaxel (Taxol) is one of the best antineoplastic drugs found from nature in the past decades. Like many other anticancer drugs, there are difficulties in its clinical administration due to its poor solubility. Therefore an adjuvant called Cremophor EL has to be employed, but this has been found to cause serious side effects. However, nanoparticles of biodegradable polymers can provide an ideal solution to the adjuvant problem and realize a controlled and targeted delivery of the drug with better efficacy and fewer side effects. A novel formulation for fabrication of nanoparticles of biodegradable polymers containing d-α-tocopheryl PEG1000 succinate (vitamin E TPGS or TPGS) was evaluated to replace the current method of clinical administration and, with further modification, to provide an innovative solution for oral chemotherapy [359]

8.10. Miscellaneous

Eeckman et al developed a novel approach of controlled drug delivery using thermo sensitive polymers. The drug release occurs at physiological temperature, at which the polymer is normally not soluble, and no medium temperature changes are required to bring about the delivery. Some anionic surfactants and poly (*N*-isopropylacrylamide) (PNIPAAm) are used in order to modify the dissolution properties of PNIPAAm and of a copolymer with *N*-vinyl-acetamide so as to induce the release of a drug contained in compression coated tablets [360].

Pulmonary surfactant is a complex mixture of lipids and several specific surfactant proteins, which together render it with unique spreading properties and a dynamic surface tension behavior. These characteristics are heralded as ideal for a carrier of choice to instil therapeutic agents into the lung, because this combination enables high local therapeutic levels while minimizing systemic side effects of the instilled agent. This review outlines the rationale to use exogenous surfactant in lung injury, including opening-up inaccessible regions of the lung to other therapeutic agents. Especially the combination of anti-microbial agents and surfactant offers an alternative for critically ill patients with pneumonia. Some caution is also indicated in combining surfactant with antibiotics without proper evaluation of possible interactions [361].

In an attempt to increase cutaneous drug delivery, ME vehicles have been more and more frequently employed over recent years. ME formulations have been shown to be superior for both transdermal and dermal delivery of particularly lipophilic compounds, but also hydrophilic compounds appear to benefit from application in MEs compared to conventional vehicles, like hydrogels, emulsions and liposomes. The favorable drug delivery properties of MEs appear to mainly be attributed to the excellent solubility properties. However, the vehicles may also act as penetration enhancers depending on the oil / surfactant constituents, which involves a risk of inducing local irritancy. The correlation between ME structure composition and drug delivery potential is not yet fully elucidated. However, a few studies have indicated that the internal structure of MEs should allow free diffusion of the drug to optimize cutaneous delivery from these vehicles [362].

Lorazepam is an anxiolytic, antidepressant agent, having suitable feature for transdermal delivery. The percutaneous permeation of lorazepam was investigated in rat skin after application of water: PEG (50:50%v/v). The enhancing effects of various surfactants (SLS, CTAB, Benzalkonium chloride or Tween 80) with different concentrations on the permeation of lorazepam were evaluated using Franz diffusion cells fitted with rat skins. The permeation profile of lorazepam in presence of the cationic surfactant, CTAB, reveals that an increase in the concentration of CTAB results in an increase in the flux of lorazepam in comparison with the control [363].

Large PC unilamellar vesicles appear to be suitable controlled and protective delivery systems of β-galactosidase. Kinetic measurements carried out on intact loaded liposomes show that most of the enzyme is entrapped inside the liposomes and its activity is latent. Nevertheless, intact liposomes also show significant activity, which can be controlled by addition of detergent. At sublytic detergent concentrations, liposome enzymatic activity reaches values two or three times greater than those of intact liposomes. This increase seems to be due to membrane structure modification that also enhances the substrate permeability across the bilayer [364].

Sorbitan monostearate, a hydrophobic nonionic surfactant, gels a number of organic solvents such as hexadecane, isopropyl myristate, and a range of vegetable oils. Gelation is achieved by dissolving/dispersing the organogelator in hot solvent to produce an organic solution/dispersion, which, on cooling sets to the gel state. Polysorbate 20 improves gel stability and alters the gel microstructure from a network of individual tubules to star-shaped "clusters" of tubules. Another solid monoester in the sorbitan ester family, sorbitan monopalmitate, also gels organic solvents to give opaque, thermoreversible semisolids. Like sorbitan monostearate gels, the microstructure of the palmitate gels comprises an interconnected network of rodlike tubules. Unlike the stearate gels, however, the addition of small amounts of a polysorbate monoester causes a large increase in tubular length instead of the "clustering effect" seen in stearate gels. The sorbitan stearate and palmitate organogels may have potential applications as delivery vehicles for drugs and antigens [365].

REFERENCES

[1] S. Schreier, S.V.P. Malheiros and E. de Paula, Biochim. Biophys. Acta, 1508 (2000) 210.

[2] M.C. Carey, D.M. Small, Arch.Intern.Med.,130(1972)506.

[3] A.T. Florence, I.G. Tuker and K.A Walters., Interaction of non-ionic alkyl and aryl ethers with membranes and other biological systems. In: Rosen, M.J. (Ed.), Structure Performance Relationships in Surfactants, ACS Symposium Series, 253, 1994 pp. 189-207.

[4] A. Lopez, F.Llinares, C. Cortell and M. Herraez, Int. J.Pharm., 202(2000)133.

[5] Z.T., Chowan and R.J. Pritchard, J. Pharm. Sci., 67(1978) 1272.

[6] B.J. Aungst, N.J. Rogers and E. Shefter, Int. J. Pharm., 33(1986) 225.

[7] L.D. Rhein, C.R. Robbins, K. Fernee and R. Cantore, J. Soc. Cosmet., 37(1986)125.

[8] P.L. Honeywell-Nguyen and J.A. Bouwstra, J. Control Release, 86(2003) 145.

[9] P.L. Honeywell-Nguyen, S. Arenja and J.A. Bouwstra, Pharm. Res., 20(2003) 1619.

[10] S. Sang-Chul and L. Hyun-Jin, Eur. J. Pharma.Biopharma., 54(2002)325.

[11] M.R. Gudipati and S. A. Stavchansky, Int. J. Pharm., 118(1995) 41.

[12] N.H. Gabboun, N.M. Najib, H.G. Ibrahim and S. Assaf, Int. J. Pharm., 212(2001) 73.

[13] A. Nokhodchi, J. Shokri, A. Dashbolaghi, D. Hassan-Zadeh, T. Ghafourian and M. Barzegar-Jalali, Int. J. Pharm., 250(2003) 359.

[14] D. Paolino, C.A. Ventura, S. Nistico, G. Puglisi and M. Fresta, Int. J. Pharm., 244(2002) 21.

[15] M.M. Jimenez, J. Pelletier, M.F. Bobin and M.C. Martini, Int. J. Pharm., 272(2004) 45.

[16] M.J. King, D. Michel and M. Foldvari, J. Pharm. Pharmacol., 55(2003)1339.

[17] E.A. Essa, M.C. Bonner and B.W. Barry, J. Control Release, 95(2000) 535.

[18] G. Cevc, Adv. Drug Deliv. Rev., 56(2004) 675.

[19] E. Touitou, Compositions for applying active substances to or through the skin. US patent 5, 540, 934, (1996).

[20] E. Touitou, Composition for applying active substances to or through the skin. US patent 5, 716, 638 (1998).

[21] B. Godin, M. Alcabez and E. Touitou, Acta Technol. Legis. Medic., 10(1999)107.

[22] G. Worral, Can. Fam. Physician, 37(1991) 92.

[23] E. Touitou, N. Dayan, L. Bergelson, B. Godin and M. Eliaz, J. Control Release, 65(2000) 403.

[24] N. Dayan and E. Touitou, Biomaterials, 21(2000) 1879.

[25] I. Dkeidek and E. Touitou, AAPS J. Pharm. Sci., 1(1999) 202.

[26] E. Touitou, B. Godin, and C. Weiss, Drug Development Research, 50(2000) 406.

[27] M. Lodzki, B. Godin, L. Rakou, R. Mechoulam, R. Gallily and E. Touitou, J. Control Release, 93(2003) 377.

[28] B. Godin and E. Touitou, J. Control Release, 94(2004) 365.

[29] B. Godin and E. Touitou, Crit. Rev. Ther. Drug Carrier Syst., 20(2003) 63.

[30] S. Lee, I. J. Danie, M.S. McAuliffe, N. Kollias, T. I. Flotte and A.G. Doukas, Lasers in Surgery and Medicine, 29(2001) 145.

[31] C. Valenta, J. Cladera, P. Oshea and J. Hadgraft, J. Pharm. Sci., 90(2001) 485.

[32] A. Tezel, A. Sens, J. Tuchscherer and S. Mitragotri1, J. Pharm. Sci., 91(2002) 91.

[33] H. Gabiga, K. Cal and S. Janicki, Int. J. Pharm., 199(2000) 1.

[34] K.A. Walters, A.T. Florence and P.H. Dugard, , Int.J..Pharm., 10 (1982) 153.

[35] Yun-Seok Rhee, Jung-Gyo Choi, Eun-Seok Park, Sang-Cheol Chi, Int.J. Pharm., 228(2001) 161.

[36] Donatella Paolino a, Cinzia Anna Ventura b, Steven Nistico` c, Giovanni Puglisi a, Massimo Fresta, , Int.J..Pharm., 244(2002) 21.

[37] M.P. Krafft, Adv. Drug Del. Rev., 47(2001) 209.

[38] H. Schreier and J. Bouwstra, J. Control. Release, 30(1994) 1.

[39] A.A. Al-Saden, A.T. Florence and T.L. Whateley, Int. J. Pharm., 5(1980) 317.

[40] B. Vora, A.J. Khopade and N.K. Jain, J. Control Release, 54(1998) 149.

[41] O.E. Philippova, L.A. Chtcheglova, N.S. Karybiants and A.R. Khokhlov, Polym. Gel Networks, 6(1998) 409.

[42] P. Hansson and B. Lindman, Curr. Opin. Colloid Interface Sci., 1(1996) 604.

[43] H. Walderhaug, B. Nystrom, F.K. Hansen and B. Lindman, Prog. Colloid. Polym. Sci., 98(1995) 51.

[44] L. Bromberg, M. Temchenko and R.H. Colby, Langmuir 16(2000) 2609.

[45] E. Kokufuta, Y.Q. Zhang, T. Tanaka and A. Mamada, Macromolecules, 26(1993) 1053

[46] R. Barreiro-Iglesias, C. Alvarez-Lorenzo and A. Concheiro, J. Control Release, 77 (2001) 59.

[47] P. Furrer, B. Plazonnet, J.M. Mayer and R. Gurny, Int. J. Pharm., 207(2000) 89.

[48] T. Wu, J.X. Guo, Q.N. Ping, F.Y Jin, X.W. Sun and Yao Xue Xue Bao, 36(2001) 381.

[49] Y.S. Rhee, J.G. Choi, E.S. Park and S.C Chi, Int. J. Pharm., 228(2001) 161.

[50] P.G Lee, R. Langer and V.P Shastri, Pharm. Res., 20(2003) 264.

[51] J.Y. Fang, H.H. Lin, H.I. Chen and Y.H. Tsai, J. Control. Release, 54(1998) 293.

[52] S.C. Shin and H.J. Lee, Eur. J. Pharma. Biopharma., 54(2002) 325.

[53] E. Peira, P. Scolari and M. R. Gasco, Int. J. Pharm.,, 226(2001) 47.

[54] H. Zhao, D.W. Park, S.K. Kim, C.H. Lee and D.D. Kim, Drug Dev. Ind. Pharm., 9(2002) 1125.

[55] M. Lodzki, B. Godin, L. Rakou, R. Mechoulam, R. Gallily and E. Touitou, J.Control Release, 93 (2003) 377.

[56] M.M. Jiménez, J. Pelletier, M.F. Bobin and M.C. Martini, Int. J. Pharm., 272(2004) 45.

[57] E. H. Lee, A. Kim, Y.K. Oh and C.K. Kim, 26(2005) 205

[58] Z. Mei, H. Chen, T. Weng, Y. Yang and X. Yang, Eur. J. Pharma. Biopharma., 56(2003)189.

[59] J.F. Woodley, Crit. Rev. Ther. Drug Carrier Syst., 11(1994) 61.

[60] F. Tang and R.T. Borchardt, Pharm. Res., 19 (2002) 765.

[61] F. Tang and R.T. Borchardt, Pharm. Res., 19 (2002) 780.

[62] E.D. Hugger, C.J. Cole, T.J. Raub, P.S. Burton and R.T. Borchardt, J. Pharm. Sci., 92(2003) 21.

[63] G. Virkel, F. Imperiale, A. Lifschitz, A. Pis, A. Alvarez, G. Merino, J. Prieto and C. Lanusse, Biopharm Drug Dispos., 24 (2003) 95.

[64] K. Kawakami, T. Yoshikawa, Y. Moroto, E. Kanaoka, K. Takahashi, Y. Nishihara and K. Masuda, J. Control Release., 81(2002)65.

[65] S. Eiamtrakarn, Y. Itoh, J. Kishimoto, Y. Yoshikawa, N. Shibata, M. Murakami and K. Takada, Biomaterials, 23 (2002) 145.

[66] G. Fontana, M. Licciardi, S. Mansueto, D. Schillaci and G. Giammona, Biomaterials, 21(2001) 2857.

[67] S.B. Ruddy, B.K. Matuszewska, Y.A. Grim, D. Ostovic and D.E. Storey, Int. J. Pharm., 182(1999) 173

[68] S. Mitragotri and J. Kost, Pharm. Res., 18(2001) 1151.

[69] Z. Hu, R. Tawa, T. Konishi, N. Shibata and K. Takada, Life Sci., 69 (2001) 2899

[70] S. Eaimtrakarn, Y.V. Rama Prasad, T. Ohno, T. Konishi, N. Y. Shibata and K. Takada, J. Drug Target, 10(2002) 255.

[71] Y.V. Rama Prasad, S.P. Puthli, S. Eaimtrakarn, M. Ishida, Y. Yoshikawa, N. Shibata and K. Takada, Int. J. Pharm., 250(2003) 181.

[72] E. Savio, L. Dominguez, A. Malangra, D. Quevedo, J. Saldana, C. Camarote, A. Ochoa and P. Fagiolino, Boll. Chim. Farm., 137(1998) 345.

[73] J. Arnold, F. Ahsan, E. Meezan and D.J. Pillion, J. Pharm. Sci., 91 (2002) 1707.

[74] S Mori, A. Matsuura, Y.V.Rama prasad, and K.Takada, Biol. Pharm. Bull., 27(2004)418.

[75] E. N. Ziv, M. Kidron, H. Bar-On, G. Friedman, E. Hyam and A. Eldor, Haemostasis, 30(2000) 225.

[76] M. Thanou, J.C. Verhoef, M.T. Nihot, J.H.M. Verheijden and H.E. Junginger, Pharm. Res., 18(2002) 1638.

[77] M. Thanou, M.T. Nihot, M. Jansen, J.C. Verhoef and H.E. Junginger, Pharm. Sci., 90(2001) 38.

[78] K. Katayama, Y. Kato, H. Onishi, T. Nagai and Y. Machida, Int. J. Pharm., 248 (2002) 93.

[79] Y. Ebato, Y. Kato, H. Onishi, T. Nagai, and Y. Machida, Drug Development Research, 58(2003) 253.

[80] S.W. Cho, J. S. Lee and S.H. Choi, J.Pharm.Sci., 93(2004)612.

[81] K. Kawakami, T. Yoshikawa, T. Hayashi, Y. Nishihara and K. Masuda, J. Control Release, 81(2002) 75.

[82] T. Kissel, T. Jung, W. Kamm and A. Breitenbach, Macromol. Symp., 172(2003)113.

[83] M. Thanou, J.C. Verhoef, P. Marbach, H.E. Junginger, J. Pharm. Sci., 89(2000) 951.

[84] M.U. Mehta, R. Venkataramanan, G.J. Burckart, R.J. Ptachcinski. B. Delamos, S. Stachak, D.H. Vanthiel, S. Iwatsuki and T.E. Starzl, Br J.Clin. Pharmacol., 25,(1988)579.

[85] R. Venkataramanan, G.J. Burckart and R.J. F'tachcinski, Semin Liver Dis., 5(1985) 357.

[86] Grevel, E. Nuesch, E. Abisch and K. Kutz, Eur.J.Clin. Pharmacol., 31(1986) 211.

[87] W.A. Ritschel, G.B. Ritschel. A. Sabouni, D. Wolochuk and T. Schroeder, Meth. Find. Expl. Clin. Pharrnacol., 11(1989) 281.

[88] J. Drewe, R. Meier, J. Vonderscher, D. Kiss, U. Posanski, T. Kissel and K. Gyr, Br. J. Clin Pharmaco.,134(1992)60.

[89] G. Fricker, A. Fahr, C. Beglinger, T. Kissel, T. Reiter and J. Drewe, Br.J. Pharrnacol 117(1996) 217.

[90] J.P. Longenecker, A.C. Moses, J.F. Flier, R.D. Silver, M.C. Carey and E.J. Dubovi, J. Pharm. Sci., 76(1987) 351.

[91] A.C. Moses, G.S. Gordon, M.C. Carey and J.F. Flier, Diabetes, 32(1983) 40.

[92] A.E. Pontioroli, A. Caldera and G. Pozza, Clin. Pharrnacokin., 17(1989) 299.

[93] W.A. Lee and J.P. Longenecker, Biopharm., 1(1988)30.

[94] T. Kissel, J. Drewe, S. Bantle, A. Rummelt and C.Beglinger, Pharrn.Res., 9(1992)52.

[95] C.S. Schasteen, M.G. Donovan and J.N. Cogburn, J. Control Re lease, 21(1992) 49.

[96] M.D. Donovan, G.L. Flynn and G.L. Amidon, Phram. Res., 7(1990) 808.

[97] M.D. Donovan, G.L. Flynn and G. L. Amidon, Pharm. Res., 8(1990) 863-868.

[98] E. Touitou, J. Control Re lease, 21(1992) 139.

[99] H. Yoshida, C.M. Lehr, W. Kok, H.E. Junginger, J.C. Verhoef and J.A. Bouwstra, J.Control Re lease, 21(1992) 145.

[100] T. Nishihata, C.S. Lee, M. Yamamoto, J.H. Rytting and T. Higuchi, J. Phram. Sci., 73(1984) 1326.

[101] G. Flemstrom and N.V.B. Marsden, Gastroenterology, 64(1976) 278.

[102] T. Nakate, H. Yoshida, A. Ohike, Y.Tokunaga, R.Ibuki and Y. Kawashima, Eur.J.Pharm.Biopharm.Sci., 55(2003) 147.

[103] M. A. Bagger, H. W. Nielsen and E.Bechgaard, Eur. J.Pharm.Sci., 14(2001) 69.

[104] C. Hascicek, N. Gonul and N. Erk, Farmaco, 58(2003) 11.

[105] A. Hussain, J. J. Arnold, M. A. Khan and F. Ahsan, J. Control Re lease, 94(2004) 15.

[106] M. Thanou, J.C. Verhoef and H.E. Junginger, Adv. Drug Del. Rev. 52(2001) 117.

[107] M. V. Margarit, I. C. Rodriguez and A. Cerezo, Int. J. Pharm., 81, (1992) 67.

[108] Y. Mrestani, B. Bretschneider, A. Hartl, R.H. Neubert, J. Pharm. Pharmacol., 55(2003) 1601.

[109] N. Yuksel, A. Karata, Y. Ozkan, A. Savaer, A. S. Ozkan and T. Baykara, Eur.J.Pharm.Biopharm.Sci., 56(2003) 453.

[110] M. Dahlback, S. Eirefelt, K. Backstrom, P. Larsson, L.O. Almer, P. Wollmer, B. Jonson, J. Aerosol Med., 15(2002) 27.

[111] J.Y. Kim and Y.S. Ku, Int. J. Pharm., 194(2000) 81.

[112] Y.V. Rama Prasad, S. Eaimtrakarn, M. Ishida, Y. Kusawake, R. Tawa, Y. Yoshikawa, N. Shibata and K.Takada, Int. J. Pharm., 268(2003) 13.

[113] Y.V. Rama Prasad, S. P. Puthli, S. Eaimtrakarn, M. Ishida, Y. Yoshikawa, N. Shibata and K.Takada, Int.J.Pharm., 250(2003) 181.

[114] S. Byrn, R. Pfeiffer, M. Ganey, C. Hoiberg and G. Poochikian, Pharm. Res., 12(1995) 9.

[115] B.Y. Shekunov and P. York, J. Cryst. Growth, 211(2000) 122.

[116] J.R. Crison, V.P. Shah, J.P. Skelly and G.L. Amidon, J. Pharm. Sci., 85(1996) 1005.

[117] J.R. Crison, N.D. Weiner and G.L. Amidon, , J. Pharm. Sci., 86(1997)384.

[118] P.E. Luner, J. Pharm. Sci., 89(2000) 382.

[119] N. Garti and H. Zour, J. Crystal Growth 172(1997) 486.

[120] K.S.N. Reddy, L.M. Salvati, P.K Dutta, P.B. Abel, K.I. Suh and R.R. Ansari, J. Phys. Chem., 100(1996) 9870.

[121] S. Luhtala, Acta. Pharm. Nord., 4(1992)271.

[122] S. Luhtala, P. Kahela and E. Kristofferson, Acta Pharm Fennica, 99(1990) 59–68.

[123] I. Weissbuch, L. Addadi, L. Leiserowitz and M. Lahav, J. Am. Chem. Soc., 110(1988) 561.

[124] J. Dugua and B. Simon, J. Crystal Growth, 44(1978) 280.

[125] Fu redi-Milhofer H, L. Tunik, N. Filipovic-Vincekovic, D. Skrtic, V. Babic-Ivancic, and N. Garti, Scanning Microscopy, 9(1995) 1061.

[126] M. Bujan, M. Sikiric, N. Filipovic-Vincekovic, N. Vdovic, N. Garti and H. Furedi-Milhofer, Langmuir, 17(2001) 6461.

[127] N. Lincopan, E.M. Mamizuka and A.M. Carmona-Ribeiro, J. Antimicrob. Chemother., 52(2003)412.

[128] A. Rodriguez, E. Junquera, P. Del Burgo and E. Aicart, J. Colloid Interface, Sci., 269(2004) 476.

[129] X. Liao and T.S Wiedmann, Pharm. Res., 20(2003) 1858.

[130] P. Li and L. Zhao, Int. J. Pharm., 249(2002) 211.

[131] Y. Ran, L. Zhao, Q. Xu and S.H. Yalkowsky, Pharm. Sci. Tech. , 2(2001)E2.

[132] P. Li and L.Zhao, J. Pharm. Sci., 92(2003) 951.

[133] R. Nagarajan, Polym. Adv. Technol., 12(2001) 2343.

[134] S. Palma, R. H. Manzo, D. Allemandi, L. Fratoni, P.L. Nostro, J.Pharm.Sci., 91(2002) 1810.

[135] I. Amar, A. Aserin, and N. Garti, J. Agric. Food Chem., 51(2003) 4775.

[136] S. J. Kim, Y.H. Cho, W. Park, D. Han, C.H. Chai and J.Y. Imm, J. Agric. Food Chem., 51(2003) 7805.

[137] Y. Saito, K. Hashizaki, H. Taguchi and N. Ogawa, Drug Dev. Ind. Pharm.., 29(2003) 345.

[138] D. Attwood, Adv. Colloid Interf. Sci., 55(1995) 271.

[139] P. Taboada, J.M. Ruso, M. Garcia and V. Mosquera, Colloids Surf. A. Physicochem. Eng. Asp., 179, 125-128 (2001).

[140] S. Tehrani, N. Brandstater, Y.D. Saito and P. Dea, Biophys. Chem., 94,87-96 (2001).

[141] M.B. Stephen, D.B. Lyle and C.M. Donald, J. Pharm. Sci., 66 (1977) 1.

[142] B.M. Michael, M.R. Susan, A.K. Christopher and, A.H. Munir, J. Pharm. Sci., 83(1994) 1418.

[143] K.Itoh, Y. Tozuka, T.Oguchi, and K. Yamamoto, Chem. Pharm. Bull., 51(2003) 40.

[144] C. Duval-Terrie, P. Cosette, G. Molle, G. Muller and De E, Protein Sci., 12(2003) 681.

[145] Gang Yang, Neera Jain and Samuel H. Yalkowsky, Int. J. Pharm., 269(2004) 141.

[146] C. Grove, W. Liebenberg, J.L. Du Preez, W. Yang and M.M De Villiers, J Cosmet Sci., 54(2003) 537.

[147] M. Sergey, Mel'nikov, W.M. Jack, S. T.Hoorn and P.A.M Astrid, Eijkelenboom, Chemistry and Physics of Lipids, 127(2004) 121.

[148] Y. Saito, K. Hashizaki, H. Taguchi and N. Ogawa, Drug Dev. Ind. Pharm., 29(2003) 345.

[149] I. Nandi, M. Bari M and H. Joshi, AAPS Pharm. Sci. Tech., 4(2003) E10.

[150] N. Hurter, P. Alexandridis and T. A. Hatton, in Solubilization Surfactant Aggregates, Eds. S.D. Christian and J. F. Scamehorn, Surfactant Science Series, vol. 55, Marcel Dekker, New York, 1995, p. 191.

[151] C. Allen, D. Maysinger and A. Eisenberg, Colloids Surf. , 16(1999) 3.

[152] X. Zhang, J. K. Jackson and H. M. Burt, Int. J. Pharm., 132(1996) 195.

[153] C.J. Rekatas, S.M. Mai, M. Crothers, M.Quinn, J. H. Collett, D. Attwood, F. Heatley, L. Martini and C. Booth, Phys. Chem. Chem. Phys. , 3(2001) 4769.

[154] T.Loftsson, A. Magnúsdóttir, M. Másson, J. F. Sigurjónsdóttir, J.Pharm.Sci., 91 (2002) 2307.

[155] L. F. Pacheco and A. M. Carmona-Ribeiro, J.Colloid Interface Sci., 258(2003) 146.

[156] E. Söderlind, M. Wollbratt and C.V. Corswant, Int.J.Pharm., 252, (2003) 61.

[157] N.M.Najib and M.S. Suleiman, Int.J..Pharm., 24(1985) 165.

[158] S. Byrn, R. Pfeiffer, M. Ganey, C. Hoiberg, G. Poochikian, Pharm. Res., 12(1995) 945.

[159] B.Y. Shekunov and P. York, J. Cryst Growth, 211 (2000) 122.

[160] J.R. Crison, V.P. Shah, J.P. Skelly and G.L. Amidon, J. Pharm. Sci., 85(1996) 1005.

[161] J.R. Crison, N.D. Weiner and G.L. Amidon, J. Pharm. Sci., 86(1997) 384.

[162] P.E. Luner, J. Pharm. Sci., 89(2000) 382.

[163] J. Dugua and B. Simon, J. Crystal Growth, 44(1978) 280.

[164] I. Weissbuch, L. Addadi, L. Leiserowitz and M. Lahav, J. Am. Chem. Soc., 110(1988) 561.

[165] S. Luhtala, P. Kahela and E. Kristofferson, Acta Pharm. Fennica, 99(1990) 59.

[166] S. Luhtala, Acta Pharm. Nord., 4(1992) 271.

[167] K.S.N. Reddy, L.M Salvati, PK. Dutta, P.B. Abel, K.I. Suh and R.R. Ansari, J. Phys. Chem., 100(1996) 9870.

[168] N. Garti and H. Zour, J. Crystal Growth 172(1997) 486.

[169] H. Furedi-Milhofer, L. Tunik, N. Filipovic-Vincekovic, D. Skrtic, V. Babic-Ivancic and N. Garti, Scanning Microscopy, 9(1995) 1061.

[170] M. Bujan, M. Sikiric, N. Filipovic-Vincekovic, N. Vdovic, N. Garti and H. Furedi-Milhofer, Langmuir, 17 (2001) 6461.

[171] N. Rodriguez-hornedo and D. Murphy, J.Pharm.Sci.,93(2004)449.

[172] W. Sun, C.K. Larive, M.J. Southard, J. Pharm. Sci., 92(2003) 424.

[173] H. Hirata and N.J. Iimura, J. Colloid Interface Sci., 268(2003) 215.

[174] S.R. Levis and P.B. Deasy, Int. J. Pharm., 230, (2001) 25.

[175] S.C. Shin and J. Kim, Int., J. Pharm., 251(2003) 79.

[176] Z. Kokot and H. Zmidzinska, Acta Pol. Pharm., 58(2001) 117.

[177] C. Olbrich and R. H. Müller, Int.J..Pharm., 180 (1999) 31.

[178] N. Rapoport, Colloids and Surfaces B: Biointerfaces, 16(1999) 93.

[179] M. V. Pattarkine and K. N. Ganesh, Biochemical and Biophysical Research Communications, 263(1999) 41.

[180] M.Wathier, A. Polidori, K. Ruiz, A.S. Fabiano and B. Pucci, Chemistry and Physics of Lipids, 115(2002) 17.

[181] G. Riess, Progress in Polymer Science, 28(2003) 1107.

[182] I. J. Castellanos, R Crespo, K Griebenow, J Control Release, 88(2003) 135.

[183] S.Tehrani, N. Brandstater, Y. D. Saito and P. Dea, Biophysical Chemistry, 94(2001) 87.

[184] M. A. Hammad and B. W. Müller, Eur. J.Pharm.Sci., 7 (1998) 49.

[185] M. Trotta, F. Pattarino and T. Ignoni, Eur.J.Pharm.Biopharm.Sci., 53 (2002) 203.

[186] S. Kantaria, G. D. Rees and M. J. Lawrence, Int.J.Pharm., 250 (2003) 65.

[187] L. F. Pacheco and A. M. Carmona-Ribeiro, J.Colloid and Interface Sci., 258 (2003) 146.

[188] C. Gadras, C. Santaella and P. Vierling, Journal of Controlled Release, 57(1999) 29.

[189] M. Wolf, M. Wirth, F. Pittner and F. Gabor, Int.J..Pharm., 256(2003) 141.

[190] R. Barreiro-Iglesias, L. Bromberg, M. Temchenko, T. A. Hatton, A.Concheiro and C. Alvarez-Lorenzo, J. Control Release, 97(2004) 537.

[191] H. Steckel, F. Eskandar and K. Witthohn, Int.J.Pharm., 257(2003) 181.

[192] M. Kursa, G. F. Walker, V. Roessler, M. Ogris, W. Roedl R. Kircheis and E. Wagner Bioconjugate Chem., 14(2003) 222.

[193] S.C. Yang, L.F. Lu, Y. Cai, J.B. Zhu, B.W. Liang and C.Z. Yang, J Control Release, 59(1999) 299.

[194] B. Lu, J.Q. Zhang and H. Yang, Drug Deliv., 10(2003) 87.

[195] R.N. Aliautdin, J. Kreuter and D.A. Kharkevich, Eksp. Klin. Farmakol. , 66(2003) 65.

[196] M. Bae, S. Cho, J. Song, G.Y. Lee, K Kim, J. Yang, K. Cho, S.Y. Kim and Y. Byun, Drugs Exp. Clin. Res., 29(2003) 15.

[197] A. Kader, P.J. Davis, M. Kara and H. Liu, J. Control Release, 55(1998) 231.

[198] P. Caliceti, M. Chinol, M. Roldo, F. M. Veronese, A. Semenzato, S. Salmaso and G. Paganelli, J. Control Release. , 83(2002) 97.

[199] H. Iinuma1, K. Maruyama, K. Okinaga1, K. Sasaki, T. Sekine1, O. Ishida, N. Ogiwara, K. Johkura And Y. Yonemura, Int. J. Cancer, 99 (2002) 130.

[200] A. Gijsens, A. Derycke, L. Missiaen, D. De Vos, J. Huwyler, A. Eberle and P. De Witte, Int. J. Cancer, 101(2002) 78.

[201] P.C. Rensen, R.L. de Vrueh, J. Kuiper, M.K. Bijsterbosch, E.A. Biessen and T.J. van Berkel, Adv. Drug Deliv. Rev., 47(2001) 251.

[202] M.S. Hong, S.J. Lim, M.K. Lee, Y.B. Kim and C.K. Kim, Drug Deliv. , 8(2001) 231.

[203] A.N. Lukyanov, Z. Gao, L. Mazzola and V.P. Torchilin, Pharm. Res., 19 (2002) 1424.

[204] A. N. Lukyanov, Z. Gao and V. P. Torchilin, J. Control Release, 91 (2003) 97.

[205] X. Boulenc, T. Breul, J. Gautier, P. Saudemon, H. Joyeux, C. Roques, Y. Berger and G. Fabre, Int.J..Pharm., 123 (1995) 71.

[206] M. Sriwongjanya and R. Bodmeier, Int.J.Pharm., 158(1997) 29.

[207] P.B. Deasy and M.P. Gouldson, Int.J..Pharm., 132(1996) 131.

[208] C. Samyn, W. Kalala, G. Van den Mooter and R. Kinget, Int.J.Pharm., 121(1995) 211.

[209] M.D. Veiga, C. Escobar and M.J. Bernad, Int.J..Pharm., 93(1993) 215.

[210] S.G. Kapsi and J.W. Ayres, Int.J..Pharm., 229(2001) 193.

[211] L.Zhu and S. Feng, Chemosphere, 53(2003) 459.

[212] J. Limpens and R.J. Scheper, Clin Exp Immunol. , 78(1989) 256.

[213] J. Afeltra, R.G. Vitale, J.W. Mouton, P E Verweij., Antimicrob. Agents Chemother., 48(2004)1335.

[214] J. Piret, S. Roy, M. Gagnon, S. Landry, A. Desormeaux, R.F. Omar and M.G. Bergeron, Antimicrob. Agents Chemother. , 46(2002) 2933.

[215] J. Davidsen, O.G. Mouritsen, K. Jorgensen, Biochim. Biophys. Acta., 1564(2002)256.

[216] C. Oelschlaeger, E. Buhler, G. Waton and S.J. Candau, Eur. Phys. J. E. Soft. Matter, 11(2003) 7.

[217] B D Andersaon, R A Conrad, K.E.Nuth and S.L. Nail, J.Pharm.Sci.,74(1985) 375.

[218] A Nokhodchi, S. Norouzi-Sani, M.R. Siahi-Shadbad, F. Lotfipoor and M Saeedi, Eur J Pharm Biopharm. , 54(2002) 349.

[219] G. Giammona, B. Carlisi, G. Pitarresi, G. Cavallaro and V. Turco Liveri J. Control Release, 22(1992) 197.

[220] M. Zacchigna, G. Di Luca, V. Maurich and E. Boccu, Farmaco. , 57(2002) 207.

[221] M. P. Krafft and J. G. Riess, Biochimie, 80(1998) 489.

[222] J.L.Murtha and H.Y.Ando, J.Pharm.Sci., 83(1994) 1222.

[223] G.B. Greenwald, A.Pendri, C.D.Comover, H.Zhao, Y.H.Choe, A. Martinez, K.Shum and S.Guan, J.Med. Chem., 42(1999) 3657.

[224] S. Supattapone, H.O. Nguyen, F.E. Cohen, S.B. Prusiner, M.R. Scott, Proc Natl Acad Sci U S A. 96(1999) 14529.

[225] Yraima Cordeiro, Luis Mauricio, T.R. Lima, P.B. Mariana, P.B. Gomes, Debora Foguel and Jerson L. Silva, J. Biol. Chem., 279(2004) 5346.

[226] D. Joester, M. Losson, R. Pugin, H. Heinzelmann, E. Walter, H.P. Merkle, F. Diederich, Angew Chem. Int. Ed. Engl., 42(2003)1486.

[227] V. Percec, A.E. Dulcey, V.S. Balagurusamy, Y. Miura, J. Smidrkal, M. Peterca, S. Nummelin, U. Edlund, S.D. Hudson, P.A. Heiney, H. Duan, S.N. Magonov and S.A. Vinogradov, Nature, 430(2004) 764.

[228] H.T. Chen, M.F. Neerman, A.R. Parrish and E.E. Simanek, J. Am. Chem. Soc., 126(2004) 1004.

[229] Z. Hyvonen, M. Ruponen, S. Ronkko and P. Suhonen, A. Urtti, Eur. J. Pharm. Sci., 15(2002) 449.

[230] I. Van Der Woude, A. Wagenaar, A.A.P. Meekel, M.B.A. Ter beest, M.H.J. Ruiters, J.B.F.N. Engberts, and D. Hoekstra, Proc. Natl. Acad. Sci. USA, 94(1997) 1160.

[231] C. L. Gebhart, S. Sriadibhatla, S. Vinogradov, P. Lemieux, V. Alakhov, and A. V. Kabanov, Bioconjugate Chem., 13(2002) 937.

[232] E. Dauty, R. Jean-Serge, G. Zuber, and B. Jean-Paul, Bioconjugate Chem., 13(2002) 831.

[233] H. Petersen, P. M. Fechner, A. L. Martin, K. Kunath, S. Clive J. Roberts, D.Fischer, M. C. Davies, and T. Kissel, Bioconjugate Chem., 13 (2002) 845.

[234] S. V. Vinogradov, T. K. Bronich, and A. V. Kabanov, Bioconjugate Chem., 91(998) 805.

[235] Y. Kakizawa, A. Harada, and K. Kataoka, Biomacromolecules, 2(2001) 491.

[236] Y. Koyama, T. Ito, T. Kimura, A. Murakami, T. Yamaoka, J. Control Release, 77 (2001) 357.

[237] H. Fritz, M. Maier and E. Bayer, J. Colloid Interface Sci., 195(1997) 272.

[238] N.Raghvachari and W.E.Fahl, J. Pharm. Sci., 91(2002) 615.

[239] S. Sung, S.H. Min, K.Y.Cho, S. Lee, Y. Min, Y.Yeom and J. Park, Biol. Pharm. Bull., 26(2003)492.

[240] M. Jumaa and B.W. Muller, Eur. J. Pharm. Sci., 9(2000) 285.

[241] B.B. Lundberg, G. Griffiths and H.J. Hansen, J. Pharm. Pharmacol., 51(1999)1099.

[242] B.B. Lundberg, J. Pharm. Pharmacol., 49(1997)16.

[243] S. Tamilvanan and S. Benita, Eur.J.Pharm.Biopharm.Sci., 58(2004) 357.

[244] S.Thomas-Gibson, A. Jawhari, P. Atlan, A.L. Brun, M. Farthing and A. Forbes, Clinical Nutrition, 23(2004) 697.

[245] E.C. Unger, T. Porter, W. Culp, R. Labell, T.Matsunaga and R. Zutshi, Advanced Drug Delivery Reviews, 56(2004) 1291.

[246] G. Dollo, P. Le Corre, F. Chevanne and R. Le Verge, Eur. J.Pharm.Sci., 22(2004) 63.

[247] J. Seki, S. Sonoke, A. Saheki, H. Fukui, H. Sasaki and T. Mayumi, Int.J.Pharm., 273(2004) 75.

[248] A. Gabizon, and D. Papahadjopoulos, Proc. Natl. Acad. Sci. U.S.A. 85(1988) 6949.

[249] T.M. Allen, C. Hansen, F. Martin, C. Redemann and A. Yau-Young, Biochim. Biophys. Acta, 1066 (1991) 29.

[250] E.B. Kullberg, N. Bergstrand, J. Carlsson, K. Edwards, M. Johnsson, S. Sjoberg, and L. Gedda, Bioconjugate Chem., 13(2002) 737.

[251] L.R. Tsuruta, W. Quintilio, M.H.B. Costa and A.M. Carmona-Ribeiro, Journal of Lipid Research, 38(2003) 1997.

[252] H. Belhaj-Tayeb, D. Briane, J. Vergote, S. Kothan, G. Leger, S.E. Bendada, M Tofighi, F Tamgac, A Cao and J.L. Moretti. , Eur J Nucl Med Mol. Imaging., 30(2003)502.

[253] K.E. Anderson, B.R. Stevenson and J.A. Rogers, J. Control Release, 60(1999) 189.

[254] D. A. Eavarone, X. Yu, and R. V. Bellamkonda, J. Biomed Mater Res., 51 (2000) 10.

[255] W.T. Phillips, R.W. Klipper, V.D. Awasthi, A.S. Rudolph, R. Cliff, V. Kwasiborski and B.A. Goins, J.Pharm.Expt.Ther., 288(1999) 665.

[256] M. Zignani, D.C. Drummond, O. Meyer, K. Hong and J.C. Leroux, Biochim. Biophys. Acta., 1463(2000)383.

[257] T.M. Allen, P. Sapra and E. Moase, Cell Mol. Biol. Lett., 7(2002)889.

[258] A. Kozubek, J. Gubernator, E. Przeworska and M. Stasiuk, Acta. Biochim. Pol., 47(2000) 639.

[259] A. Paavola, I. Kilpelainen, J. Yliruusi and P. Rosenberg, Int. J. Pharm., 199(2000) 85.

[260] V. DiTizio, G.W. Ferguson, M.W. Mittelman, A.E. Khoury, A.W. Bruce and F. DiCosmo, Biomaterials, 9(1998) 1877.

[261] R.M. Schiffelers, G.A. Koning, T.L.M.T. Hagen, M.H.A.M. Fens, A. J. Schraa, A. P.C.A. Janssen, R.J. Kok, G. Molema and G. Storm, J. Control Release, 91 (2003) 115.

[262] S. Dagar, A. Krishnadas, I. Rubinstein, M.J. Blend and H. Onyuksel, J. Control Release, 91(2003) 123.

[263] P. Sapra and T.M. Allen, Progress in Lipid Research, 42 (2003) 439.

[264] G. F.A. Kersten and D. J.A. Crommelin, Vaccine, 21 (2003) 915.

[265] R. Scheuermann, I.M. Tucker, A.M. Creeth, H. Dilger, B. Beck, E. Roduner, Phys. Chem. Chem. Phys., 4 (2002) 1510.

[266] E. Roduner Physica B 326 (2003) 19.

[267] B.J. Lestini, S.M. Sagnella, Z. Xu, M.S. Shive, N.J. Richter, J. Jayaseharan, A.J. Case, K. Kottke-Marchant, J.M. Anderson and R.E. Marchant, J. Control Release, 78 (2002) 235.

[268] J.K. Kim, S.H. Choi, C.O. Kim, J.S. Park, W.S. Ahn and C.K. Kim, J. Pharm. Pharmacol. , 55(2003) 453.

[269] M.S. Hong, S.J. Lim, Y.K. Oh and C.K. Kim, J. Pharm. Pharmacol. , 54(2002) 51.

[270] C. Managit, S. Kawakami, M. Nishikawa, F. Yamashita and M. Hashida, Int.J..Pharm., 266 (2003) 77.

[271] K. Maruyama, Biosci. Rep., 22(2002) 251.

[272] A. C. Sintov and L.Shapiro, J.Controlled Release, 95 (2004) 173.

[273] T. Chandy, G.H. Rao, R.F. Wilson and G.S. Das, Drug Deliv., 9(2002) 87.

[274] R.A. Jain, C.T. Rhodes, A.M. Railkar, A.W. Malick and N.H. Shah, Eur. J. Pharm. Biopharm., 50(2000)257.

[275] A. Shukla, A. Krause and R.H. Neubert, J. Pharm. Pharmacol., 55(2003) 741.

[276] C. Mallard, J. Coudane, I. Rault and M. Vert, J. Microencapsul., 17(2000)13.

[277] K.M. Park and C.K. Kim, Int. J. Pharm., 181(1999) 173.

[278] L. Li, I. Nandi and K.H. Kim, Int. J. Pharm., 237(2002) 77.

[279] B.A. Brime, M. A. Moreno, G. Frutos, M. P. Ballesteros and P. Frutos, J.Pharm.Sci.,91(2002) 1178

[280] T. F. Vandamme, Progress in Retinal and Eye Research, 21 (2002) 15.

[281] S. Kantaria, G.D. Rees and M. Lawrence, J. Int. J. Pharm., 250(2003) 65.

[282] Y.Y. Huang and T.W. Chung, J. Microencapsul., 18(2001)457.

[283] M. Genty, G. Couarraze, R. Laversanne, C. Degert, J. Maccario and J.L. Grossiord, J. Control Release., 90(2003)119.

[284] M. Iwata, Y. Nakamura and J.W. McGinity, J. Microencapsul., 16(1999)49.

[285] P. Kan, Z.B. Chen, C.J. Lee and I.M Chu, J. Control Release, 58(1999) 271.

[286] I. Gutierro, R.M Hernandez, M. Igartua, A.R. Gascon and J.L Pedraz, J. Microencapsul., 20(2003)525.

[287] P.L. Honeywell-Nguyen and J.A. Bouwstra, J. Control Release, 86(2003) 145.

[288] E. Memisoglu, A. Bochot, M. Sen, D. Charon, D. Duchene and A.A. Hincal, J. Pharm. Sci., 91(2002) 1214.

[289] S. Miyazaki, A. Takahashi, W. Kubo, J. Bachynsky and R. Loebenberg, AAPS J. Pharm. Sci., 6(2003) 238.

[290] A. Lamprecht, N. Ubrich, H. Yamamoto, U. Schafer, H. Takeuchi, C.M. Lehr, P. Maincent, and Y. Kawashima, J. Control Release, 71(2001) 297.

[291] M. Trotta, M. Gallarate, M.E. Carlotti and S. Morel, Int. J. Pharm., 254(2003) 235.

[292] L. Araujo, R. Lobenberg and J. Kreuter, J. Drug Target, 6(1999) 373.

[293] A. Friese, E. Seiller, G. Quack, B. Lorenz and J. Kreuter, Eur. J. Pharm Biopharm., 49(2000)103.

[294] P. Calvo, B. Gouritin, H. Chacun, D. Desmaele, J. D. Angelo, J.P Noel, D. Georgin, E. Fattal, J.P Andreux and P. Couvreur, Pharm. Res., 18(2001) 1157.

[295] G. Gonzalez-Martin, I. Merino, M.N. Rodriguez-Cabezas, M. Torres, R. Nunez and A. Osuna, J. Pharm. Pharmacol., 50(1998)29.

[296] J. Kreuter, Adv. Drug Deliv. Rev., 47(2001) 65.

[297] A. Mitra and S. Lin, J. Pharm. Pharmacol., 55(2003)895.

[298] M. Tobio, R. Gref, A. Sanchez, R. Langer and M. Alonso, J. Pharm. Res., 15(1998) 270.

[299] S. Barbault-Foucher, R. Gref, P. Russo, J. Guechot and A. Bochot, J. Control Release, 83(2002) 365.

[300] I. Oh, K. Lee, H.Y Kwon, Y.B. Lee, S.C. Shin, C.S. Cho and C.K. Kim, Int. J. Pharm., 181(1999) 107.

[301] J. Kreuter, D. Shamenkov, V. Petrov, P. Ramge, K. Cychutek, C. Koch-Brandt and R.J. Alyautdin, Drug Target, 10(2002) 317.

[302] S.P. Vyas, N. Mysore, V. Jaitely, N. Venkatesan, Pharmazie, 53(1998) 466.

[303] G.K Pillai and M.L. Salim, Int. J. Pharm., 193(1999) 123.

[304] P. Arunothayanun, T. Sooksawate and A.T. Florence, J. Control Release, 60(1999) 391.

[305] C.O. Rentel, J.A. Bouwstra, B. Naisbett and H.E Junginger, Int. J. Pharm., 186(1999) 161.

[306] V. Sihorkar and S.P Vyas, Pharmazie, 55(2000) 107.

[307] A. Manosroi, P. Wongtrakul, J. Manosroi, H. Sakai, F. Sugawara, M.Yuasa and M. Abe, Colloids and Surfaces B: Biointerfaces, 30 (2003) 129.

[308] B.W. Barry, Eur. J.Pharm.Sci., 14 (2001) 101.

[309] J.Y. Fang, Chi-Tzong Hong, W.T. Chiu and Y.Y.Wang, Int.J.Pharm., 219 (2001) 61.

[310] H. Chengjiu and D.G. Rhodes, Int.J.Pharm. , 185 (1999) 23.

[311] Y. Hao, F. Zhao, N. Li, Y. Yang and K. Li , Int.J..Pharm., 244 (2002) 73.

[312] T.R. Desai and W.H. Finlay, Int.J..Pharm., 241 (2002) 311.

[313] M. L. Adams, A. Lavasanifar and G. S. Kwon, J Pharm Sci., 92(2003) 1343.

[314] C.S.Chern, H.C.Chiu and Y.C.Chuang, Polym. Int., 53(2004) 420.

[315] C. Scholz, M. Iijima, Y. Nagasaki and K. Kataoka, Polym. Adv. Technol., 9(1998) 768.

[316] J.H.Kim, K. Emoto, M. Iijima, Y. Nagasaki, T. Aoyagi, T. Okano, Y. Sakurai and K.Kataoka, Polym. Adv. Technol., 10(1999) 647.

[317] S.B.La, T. Okano, and K. Kataoka, J.Pharm.Sci., 85 (1996) 85.

[318] T. Ouchi, H. Miyazaki, H. Arimura, F. Tasaka, A. Hamada and Y. Ohya, J.Polymer Science: Part A: Polymer Chemistry, 40(2002) 1426.

[319] A. V. Kabanov, E. V. Batrakova and V. Y. Alakhov, J. Control Release, 82 (2002) 189.

[320] A.V. Kabanov, E.V. Batrakova and V.Y. Alakhov, J. Control Release, 82(2002) 189.

[321] A. Marin, H. Sun, G.A. Husseini, W.G. Pitt, D.A. Christensen and N.Y. Rapoport, J. Control Release, 84(2002) 39.

[322] A.H El-Kamel, Int. J. Pharm., 241(2002) 47.

[323] J.R. DesNoyer and A.J. McHugh, J. Control Release, 86(2003) 15.

[324] K.A. Mehta, M.S. Kislalioglu, W. Phuapradit, A.W. Malick and N.H. Shah, Drug Dev. Ind. Pharm.., 28(2002)275.

[325] B. Ochietti, N. Guerin, S.V. Vinogradov, Y. St-Pierre, P. Lemieux, A.V. Kabanov, and V.Y Alakhov, J. Drug Target, 10(2002) 113.

[326] A.V. Kabanov, E.V. Batrakova and V.Y. Alakhov, Adv. Drug Deliv. Rev., 54(2002) 759.

[327] E.V. Batrakova, H.Y. Han, V.Yu. Alakhov, D.W. Miller and A.V. Kabanov, Pharm. Res., 15 (1998) 850.

[328] A.V. Kabanov, P. Lemieux, S. Vinogradova and V. Alakhov, Adv. Drug Deliv. Rev., 54 (2002) 223.

[329] K. Kataoka, T. Matsumoto, M. Yokoyama, T. Okano, Y. Sakurai, S. Fukushima, K. Okamoto, G.S. Kwon, J Control Release, 64(2000) 143.

[330] J. Fu, J. F. Kraul and E. Hanes, J.Biomaterials, 23(2002) 4425.

[331] N. Rapoport, W.G. Pitt, H. Sun and J.L Nelson, J. Control Release., 91(2003)85.

[332] A. Harada, H. Togawa and K.Kataoka, Eur. J. Pharm. Sci., 13(2001) 35.

[333] P. N. Jalenjak and I. J. enjak, Int.J.Pharma., 272(2004)57.

[334] M.C.Jones and J.C.Leroux, Eur.J.Pharma. Biopharm., 48 (1999) 101.

[335] R. B. Greenwald, H. Zhao, J. Xia, and A. Martinez, J. Med. Chem., 46(2003) 5021.

[336] Z. Yang, Y. Zhang, P. Markland and V. C. Yang, J. Biomed. Mater. Res., 62(2002) 14.

[337] K. Moriyama, T. Ooya and N .Yui, J Control Release, 59(1999) 77.

[338] R.Cortesi, D.C.Mischiati, M.Brogatti, L.Bredo, A.Romancelli, M.Savino, C.Pedone, R.Gambari and C. Nastruzzi, AAPS J Pharma Sci., 6(2004) 1.

[339] B. Rajashekhar and E.T. Kaiser, J Biol Chem., 261(1986) 13617.

[340] M. Tobio, S.P. Schwendeman, Y. Guo, J. McIver, R. Langer and M.J. Alonso, Vaccine, 18(1999) 618.

[341] J. Chen and E. Dickinson, J. Agric. Food Chem., 46(1998) 91.

[342] I. Toth, J.P. Malkinson, N. S. Flinn, B. Drouillat, A. Horvath, J. EArchegyi, M. Idei, A. Venetianer, P. Artursson, L. Lazorova, B. Szendę and G. Keri, J. Med. Chem., 42(1999) 4010.

[343] M. Torres-Lugo, M. Garcia, R. Record, N.A. Peppas, J Control Release, 80(2002) 197.

[344] B .Qiu, S. Stefanos, J. Ma, A. Lalloo, B.A. Perry, M.J. Leibowitz, P.J. Sinko and S. Stein, Biomaterials, 24(2003) 11.

[345] Y. Ke, C.L. McGraw, R.L. Hunter and J.A. Kapp, Cell Immunol., 176(1997)113.

[346] T.K. Mandal and L.A. Bostanian, Pharm. Dev. Technol., 5(2000) 555.

[347] P. Bouillot, N. Ubrich, F. Sommer, T.M. Duc, J.P. Loeffler, E. Dellacherie, Int. J. Pharm. , 181(1999) 159.

[348] H. Li, J.H. Song, J.S. Park and K. Han, Int J Pharm., 258(2003) 11.

[349] R.B. Greenwald, K. Yang, H. Zhao, C. D. Conover, S. Lee, and D. Filpula Bioconjugate Chem., 14(2003) 395.

[350] A. Kim, M.O. Yun, Y.K. Oh, W.S. Ahn and C.K. Kim, Int. J. Pharm. , 180(1999) 75.

[351] S. Zhou, X. Liao, X. Li, X. Deng and H. Li, J Control Release, 86(2003) 195.

[352] R.N. Gursoy and S. Benita, Biomedecine and Pharmacotherapy, 58(2004) 173.

[353] C.M. O'Driscoll, Eur. J. Pharm. Sci., 15 (2002) 405–415

[354] C.W. Pouton, Eur. J. Pharm. Sci., 11 (2000) S93.

[355] T.R. Kommuru B. G. M.A. Khan, I.K. Reddy, Int. J. Pharm., 212 (2001) 233.

[356] G. Krishna and B.B. Sheth, PDA J Pharm Sci Technol. , 53(1999) 168.

[357] T. Gershanik and S. Benita, Eur. J. Pharm. Biopharm.,50(2000) 179.

[358] H.J. Kim, K.A. Yoon, M. Hahn, E.S. Park and S.C. Chi, Drug Dev. Ind. Pharm., (2000) 523.

[359] L. Mu and S.S. Feng, J.Controlled Release, 86 (2003) 33.

[360] F. Eeckman, A.J. Moes, K. Amighi, J.Controlled Release, 88 (2003) 105.

[361] J. J. Haitsma, U. Lachmann and B. Lachmann, Adv. Drug Del. Rev., 47 (2001) 197.

[362] M. Kreilgaard, Adv. Drug Del. Rev., 54 (2002) S77.

[363] A. Nokhodchi J. Shokri, A. Dashbolaghi, D. Hassan-Zadeh, T. Ghafourian, M. Barzegar-Jalali, Int J Pharm., 250 (2003) 359.

[364] M.C. Annesini, C. M. Braguglia, A. Memoli, L. G. Palermiti and S.D. Sario, Biotechnol. Bioeng., 55(1997)261.

[365] S. Murdan, G. Gregoriadis and A. T. Florence, J.Pharm. Sci., 88 (1999) 608.

EPILOGUE

The liquid membrane hypothesis of drug action described in the monograph is a new facet of drug action, which had hitherto gone unnoticed. The reason why it had gone unnoticed so far appears to be the fact that passive transport has been traditionally considered unimportant for biological action; the transport through liquid membrane is, of course, passive in nature. It has been emphasized earlier also, and we would like to emphasize once again that the liquid membrane hypothesis of drug action in no way contradicts or dispenses with the active interaction of drugs with receptors. It relates to the access of relevant permeants to the active sites. It is an event that must precede the active interaction. Thus, the liquid membrane hypothesis of drug action is a missing episode in the whole story of drug action. When this hypothesis is viewed in the light of existing theories, a more rational picture emerges. Before we close we would like to admit that the opinions expressed or conclusions drawn in this monograph or for that matter in any work of this kind are never final. The purpose always is to arouse curiosity, raise more and more questions while answering the existing ones so that the act of enquiry goes on and on. If the present text is able to achieve this purpose even to a small extent, the authors will consider themselves amply rewarded.

Author Index

Numbers indicate the pages on which complete reference is listed

E

G

L

T

Subject Index